International and European Environmental Policy Series

Edited by
Alexander Carius
R. Andreas Kraemer

Springer-Verlag Berlin Heidelberg GmbH

Sebastian Oberthür Hermann E. Ott

The Kyoto Protocol

International Climate Policy for the 21st Century

In Collaboration with Richard G. Tarasofsky

Foreword by Ernst Ulrich von Weizsäcker

With 14 Figures and 15 Tables

Springer

Dr. Sebastian Oberthür
Ecologic
Pfalzburger Str. 43/44
10717 Berlin
Germany
E-mail: *oberthuer@ecologic.de*

Dr. Hermann E. Ott
Wuppertal Institute for Climate, Environment and Energy
Doeppersberg 19
42103 Wuppertal
Germany
E-mail: *hermann.ott@wupperinst.org*
URL: *http://www.wupperinst.org*

ISBN 978-3-642-08575-8

Library of Congress Cataloging-in-Publication Data Applied For

Die Deutsche Bibliothek - CIP Einheitsaufnahme
Oberthür, Sebastian:
The Kyoto protocol: international climate policy for the 21st century; with 15 tables/ Sebastian Oberthür; Hermann E. Ott.

ISBN 978-3-642-08575-8 ISBN 978-3-662-03925-0 (eBook)
DOI 10.1007/978-3-662-03925-0

Originally published by Springer-Verlag Berlin Heidelberg New York in 1999
Softcover reprint of the hardcover 1st edition 1999

Cover Design: Erich Kirchner
Typesetting: Camera-ready by Jan Peter Schemmel, Ecologic, Berlin
SPIN: 10776603 30/3111 - 5 4 3 2 1 – Printed on acid free paper

About Ecologic

Ecologic is a not-for-profit institution for applied environmental research and policy consultancy, seated in Berlin. Ecologic conducts implementation-focused research projects and prepares expert opinions on a range of issues relating to nature conservation and environmental protection. These include new approaches in environmental policy, ecologically sustainable resource policy, international environmental agreements and institutions, environmental planning and the integration of environmental concerns into other policy realms. Ecologic's work focuses in particular on analyzing the environmental policy of the European Union and its member states and enhancing the effectiveness of international environmental regimes.

Ecologic is part of the network of Institutes for European Environmental Policy with offices in a number of principal European cities, and is part of numerous other issue-focused and project-related international networks. Ecologic works primarily for international and supranational organizations, parliaments, governments, national agencies and local authorities, industry federations, trade unions, and environmental or conservation NGOs. Ecologic pursues exclusively academic and non-profit aims. It is an economically and academically independent, nonpartisan institute.

Ecologic publishes current research reports and studies in its book series "Contributions to international and European environmental policy".

About the Wuppertal Institute

The Wuppertal Institute for Climate, Environment and Energy was established with the aim of systematically addressing both the global environmental challenges and the complex task of ecological structural change. From the optimistic perspective of "more for less" it works towards overcoming what are often perceived as the conflicting demands of economy and ecology. Increased prosperity with a reduction in consumption of natural resources – this is the paradigm for eco-efficient innovation and for a new generation of technology that has to be more energy efficient, uses less space and absorbs fewer raw materials. To help achieve those aims, the Wuppertal Institute draws up concrete concepts in almost all fields of the environment and provides tangible visions for new models of prosperity.

As part of the Science Centre of North-Rhine Westfalia, the Wuppertal Institute has two sister institutes, the Institute for Cultural Studies in Essen and the Institute for Work and Technology in Gelsenkirchen. Since its foundation in 1991, the Institute has expanded to about 120 staff, organised in five divisions: Climate Policy, Energy, Transport, Material Flow/Ecological Accounting, and New Models of Wealth. It is headed by Prof. Ernst Ulrich von Weizsäcker, Member of the Club of Rome and, since October 1998, Member of the German Parliament. Research and strategy are guided by an International Advisory Board chaired by Prof. Hartmut Grassl, Director of the Climate Programme of the World Meteorological Organisation.

Foreword

This is a timely and most welcome book. Shortly before the turn of the century, two experts on the political and legal dimensions of international environmental negotiations present a comprehensive and concise commentary on the Kyoto Protocol and the prospects for international climate policy. This treaty, adopted in December 1997, attempts to reverse the trend towards rising emissions of greenhouse gases and presents one of the most astounding achievements in international environmental policy.

Let us remember a few facts. Since the discovery of the man-made greenhouse effect more than a hundred years ago, the world has embarked on a large-scale "experiment" with nature and the global climate. This experiment is irreversible in human time-scales and is certainly one of the biggest threats to civilisation and the lives of many million people. Our civilisation can be traced back by about 10,000 years - exactly those 10,000 years that were characterised by a pretty stable global climate. Now we are endangering this stability, with unpredictable consequences for the lives of our children and grandchildren.

The immediate danger has been apparent for some 20 years now. The challenge: to reduce the world-wide emissions of greenhouse gases from the burning of fossil fuels and other sources by about 60% until the middle of the next century. A tremendous task! But it can be done, as has been shown by numerous scientists and engineers around the world. The solution: the switch to clean energy derived mainly from renewable energy sources like solar, wind, bio-mass and hydropower, coupled with a revolution in energy and resource efficiency by a factor of four and more. This is what I call the "efficiency revolution". In our Book "Factor Four", Amory Lovins, L. Hunter Lovins and I have provided some 50 examples of how this could be achieved even today.

At present, many factors impede the efficiency revolution, although it would come at no cost and, in many cases, be quite profitable. Some of the main reasons are the structural deficits of our energy systems, our infrastructure (transportation, for example!) and our predominant production and consumption patterns. To change these deficiencies is not easy, because it requires foresight and long-term planning. Politicians are usually not well equipped to deal with challenges of this kind, because they are more interested in short-term results.

Therefore, it is all the more remarkable that the world community unanimously adopted the Kyoto Protocol in December 1997. This is an attempt at long-term planning by politicians – very long-term, that is to say, as it will occupy humankind for the greatest part of the next century! Everybody involved in the negotiations deserves praise for the relentless efforts to put together what has been termed "the most complicated non-military agreement ever". World leaders like Bill Clinton, Boris Yeltsin and the former Chancellor of Germany Helmut Kohl were directly involved and pushed for compromise.

In its first part, this book helps us to understand all the intricacies of the negotiations, describes the actors and carefully unveils the many compromises that led to agreement. It is quite fascinating to watch the picture unfold. I am sure that even those who were directly involved will enjoy refreshing their memories and might learn a lot. The authors were present at the formal as well as at many informal meetings and have gained a wealth of insights which they present in a comprehensive manner.

The second part of the book is devoted to explaining the provisions of the Kyoto Protocol itself. Everyone interested in climate policy will find important insights in this comprehensive commentary, whether an interested layman, an academic or directly involved in the upcoming negotiations. The authors shed light on the ambiguities of many provisions, offer guidance for their interpretation and provide a far-reaching outlook on the "unfinished business" that Kyoto left for future rounds of negotiations.

Lastly, the authors take us into the next century. They draw important lessons from the factors that shaped the outcome in Kyoto, undertake a careful evaluation of the Protocol and present a picture of the political landscape after the adoption of the treaty. Most importantly, however, the authors make an attempt to explore the possibilities for a Leadership Initiative that may form the basis for a renewed effort to breath life into the Kyoto Protocol.

This book is characterised by a unique combination of diligence, comprehensiveness and creativity. Were it on a different topic, I would say it belongs in every household. As it stands, it certainly belongs in the hands of everyone interested in climate policy and, indeed, in the fate of our planet at the turn of the century.

Ernst Ulrich von Weizsäcker
President, Wuppertal Institute
July 1999

Preface

Like the Kyoto Protocol, which represents the culmination of 10 years of negotiations, this book was written after many years of involvement in this process by the two authors. Both of us have been duly following the diplomatic circus around the world and have written extensively on various policy-related or legal aspects of the emerging climate regime. We have also worked on other environmental regimes. Nevertheless, climate change has never stopped to attract our intellectual and emotional attention and, indeed, to fascinate us.

Maybe the root of this fascination lies in the unique challenge to human civilisation posed by the increasing concentrations of greenhouse gases in the Earth's atmosphere. It is a challenge to mankind's ingenuity, ability to adapt to changing living conditions and, most importantly, capability to act today in response to a threat that will only affect our children and grandchildren. The adoption of the Kyoto Protocol, a treaty between sovereign states, is a first serious attempt to tackle the problem. And despite the many obstacles and the powerful forces threatening the fledgling treaty, we have not lost our belief that mankind will live up to this challenge.

We are proud to present this book and sincerely hope that it will contribute to find mankind's path into the future. Our work should help to understand the issues at stake and to provide a guide to the further development of the climate regime. We neither claim exhaustiveness nor the right judgement at all times. Therefore, we welcome any comments and suggestions.

Acknowledgements

We are indebted to many friends and colleagues who provided advice and support. Due to our long participation in the climate process, we discussed the issues with many experts, who enhanced our understanding and often directed our attention to problems that otherwise would have gone unnoticed. Some of our ideas have their origins in these fruitful exchanges. We apologise for not being able to mention all of them here, but we are sure they are aware of their contribution.

Special thanks go to those who have reviewed parts of the book: Matthias Buck, Anke Herold, R. Andreas Kraemer, John Lanchbery, Jürgen Lefevere, Fanny Mißfeldt, Kelly Sims, Stefan Singer, Azza Talaab, Jake Werksman, André Witthöft-Mühlmann and Farhana Yamin. Their comments helped us to avoid many mistakes. Responsibility for any remaining errors and for other shortcomings remains, of course, with the authors.

Many colleagues and assistants in our organisations have helped with the research, offered comments on drafts and provided support in finalising the book. These are Thorsten Brinkmann, Matthias Duwe, Markus Kaplan, Malte Meinshausen, Henrike Peichert, Neeta Sharma-Höfelein and Hauke von Seht. Thomas Langrock, Jan-Peter Schemmel and Dennis Tänzler were extremely helpful during the last stages and we would not have been able to finish the book in time without them.

Richard G. Tarasofsky proved to be a very special collaborator. He vastly improved the language and offered many helpful comments as an expert in international environmental law. He furthermore provided substantial input into Chapter 23. Without him, the book would certainly have a different flavour. In the final days before submitting the manuscript to the publisher, editing by Darla Nickel improved many of the last pages we wrote.

Furthermore, we gratefully acknowledge the financial support granted by the European Commission, DG XI, the German Federal Ministry for the Environment, Nature Conservation and Nuclear Safety, the Norwegian Ministry of Environment, the Danish Ministry of Environment and Energy and the Danish Environmental Protection Agency. These contributions allowed us to prepare, in collaboration with Farhana Yamin of FIELD, an earlier analysis of the Kyoto Protocol.

Before concluding, we wish to express our gratitude to our institutions, Ecologic and the Wuppertal Institute. They provided an ideal background for our undertaking, encouraged us to go ahead and allowed us to continue the project on the Kyoto Protocol beyond the financial means granted by our funders.

And finally, we would like to express our deep gratitude to our families for the tolerance, understanding and support they provided throughout the making of this book – many, many thanks to Sigrid, Ute, Pablo and Yassine.

Hermann E. Ott
Sebastian Oberthür
Wuppertal/Berlin, in July 1999

Contents

List of Figures

List of Tables

List of Boxes

Abbreviations

AG 13	Ad Hoc Group on Article 13
AGBM	Ad Hoc Group on the Berlin Mandate
AIJ	Activities Implemented Jointly
AOSIS	Alliance of Small Island States
APEC	Asia-Pacific Economic Cooperation
CAN	Climate Action Network
CDM	Clean Development Mechanism
CEITs	Countries with Economies in Transition
CERs	Certified Emission Reductions
CFCs	Chlorofluorocarbons
CH_4	Methane
CNE	Climate Network Europe
CO_2	Carbon Dioxide
COP	Conference of the Parties (to the FCCC)
COP/MOP	Conference of the Parties serving as the meeting of the Parties (to the Kyoto Protocol)
COW	Committee of the Whole
EC	European Community
ERUs	Emission Reduction Units
EU	European Union
FCCC	Framework Convention on Climate Change
G-7	Group of Seven Industrialised Countries (Canada, France, Germany, Italy, Japan, United Kingdom, United States)
G-77 (and China)	Group of 77 (and China)
Gt	Gigatonnes
GATT	General Agreement on Tariffs and Trade
GEF	Global Environment Facility
GHGs	Greenhouse Gases
GWP	Global Warming Potential
HCFCs	Hydrochlorofluorocarbons
HFCs	Hydrofluorocarbons
ICAO	International Civil Aviation Organization
ICLEI	International Council of Local Environmental Initiatives

IMO	International Maritime Organization
INC	Intergovernmental Negotiating Committee (for a Framework Convention on Climate Change)
IPCC	Intergovernmental Panel on Climate Change
JI	Joint Implementation
JUSSCANNZ	A Group of Countries Comprising Japan, Unites States, Switzerland, Canada, Australia, Norway and New Zealand
LUCF	Land-use Change and Forestry
Mt	Megatonnes
N_2O	Nitrous Oxide
O_3	Ozone
ODA	Official Development Assistance
QELROs	Quantified Emission Limitation and Reduction Objectives
QELRCs	Quantified Emission Limitation and Reduction Commitments
OECD	Organization for Economic Co-operation and Development
OPEC	Organization of Petroleum Exporting Countries
PAMs	Policies and Measures
PFCs	Perfluorocarbons
SBI	Subsidiary Body for Implementation
SBSTA	Subsidiary Body for Scientific and Technological Advice
SF_6	Sulphur Hexafluoride
UNCED	United Nations Conference on Environment and Development (Earth Summit 1992)
UNCTAD	United Nations Conference on Trade and Development
UNDP	United Nations Development Programme
UNEP	United Nations Environment Programme
UNFCCC	United Nations Framework Convention on Climate Change
UNGASS	United Nations General Assembly Special Session (on AGENDA 21) (New York, June 1997)
UNICEF	United Nations Children's Fund
UNIDO	United Nations Industrial Development Organization
WBCSD	World Business Council for Sustainable Development
WBGU	German Advisory Council on Global Change
WMO	World Meteorological Organization
WTO	World Trade Organization
WRI	World Resources Institute
WWF	World Wide Fund for Nature

Part I
Building Blocks and Negotiating History

The third session of the Conference of the Parties to the UN Framework Convention on Climate Change (UNFCCC), which took place in Kyoto in December 1997, produced a landmark treaty. This was the culmination of a negotiating process without precedent in international environmental affairs. The Kyoto Protocol was the result of a complex multiyear negotiating "game", extending far beyond the 12-day gathering in Kyoto. Like chess-players, governments had to act strategically, taking into account the near- and long-term consequences of their actions and the possible reactions of their counterparts. Unlike chess, the "rules of the game" evolved during the play. But of course, Kyoto was more than a game. Indeed, the involvement of Foreign Ministers, Prime Ministers and even Heads of State marked the elevation of climate change into the sphere of "high politics".

To understand the Kyoto Protocol and the future prospects of international climate policy, one has to grasp the building blocks upon which the Kyoto process was founded and the accompanying political dynamics that have developed. Some of the most important building blocks shall be explored in the following chapters: the science of climate change (Chapter 1), the key players and their interests (Chapter 2), and the UNFCCC of 1992 (Chapter 3). Moreover, the Kyoto Protocol is incomprehensible without some familiarity with the Berlin Mandate that led to more than 30 months of international negotiations, the so-called "Kyoto process" (Chapter 4). During that period, the key players used international and regional political events, as well as bilateral activities outside the climate policy arena, to further their cause and evaluate the other players' positions and room for manoeuvre (Chapter 5). International climate politics were also influenced by political developments within countries, including non-governmental and business developments (Chapter 6). Finally, the endgame at Kyoto and its special characteristics proved to be of critical importance (Chapter 7).

1. The Science of Climate Change

Scientists have known about the greenhouse effect for over a century. Swedish chemist Svante Arrhenius described the basic mechanism as early as 1896.[1] The greenhouse effect is caused by the Sun's radiation that is reflected off the Earth's surface and trapped by carbon dioxide (CO_2) and other greenhouse gases (GHGs) in the atmosphere. This natural greenhouse effect increases the global mean temperature to about 15 degrees Celsius, warm enough to sustain life on Earth. By burning fossil fuels and releasing more CO_2 into the atmosphere, humans have altered this basic mechanism leading to an additional human induced greenhouse effect also known as "global warming".

Since the times of Svante Arrhenius, our understanding of how human activities influence the climate and the possible effects of this has improved considerably. The authoritative assessment of the state of the science of climate change, its impacts and its economic and social dimension is provided by the Intergovernmental Panel on Climate Change (IPCC). In this chapter, the role of the IPCC as a body that assesses the scientific consensus in the area of climate change is highlighted first. Second, the causes and potential impacts of climate change are summarised. This is mainly based on the scientific knowledge available to negotiators in the Kyoto process (which has, however, not changed much since). After a brief evaluation of the role played by the "climate sceptics" in the debate, the rationale for acting on climate change under the existing scientific uncertainty is explored.

1.1 The Intergovernmental Panel on Climate Change

The Intergovernmental Panel on Climate Change (IPCC), which was set up in 1988 by the United Nations Environment Programme (UNEP) and the World Meteorological Organization (WMO), regularly reviews the state of scientific knowledge on climate change. All states that are members of the United Nations and of the WMO are therefore members of the IPCC and its three working groups: Working Group I on the science of climate change, Working Group II on scientific-technical analyses of impacts, adaptations and mitigation of climate change, and Working Group III on the economic and social dimensions of climate change. The IPCC, thus, is a hybrid body, partly political-intergovernmental and partly scientific-technical.[2]

The reports of the IPCC working groups, including summaries for policy-makers, are drafted and reviewed by hundreds of experts from all over the world. These experts are nominated by governments as well as intergovernmental and

---

1 Arrhenius 1896.

2 On the history of the IPCC see in detail Lanchbery/Victor 1995; the tasks of the IPCC working groups have been subject to slight adaptations over the years.

non-governmental organisations, but they act in their personal capacity. About 2,500 scientists from 100 different countries participated in the IPCC Second Assessment Report of 1995. The drafts are regularly sent to governments for comment. Finally, the IPCC working groups and the IPCC plenary – which includes the political representatives of the member states – approve, line by line, the summaries for policy-makers and the synthesis of the assessment, while the full reports are "accepted" without discussion. As such, the full reports of the IPCC represent a broad international scientific consensus that has been "filtered" politically. The summaries for policy-makers and the syntheses by contrast, which provide the main input of the IPCC to the international climate negotiations, consist of politically negotiated text.

The IPCC presented its first assessment report in 1990 to the Second World Climate Conference held in Geneva.[3] This report, along with a Supplementary Report of the IPCC,[4] provided the scientific basis for the negotiations on the Climate Change Convention of 1992. The Second Assessment Report, adopted by the IPCC in December 1995, established the scientific basis of the Kyoto process. The IPCC found, for the first time in the history of international discussions on climate change, not only that "the balance of evidence suggests that there is a discernible human influence on global climate".[5] It also identified a large potential for "no-regrets" measures (i.e. those measures whose benefits equal or exceed their costs to society) to limit and reduce GHG emissions.

1.2 Causes and Effects

Since the late 19th century, global mean temperature on the Earth's surface has risen by about 0.3–0.6°C (see Figure 1.1). By 2100, temperature increases between 1 and 3.5°C are expected to occur due to global warming (best estimate: 2°C). While change will probably occur on a global scale, the impacts on local and regional climates are expected to vary considerably. Temperature increases will be felt more strongly in the higher latitudes than in areas close to the equator. Furthermore, the frequency of severe storms and other extreme weather events like droughts and floods are likely to increase. The El Niño phenomenon – a change in the Southern Ocean oscillation that usually occurs once every nine years and leads to severe changes in weather patterns world-wide – might become more serious and frequent in a warmer world.[6]

The expected sea-level rise of 15 to 95 cm by 2100 could endanger coastal areas and some low-lying island nation states. Shifting climatic zones might be fatal for forests and bring about an upsurge of diseases, such as malaria, to

---

3 Jaeger/Ferguson 1991; IPCC 1990.

4 IPCC 1992.

5 IPCC 1996a; see also Houghton 1997 and World Energy Council (WEC) 1996.

6 See IPCC 1996b for a description of the probable effects and impacts of global climate change.

regions where these were previously unknown. Drastic climate change might also endanger fragile ecosystems of mountains and wetlands. Desertification is likely to increase, while the cryosphere (the area which is covered by ice and snow) is likely to shrink. Both effects might reinforce shortages of water supply in some areas. Agricultural output and fisheries could be affected, increasing the risk of hunger and famine in several regions. Furthermore, areas prone to flooding are likely to experience extensive damage to infrastructure as a result of climate change.[7]

Figure 1.1: Global Average Surface Temperature, 1860–1998

Note: While the bars represent the annual temperatures the curve indicates the five-year average.

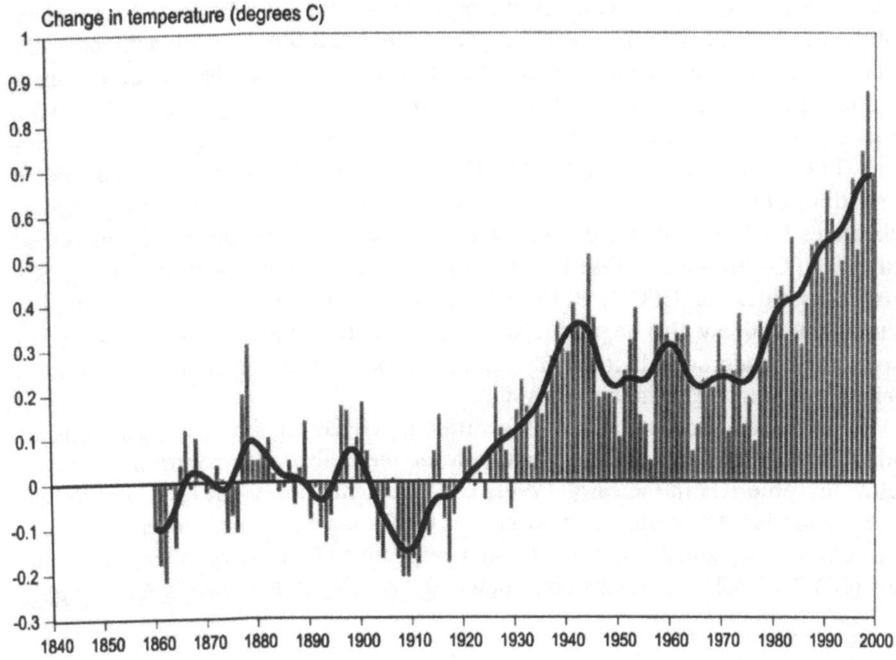

Source: Hadley Centre 1998

Contrary to conventional wisdom, global warming could also result in the cooling of certain regions. For example, the Gulf Stream that currently warms the waters of the North Atlantic Ocean and tempers the Western-European climate might take another course – as was the case in earlier times when icebergs appeared regularly in the North Sea. This could happen within a few years or dec-

7 IPCC 1996b.

ades and could provide one of the future climate change surprises.[8] As a further hazard, the West Antarctic ice sheet might "surge", causing a sudden rise in the sea level.[9] Because of the varied impacts of the greenhouse effect, the term "climate change" has largely replaced the more popular "global warming" in the international discussions.

Human induced emissions of a number of gases have contributed to the warming of the atmosphere. The major GHGs are CO_2, methane (CH_4), nitrous oxide (N_2O), and halocarbons such as fully and partially halogenated chlorofluorocarbons (CFCs and HCFCs), halons etc. Among the principal GHGs that were subject of the Kyoto process, CO_2 is still the major culprit, accounting for 70–72% of the additional greenhouse effect accumulated since industrialisation. In comparison, CH_4 and N_2O have contributed 21–22% and 6–7% respectively (Table 1.1). Historically, halocarbons have contributed about 10% to the additional greenhouse effect.[10] Many of these chemicals are regulated under the Montreal Protocol on Substances that Deplete the Ozone Layer of 1987 and thus are not subject to the international climate regime.[11] However, non-ozone depleting halocarbons cannot be controlled under the Montreal Protocol, but were on the agenda of the Kyoto process. The most important of these, hydrofluorocarbons (HFCs), perfluorocarbons (PFCs) and sulphur hexafluoride (SF_6), have accounted for less than 1% of the additional greenhouse effect since industrialisation. HFCs are generally used as substitutes for CFCs that have been consumed for diverse purposes (e.g. as refrigerants). HFCs are also released as a by-product of the manufacturing of HCFCs (also substitutes for CFCs). PFCs are mainly emitted during the production of aluminium, but may also be produced for use in several applications. SF_6 is used in producing soundproof windows, as a substitute for air in filling automobile tyres as well as in a few other minor applications.

The current apportionment of the different GHGs in the total emissions of industrialised countries differs slightly from their historical contribution given below in Table 1.1. In the early 1990s, CO_2 accounted for about 82% of industrialised countries' emissions of the principal GHGs not controlled under the Montreal Protocol, weighted according to their respective Greenhouse Warming Potentials (GWPs).[12] CH_4 emissions contributed about 12%, and N_2O emissions roughly

---

8 Rahmstorf 1997.

9 IPCC 1996b.

10 IPCC 1996a.

11 On the Montreal Protocol see especially Benedick 1998a; Parson/Greene 1995; Oberthür 1997; Ott 1998a.

12 The Greenhouse Warming Potential (GWP) is a measure that allows a molecule-by-molecule comparison of the greenhouse forcing potential of different GHGs. Because of the different lifetimes of the gases, the GWP varies with the time horizon selected. It has become common to select a time horizon of 100 years. To compare the climate relevance of different GHGs, their emissions can be multiplied by the GWP value to calculate CO_2 equivalents (CO_2 has the GWP value 1). One molecule of CH_4 thus equals 21 molecules of CO_2 equivalents; one molecule N_2O equals 310 CO_2 equivalents.

7%. Emissions of halocarbons (HFCs, PFCs, SF_6), which have high GWPs and the atmospheric concentrations of which grow particularly dynamically, accounted for about 2% (Tables 1.1 and 1.2).[13] While halocarbons were non-existent in the atmosphere before human production, atmospheric concentrations of CO_2, CH_4 and N_2O have grown by nearly 30%, 150% and 15% respectively from pre-industrial levels (Table 1.2). This increase has been measurable in all parts of the world, since these GHGs mix in the atmosphere and enhance greenhouse forcing irrespective of their place of release.

Table 1.1: The Contribution of Principal GHGs not Controlled under the Montreal Protocol to Anthropogenic Climate Change

Gas	Main sources and their share in emissions in industrialised countries in the 1990s	Contribution to Radiative Forcing	Share of GHG emissions of industrialised countries in early 1990s
CO_2	fossil fuel combustion: >95% industrial processes: 2–3%	70–72%	ca. 82%
CH_4	fossil fuel production, distribution and combustion: ca. 1/3 agriculture: ca. 1/3 waste: ca. 1/3	21–22%	ca. 12%
N_2O	agriculture: 40% fossil fuel combustion: 20–25% industrial processes: ca. 1/3	6–7%	ca. 4%
HFCs, PFCs, SF_6	Industrial processes; industry; consumers (shares not available)	< 1%	ca. 2%

Source: own calculation on basis of IPCC 1996a, FCCC/SBI/1997/19 and Add. 1, and FCCC/CP/1998/11; only GHGs not controlled under the Montreal Protocol have been included in the total.

The largest amount of GHG emissions are caused by the combustion of fossil fuels to generate electricity and power cars, aircraft and other engines. In addition to the energy sector, other major contributing sectors include agriculture (CH_4, N_2O), waste (CH_4), and industry (industrial processes: CO_2, N_2O). Table 1.1 gives

---

13 For figures on the rapid development of production of the most common HFC, HFC 134a, see AFEAS 1998.

the approximate shares of these sectors in the emissions of the principal GHGs of industrialised countries in the 1990s. An additional estimated 3% of total CO_2 emissions occur from the use of international bunker-fuels, i.e. the consumption of fuel by aircraft and ships in international transport.[14]

Table 1.2: Atmospheric Concentrations, Lifetimes, and Greenhouse Warming Potentials of Selected GHGs

Gas	Concentrations				Lifetime	GWP
	pre-industrial	current (1994)	change	recent rate of change (mid-1990s)	(years)	
CO_2	ca. 280 ppmv	358 ppmv	ca. 30%	0.40%	50–200.0	1
CH_4	700 ppbv	1,721 ppbv	ca. 150%	0.50%	12.2 (±3.0)	21
N_2O	275 ppbv	311 ppbv [a]	ca. 15%	0.25%	120.0	310
HCFC-22	0	110 pptv	–	5.00%	12.1 (±2.4)	1,500
HFC-23	0	–	–	–	264.0	11,700
HFC-134a	0	–	–	–	14.6	1,300
CF_4 (a PFC)	0	70 pptv [a]	–	2.00%	50,000.0	6,500
SF_6	0	3–4 pptv	–	5–7.00%	3,200.0	23,900

[a] in 1992

Source: IPCC 1996a; the GWP given refers to a time horizon of 100 years. The GWP of CH_4 includes indirect effects (tropospheric ozone production, stratospheric water vapour production), the GWP of HCFC 22 only takes into account direct effects. ppmv = parts per million by volume; ppbv = parts per billion by volume; pptv = parts per trillion by volume.

Human activities also influence atmospheric concentrations of ozone (O_3), another potent GHG which was, however, not a subject of the Kyoto process. In nature, ozone concentrations are negligible in the troposphere, but higher in the stratosphere. Humans influence this distribution in two ways. First, by emitting ozone-depleting substances such as CFCs and halons, they have destroyed ozone in the stratosphere. This has a cooling effect that offsets the greenhouse effect to a limited extent. Second, and more important, the release of certain pollutants (NO_x,CO, non-methane hydrocarbons and others) results in the regional build-up of tropospheric ozone, which aggravates the additional greenhouse forcing.

14 See FCCC/CP/1996/12/Add.2 or <http://www.unfccc.de/emiss/ta10.pdf>; IPCC 1999.

Uncertainties of these processes have, however, not allowed for their exact quantification.

Also not considered in the Kyoto negotiations were emissions of aerosols (microscopic particles and droplets resulting especially from the burning of fossil fuels and biomass). These have a regionally differentiated cooling effect on greenhouse forcing, by scattering and absorbing solar radiation and influencing the properties, amount and lifetime of clouds. While also contributing to acid rain, such aerosols could have masked about 20–40% of the greenhouse forcing of GHG emissions to date. Locally, it could have even led to net cooling.[15] However, significant progress has been made in curbing these aerosol emissions. As GHG emissions continue to rise at the same time, the positive greenhouse forcing will increasingly outweigh this cooling effect.

A certain part of CO_2 in the biosphere is stored in forests and other biomass. Burning of carbon reservoirs, e.g. forests and woodlands, thus causes additional CO_2 emissions. Conversely, afforestation enhances these so-called "sinks" that absorb CO_2 from the atmosphere (see Chapter 11.3). However, significant scientific uncertainties exist as regards the calculation of a sink's absorption capacity.[16] According to reports from industrialised countries in the early 1990s, the net effect of their land-use change and forestry sectors was an absorption of atmospheric carbon equalling about 10–15% of their total CO_2 emissions.[17]

For developing countries, the sources of GHG emissions are slightly different. In particular, tropical deforestation has been a net source of CO_2 in a number of developing countries. In the early 1990s, CO_2 emissions from deforestation accounted for an estimated 15–25% of global CO_2 emissions. Due to the less intensive use of nitrogen fertiliser than in industrialised countries, N_2O emissions have contributed a smaller share to developing countries' GHG emissions. By contrast, CH_4 emissions are relatively higher due to widespread agricultural sources (especially rice paddies in Asia). Nevertheless, CO_2 from fossil fuel combustion has been responsible for the largest amount of GHG emissions even from developing countries. The per-capita CO_2 emissions from developing countries were, however, less than half of the industrialised country average in 1990 (see Figure 2.3). In historical perspective, developing countries have been responsible for less than one-third of the CO_2 and CH_4 accumulated in the atmosphere, and less than 20% of all industrial CO_2 emissions.[18]

---

15 IPCC 1995.
16 WBGU 1998.
17 See FCCC/TP/1997/5.
18 See IPCC 1996c, p. 94–95.

1.3 Climate Sceptics

A small group of so-called "climate sceptics"[19] has argued that the IPCC's measurement methods are faulty and that the available data do not show any significant warming trend. This criticism has neglected that the IPCC is not conducting its own research programme. Rather, the IPCC assesses the state of scientific knowledge taking into account only peer-reviewed work. It is this limitation that has probably allowed the IPCC process to arrive at consensual conclusions. As a result, the IPCC's findings represent a broad international scientific consensus.

Because the IPCC confines itself to the assessment of peer-reviewed work, its ability to respond to the climate sceptics has been limited. This is because almost all the literature of the "climate sceptics" has itself not been subject to scientific peer-review, putting into doubt their scientific credibility. However, the vast majority of scientists organised in the IPCC have rejected the claims of the sceptics. As former IPCC Chairman Bert Bolin has pointed out, "many of such comments have nonetheless been considered and most of them rejected in the IPCC process because of inadequate scientific bases".[20]

The credibility of the climate sceptics is also impaired by the fact that they are largely funded by entities explicitly opposed to any effective measures to counter human induced climate change. In 1995, a report on the funding of the main sceptics, such as Professors Richard Lindzen and Fred Singer, disclosed that fossil fuel groups, e.g. the Western Fuels Association, heavily support their public appearances as well as their studies.[21] Since then, evidence of the funding of sceptics by Exxon, Mobil Oil and other fossil fuel industries has further undermined the credibility of these scientists. Their theories have received the warmest reception in the United States, the site of many fossil fuel multinationals. Moreover, a Republican-controlled Congress helped to raise the profile of the sceptics in the public debate prior to Kyoto, when the sceptics also intensified their efforts in Europe – funded by the same sources.[22]

1.4 Acting under Uncertainty

Despite the broad consensus, uncertainties remain, not only with respect to the impacts of climate change, but also regarding sources and sinks of GHG emissions. Some renowned scientists have criticised the IPCC findings by pointing

19 While these scientists have in fact denounced the science of climate *change*, they are generally referred to as the "climate sceptics".

20 Bert Bolin, Report to the Second Session of the Conference of the Parties to the UN Framework Convention on Climate Change, Geneva, 8 July 1996 (on file with authors).

21 Gelbspan 1995.

22 Gelbspan 1997; Bals 1997; Greenpeace International/Stockholm Institute 1997.

to the scientific uncertainties that are still significant.[23] These are valid comments, since knowledge can not be increased and sustained without criticism. Indeed, such criticism has largely driven the scientific progress achieved in climate change. But given the vast environmental, economic and social risks involved, policy-makers have to follow the best available science, i.e. the scientific knowledge that best stands up against criticism. If, after a thorough examination of claims and counterclaims, the vast majority of scientists support the conclusions of the IPCC, a sound basis for policy-making is provided as acknowledged by most policy-makers themselves (see Chapter 4).

Acting under uncertainty is, however, nothing new in international environmental policy. The Montreal Protocol on Substances that Deplete the Ozone Layer was adopted in 1987 under conditions of significant scientific uncertainty. The ozone depleting effect of CFCs, which is accepted today, was then still disputed. The ozone hole had been detected, but its cause was unknown. Nevertheless, governments decided that it would be prudent to limit the production and use of CFCs and halons on the basis of the available evidence rather than to wait for the "smoking gun" to be found.[24]

The science of climate change in the mid-90s resembled the science of the ozone layer a decade ago. First signs of a changing climate had been detected and, according to the IPCC, the balance of evidence suggested that human behaviour was contributing to this change. The ultimate proof that the observed warming trend was due to anthropogenic GHG emissions was (and is), however, still missing. The IPCC in its Second Assessment Report emphasised that considerable scientific uncertainties still existed regarding the causes and possible impacts of climate change. In particular, the timing, scope and regional distribution of impacts are not well-known and knowledge about the effects are not rigorously linked to empirical observation. Many variables intervene and influence each other, making it difficult to disentangle causes from effects. The human influence on climate is also difficult to detect against the "noise" of natural variability.[25]

As in all cases of scientific uncertainty, the question arises whether precautionary action should be taken. From the IPCC's analyses of future emission scenarios, it has become obvious that significant reductions of GHG emissions would be needed to stabilise atmospheric GHG concentrations at comparatively safe levels. Stabilisation at current levels would require reductions in world-wide emissions of CO_2, CH_4 and N_2O of 50–70%, 8% and 50% respectively, with even bigger cuts in industrialised countries.[26] Based on scientific progress and the well-established precautionary principle, an international consensus has thus emerged that acknowledges the need for action. Deferring action to some future date, when some miracle technology might allow the world to reduce emissions rapidly, risks

23 For several see Jefferson 1996.

24 Parson 1993; Oberthür 1997; Ott 1998a.

25 IPCC 1996a; see also Jefferson 1996.

26 IPCC 1995; IPCC 1997; WBGU 1997.

losing the precious time needed to redirect social development patterns and committing Earth to irreversible climate change.

In its Second Assessment Report, the IPCC therefore recommended a number of policy options for mitigation (as well as adaptation and improvement of knowledge) in order to prudently deal with climate change. Even action beyond no-regrets measures could be justified in such a risk aversion strategy. In this respect, the IPCC listed a portfolio of possible actions, ranging from energy efficiency measures to research.[27] With proof that feasible options for mitigating climate change were available, it was now up to politics to decide on their implementation: the ball was thus in the field of policy-makers.

---

27 IPCC 1996c.

2. Key Players and Interests

Governments were the decisive players in the Kyoto process, as only they were competent to adopt the Kyoto Protocol. Their negotiating behaviour and positioning was not only influenced by the state of scientific knowledge, but also by their perceived or real interests. For example, dependence on production and use of fossil fuels varies greatly, building the basis of differing "polluter interests". Also, countries' vulnerability to the impacts of climate change varies, as does their interest in climate change mitigation and adaptation. The availability of affordable options to reduce GHG emissions and adapt to climate change is another crucial factor ("helper interest"). Finally, the players' goals are also influenced by factors related only indirectly to climate change, e.g. cultural pre-determinations and institutional structures.[1] These are the main elements of a country's constellation of interests. A country dominated by polluter interests can be expected to be less enthusiastic about elaborating stringent action to curb GHG emissions. Countries greatly affected by the impacts of climate change and/or strong helper interests will lean more towards supporting such action, while others might be positioned in the middle ground.[2]

More than 170 countries and several hundred non-governmental and inter-governmental organisations were involved in the international negotiations on the Kyoto Protocol. Of course, not all the participating countries had an equal impact on the Kyoto process. Based on the respective special interest, a number of groupings of countries emerged in the process and can be distinguished:

- the European Union and its Member States,
- the US, Japan and other non-EU OECD countries,
- Russia and other countries in transition to a market economy (CEITs),
- oil-exporting developing countries of the Organization of Petroleum Exporting Countries (OPEC),
- the developing countries bound together in the Alliance of Small Island States (AOSIS), and
- the majority of developing countries (including China and India).

In addition, a number of non-governmental and international organisations have invested considerable resources to influence the process and deserve further analyses.

---

1 See in general on different policy styles and cultures in environmental policy Vogel 1986; Jänicke/Weidner 1997.

2 On the categorisation of interests in international climate policy see Oberthür 1993; see also Prittwitz 1990.

2.1 Industrialised Leader: The European Union

The major industrialised leader in promoting international climate policy during the 1990s has been the European Union (EU).[3] Since the accession of Austria, Sweden and Finland in 1995, the EU consists of 15 Member States. In the absence of a European "government", the European Commission has also been represented at the international negotiations. In contrast to the strong role of the Commission in other international environmental negotiations,[4] the competences of the Union, and thus of the European Commission, in the field of climate policy have remained limited (despite attempts to broaden them). Therefore, the Commission has not played a major role in international climate negotiations. Instead, common positions have been agreed by the Member States, with the participation of the Commission. The country holding the Presidency of the Union – a position that rotates every six months – co-ordinates the members and presents the EU position at the international negotiations. Member States and the Commission have increasingly refrained from making interventions additional to those of the Presidency, thereby improving the public appearance of the Union as a unitary actor.[5]

Among industrialised countries, the EU held a share of 24.3% of CO_2 emissions in 1990. In absolute terms, this amounted to 3,300 Mt CO_2. It equalled roughly 15–16% of global CO_2 emissions.[6] Total emissions of GHGs not controlled under the Montreal Protocol amounted to roughly 4,100 Mt of CO_2 equivalent.[7] Per-capita emissions within the EU were 8.7 t which was double the world average, but lower than the OECD average of 13.1 t. Although EU Member States belong to the high-emitting countries, their performance has been better than the other OECD countries, such as the US, Canada and Australia (see Figures 2.1–2.3).

---

3 With the entry into force of the Maastricht Treaty in November 1993, the European Economic Community (EEC) was renamed the "European Community" (EC). Moreover, rules on a Common Foreign and Security Policy and co-operation in the area of justice and home affairs were agreed separately from the "Treaty Establishing the European Community" and the political entity thus created was referred to as the European Union (EU). While the EC Treaty has remained the legal basis for the participation of the EC/EU in international environmental agreements, it has become common to refer to the EU as the political actor. We follow that custom in the following.

4 For the Montreal Protocol see, for example, Oberthür 1999a.

5 On the division of competence on climate policy of the Union see generally Macrory/Hession 1996; on the development of internal European climate policy see Haigh 1996; Oberthür/Bär 1998; Michaelowa 1998.

6 Global CO_2 emission data are uncertain and different sources provide differing figures. Official emission data have by now only been reported and compiled under the UNFCCC for industrialised countries. We therefore refer to these data in the following.

7 European Commission 1996a; from hereon, any reference to GHGs will refer to GHGs not controlled under the Montreal Protocol (CO_2, CH_4, N_2O, HFCs, PFCs, SF_6), unless indicated otherwise.

Figure 2.1: Contribution of Major Players to Total CO_2 Emissions of Industrialised Countries in 1990

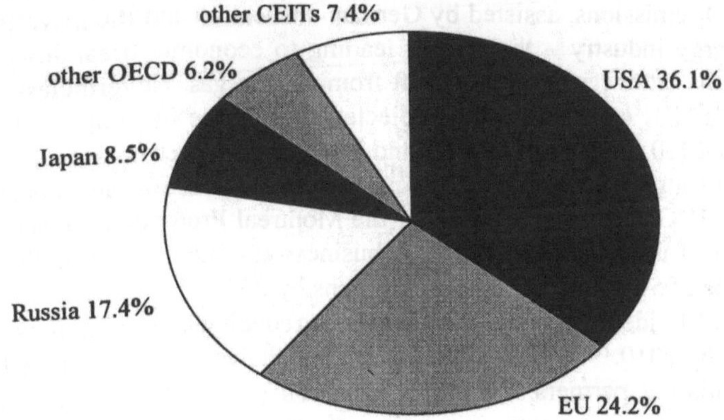

Note: Without Belarus, Lithuania and Ukraine

Source: FCCC/CP/1997/7/Add.1, Annex; aggregate figures based on own calculation, percentages do not add up to 100 because of rounding.

The EU is a large importer of fossil fuels. Currently, just over half of the EU's energy demand is met by its own production. Without policy changes, dependence on energy imports could rise to 55–70% of domestic demand by 2020. Traditionally, several Member States have produced coal. In the 1980s, the United Kingdom became a major producer of oil and gas, when it started large-scale exploitation of the fossil fuel reserves of the North Sea. However, proven reserves of fossil fuels in the EU are fairly limited and production costs are comparatively high. Consequently, coal production within the Union has decreased, and oil and gas production are set to decline in the future due to resource depletion.[8]

Hence, the European Union has a vested interest in reducing its energy consumption and especially its fossil fuel consumption. This interest had been hightened by the oil price "shocks" of the 1970s and early 1980s, which drastically increased Europe's energy bill. In the mid-1990s, the EU's energy policy objectives were to promote the sustainable development of the energy sector and enhance energy saving in the region. The EU's aim of improving energy efficiency by 20% between 1985 and 1995, however, was not achieved, mainly on account of declining world energy prices. Annual improvements in energy intensity fell from 2%

8 European Commission 1996c; see also IEA 1996; 1998.

between 1985–1990 to just 0.6% between 1990–1995, resulting in an overall improvement of only 12% between 1985 and 1995.[9]

Nevertheless, since 1990 the European Union has been relatively successful in reducing its CO_2 emissions, assisted by German unification and the privatisation of the British energy industry – the former leading to economic breakdown in East Germany and the latter resulting in a shift from coal to gas. Nevertheless, prior to Kyoto the EU's CO_2 emissions were projected to increase by roughly 8% above 1990 levels by 2010. Emissions of CH_4 and N_2O, on the other hand, were forecast to decrease by roughly 15% and 30%, respectively. Production and consumption of fluorinated GHGs not controlled under the Montreal Protocol were set to surge by up to 50%. Taken together, under a business-as-usual scenario, this would amount to a rise of 5–6% of total GHG emissions by 2010.[10]

In 1997, the EU identified low-cost means for reducing CO_2 emissions by 15% of 1990 levels by 2010.[11] Although the underlying analysis was considered flawed by some negotiating partners, the large reduction potential achievable through energy efficiency improvements, promotion of renewables and increased use of cogeneration technology, among other things, have also been identified by other researchers.[12] Europe holds significant shares of the world markets in renewable energy and energy-efficient technology, which could grow as international climate policy develops. Finally, many options for further reducing the emissions of other GHGs have also been identified.[13]

Studies about possible impacts of climate change on Europe have received considerable public attention. Europe might not only be hit by the projected sea-level rise, but also by shifting vegetation zones. In addition, a further warming may worsen problems of desertification already affecting the Southern Member States (Spain, Portugal, Italy, and Greece). Of particular public concern has been the danger of a shift in the Gulf Stream's course, which brings Western Europe its moderate climate.

In the early 1990s, many European environmental groups made climate change one of their top priorities. They set up the Climate Network Europe (CNE) in 1989, a regional sub-network of the global Climate Action Network (CAN). In June 1998, CNE had 76 members in 19 European countries. The climate change issue also found its way into parliamentary politics, as a result of Green Parties entering parliaments in more than two-thirds of the EU Member States, including Germany, Austria, Sweden, Belgium, Finland, Italy, Ireland, France and the Netherlands.[14] The European Parliament, which is elected directly by the citizens of the

---

9 ENDS Daily 30 April 1998; European Commission 1997a; 1997b; 1998a.

10 European Commission 1997c; 1999; Oberthür/Bär 1998.

11 Blok et al. 1997 .

12 E.g. WWF 1996.

13 European Commission 1996b; 1997c; see also WWF 1996; CEPS 1997; Oberthür/Bär 1998.

14 In the UK, Greens have consistently failed to enter the Parliament due to the British "first past the post" electoral system; in some EC Member States determination of parties as

European Union, has also generally supported stringent climate policy. The result was that climate sceptics were less influential in the EU than in the US.[15]

Nevertheless, there is a wide gap between the ends and the means – the plans for reducing GHG emissions and their actual implementation. Before Kyoto, promising measures like energy taxes, were proposed at the EU level by the European Commission but did not receive approval by Member States mainly due to fears of competitive disadvantages and strong opposition from industry. Member State governments have also been hesitant to adopt common policies and measures because this would imply shifting competences to the EU, thus changing the balance of powers within the Union. This, however, is a consequence that might eventually be unavoidable in the interest of effective action on climate change.[16]

Although the EU largely succeeded in appearing as a unitary actor during the Kyoto process, severe internal differences persisted. The most obvious among them was the North-South divide. The poorer Southern ("cohesion"-) countries Portugal, Spain and Greece (along with Ireland) were not very enthusiastic about limiting GHG emissions and have requested increases in emissions, which would be compensated by Northern Member States. The Netherlands, Denmark, Austria, Germany and, more recently, the United Kingdom, have acted as protagonists of climate policy, but even among them, consensus has proven difficult to achieve due to different perceptions and national circumstances.[17] Nevertheless, the EU has managed to be well ahead of other industrialised countries, most notably the JUSSCANNZ group, in pushing effective international climate policy.

2.2 Industrialised Laggards: "JUSSCANNZ"

The major opponent of the EU in the Kyoto process was an informal coalition known as JUSSCANNZ. JUSSCANNZ is an acronym that denotes Japan, the United States, Switzerland Canada, Australia, Norway and New Zealand.[18] In addition, Iceland and South Korea sometimes participated in the JUSSCANNZ meetings. Although JUSSCANNZ members held differing positions on single

---

"Green" is not straightforward (e.g. Portugal); see Müller-Rommel 1993; Richardson/Rootes 1995.

15 There have, however, been attempts to export the sceptics view to Europe, see Bals 1997; on the regional climate change impacts in Europe see IPCC 1997; Nick Nuttall, "Warning on greenhouse effect: Hotter-than-ever world adds to fear of climate change", in: The Times, Saturday, 6 January 1996.

16 See Macrory/Hession 1996; Haigh 1996; Collier 1997; the state of the policy initiatives of the European Commission on climate change prior to Kyoto is summarised in European Commission 1997c.

17 O'Riordan/Jäger 1996; Brauch 1996; Collier 1996; 1997.

18 In the run up to Kyoto the five "traditional" JUSCANZ countries (Japan, USA, Canada, Australia and New Zealand) were joined by Switzerland and Norway; reference to this group before Kyoto is thus made using the acronym JUSSCANNZ.

issues, their common ground was a general opposition to stringent commitments for GHG emission reduction.

The United States as the most powerful member of JUSSCANNZ is the largest emitter of CO_2 and GHGs, in absolute as well as relative terms. In 1990, its per-capita emissions of CO_2 amounted to roughly 20 t, nearly five times the world average (Figure 2.3). Its CO_2 emissions of about 5,000 Mt represented more than 35% of the total industrialised countries' emissions. Including other GHGs increases this total to 5,800 Mt. Between 1990 and 1995, CO_2 emissions in the US rose by 5%, while emissions of all GHGs increased by 6%. Total GHG emissions were projected to increase by 26% by 2020, as compared to 1990 (2010: 23%).[19]

Figure 2.2: Total CO_2 Emissions of Selected Players in 1990

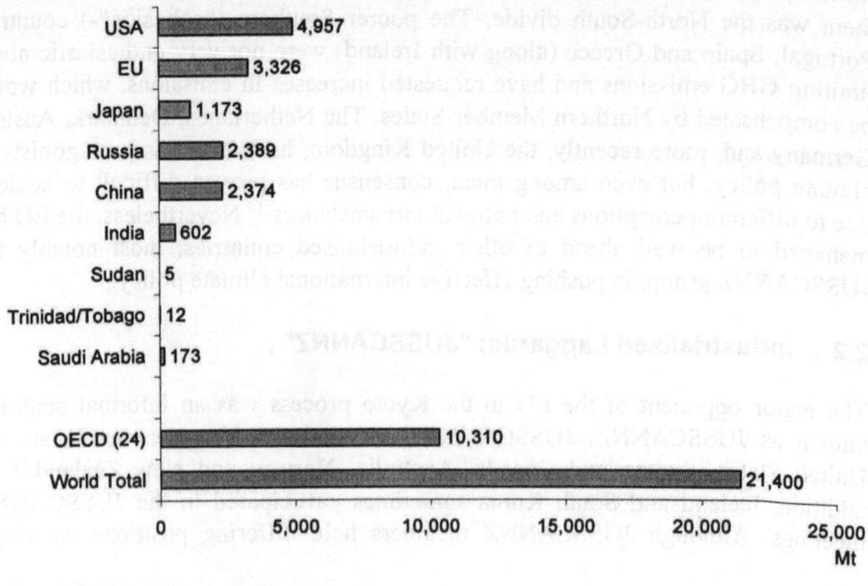

Sources: UNFCCC Secretariat and IEA 1997.

The US stance on climate change has been largely determined by its position as the world's largest producer of coal, oil and gas, although it is a net importer of energy products.[20] As regards vulnerability to the effects of climate change, the US has a long coastline (sea-level rise) and parts of the country are prone to extreme weather events, such as tropical cyclones and droughts, whose occurrences are

[19] FCCC/SBI/1997/19 and Add.1.

[20] For data on energy production and consumption see for example WRI 1996; IEA 1996; 1998.

likely to increase with global warming. Other impacts of climate change, including the northward shift of certain vector-borne diseases such as malaria, will also be felt in North America.[21]

Environmental NGOs have been active and well organised as the Climate Action Network (CAN) in the US. Public support for a global climate treaty was also high before Kyoto (67%), as revealed by a national survey conducted in February 1997.[22] Albert Gore, before becoming Vice President of the United States, had written his book "Earth in the Balance" that strongly supported global environmental protection.[23] In late 1991, he criticised President Bush's stance in the international climate negotiations as "the single worst abdication of leadership ever".[24]

Outweighing these positive forces is the country's pattern of economic and societal development, which has relied heavily on the availability of low-price energy, making the US one of the most energy intensive economies in the OECD. As Steve Rayner put it, "the history of US energy demand and the existing resources, infrastructure and institutions make the US economy as dependent upon fossil fuel as a heroin addict is on the needle".[25] Because of the very high energy intensity associated with American technology and lifestyles, low-costs means of saving energy and reducing GHG emissions are in fact abundant.[26] Nevertheless, the perception (furthered by some for obvious political reasons) that reducing CO_2 emissions would be exorbitantly costly has been comparatively widespread in the US.[27]

In addition to the perceived costs of an effective climate policy, well-organised lobby groups opposing climate policy, such as the Global Climate Coalition (see below, Section 2.5) have been particularly influential in the United States. There are several reasons for this. First, the fossil fuel market is dominated by a small (and still decreasing number) of big companies (such as Exxon, Mobil, Sun Oil, etc.), most with headquarters in the US. This has facilitated effective organisation of industry interests. Second, the "open" US political system has provided lobby groups excellent access to policy-makers. And finally, dependence of US politicians on financial support in order to run election campaigns has provided the fossil fuel lobby with open ears especially in the Congress. In 1994, a land-slide victory for the Republican Party further strengthened anti-environmental forces in the US.

---

21 IPCC 1997.

22 The Charlton Report 1998: National Issues 1/98 – Environmental Values: Global Climate Change Issue, at <http://www.charltonresearch.com/GCC_398.html> as of 6 May 1998.

23 Gore 1992.

24 ECO-Newsletter, 18 December 1991, p. 1.

25 Rayner 1991.

26 WWF 1997a; IPCC 1996b; CNE/USCAN 1997.

27 See Grubb 1996.

Japan, the host of the Kyoto Conference (see Section 7.1), is the second largest emitter of GHGs among JUSSCANNZ. It added more than 1,000 Mt of CO_2 (total GHGs: about 1,200 Mt CO_2 equivalents) to the atmosphere in 1990 (more than 8% of the industrialised country total). Japanese CO_2 emissions increased by 8% between 1990 and 1995 and a further increase of 12 percentage points by 2010 was projected prior to Kyoto. Despite similar emission trends and vulnerability, the US and Japan differed considerably in many respects. Due to Japan's lack of own fossil fuel reserves and thus high dependence on fossil fuel imports, especially oil, it has had a vested interest in reducing its fossil fuel consumption (as evident from its large nuclear energy sector). Moreover, as the most energy-efficient industrialised economy, Japanese per-capita CO_2 emissions are less than half of those of the US (Figures 2.1–2.3).[28]

However, it is exactly the relatively high energy efficiency of its economy that qualified Japan as a member of JUSSCANNZ. Because Japan made special efforts to reduce its energy use in the past, the cost-effective potentials for further cutting its emissions of CO_2 and other GHGs in the future are more limited. However, many options for limiting and reducing GHG emissions still do exist.[29] In addition, Japan's energy efficient technology might experience a higher demand on the world markets if international controls on GHG emissions are implemented.[30] Nonetheless, this has not induced Japan to support strong international action on climate change.

GHG emissions from the other members of JUSSCANNZ have been far less significant (Figure 2.1). All other JUSSCANNZ members, except Switzerland, which expected emissions reductions of 5% by 2010, experienced an increase in their GHG emissions between 1990 and 1995 – and further increases were projected prior to Kyoto.[31] However, most of these countries are also vulnerable to climate change. Australia and New Zealand, in particular, have already experienced the negative effects of human induced atmospheric changes in the case of the Antarctic ozone hole, and could experience severe droughts in the future. Environmental movements are also quite active in these countries.

Other interests that influence positions on climate change, however, varied among the "other" JUSSCANNZ countries. Although none of them reached the US level of per-capita CO_2 emissions (19.8 t) in 1990, Australia and Canada came relatively close (Australia 16.9 t and Canada 17.4 t). Both the Australian and Canadian economies are resource-intensive. Australia produces coal, while Canada produces oil and gas, and, unlike the US, both countries export fossil

---

28 Japan 1997.

29 WWF 1997b; CNE/USCAN 1997.

30 Grubb 1996.

31 FCCC/SBI/1997/19 and Add.1.

fuels.[32] For example, in 1995/96 Australia exported about 70% of its total energy production (valued at around 18% of Australia's export earnings) with black coal (53%) and uranium (33%) as the major exports.[33]

Among the other countries, New Zealand could nearly meet its energy consumption by domestic production, Switzerland imported about two-thirds of its demand, and Norway is a net exporter of fossil fuels because of its limited gas and oil reserves in the North Sea. New Zealand also plans to greatly increase forest growth to limit increases of net GHG emissions.[34]

Norway's situation is special in two respects. First, with almost three-quarters of energy consumption in 1990 supplied by hydropower, it has (like Iceland) almost exhausted its potential for emissions reductions in the energy sector. Any increase in demand for electricity would thus have to be met by fossil-fuel powered capacity, which would lead to increased emissions. Second, emissions from Norway are bound to rise further because of the production of oil and gas that leads to GHG emissions (with limited reduction potential). This source accounted for 23% of Norwegian CO_2 emissions in 1993. A growing demand for relatively carbon-efficient natural gas by other countries, aimed at limiting their own emissions, might thus even lead to rising Norwegian GHG emissions.[35]

2.3 Russia and the "Countries with Economies in Transition"

Following the breakdown of the former "Soviet block", the successor states have become known as "countries with economies in transition" (CEITs). Russia, the most important CEIT, accounted for 17.4% of all CO_2 emissions of industrialised countries in 1990 (nearly 2,400 Mt; Figures 1 and 2). The other successor states of the Soviet Union, most importantly Belarus and the Ukraine, as well as the former communist countries of Central and Eastern Europe (among them Poland, the Czech Republic, Slovakia, Hungary, Bulgaria, Romania, Slovenia, Croatia) emit far less CO_2 and other GHGs in absolute terms. Ukraine, the second largest emitter among the CEITs, was responsible for CO_2 emissions in the order of about 30% of the share of Russia (about 700 Mt). Overall energy efficiency, however, was equally low in all these countries.

Russian per-capita CO_2 emissions of more than 16 t in 1990 were among the highest in the world (see Figure 2.3), mainly due to the very inefficient use of energy resources. With the economic collapse, Russian CO_2 emissions have fallen by about 30% since 1990. Accordingly, per-capita emissions were down to

32 See for CO_2 per-capita emissions FCCC/CP/1996/12/Add. 2 <http://www.unfccc.de/emiss/ta9.pdf>; for data on energy consumption and production see WRI 1996; IEA 1996; 1998.

33 Australia 1997.

34 On the development of New Zealand's climate policy see Gillespie 1998a.

35 On Norway's climate policies see Sydnes 1996.

10–11 t in the mid-1990s, which was lower than the OECD average.[36] Comparable developments have occurred in other CEITs. Even with steady economic growth, which would lead to rising GHG emissions, emissions are unlikely to reach "communist" levels again by 2010.[37]

Figure 2.3: Per-capita CO_2 Emissions of Selected Players in 1990

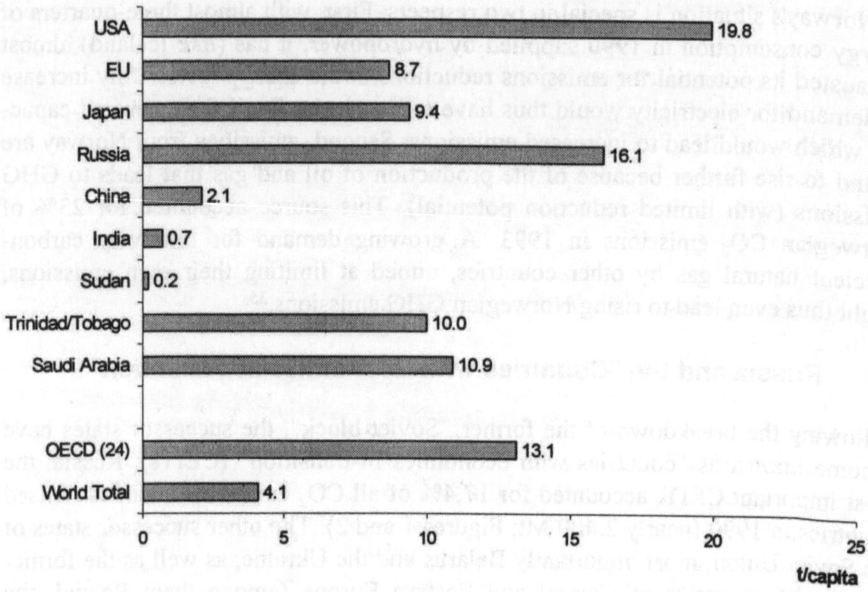

Source: UNFCCC Secretariat and IEA 1997.

This is because competition and energy prices will force some efficiency improvements. Further advances, however, will depend on the strategy chosen to rebuild the energy industries in Russia and the other CEITs.[38] If new energy infrastructures were designed in a way that exhausts the existing potential for further

---

36 IEA 1997.

37 See e.g. Simeonova/Mißfeldt 1997. This has been the root of the so-called "hot-air" problem: Russia and Ukraine have, upon their insistence, been granted the right to emit at their 1990 levels in the commitment period of the Kyoto Protocol 2008–2012. As they are unlikely to reach such emission levels by that date, they may sell the surplus emission reductions (which have not come about as the result of any climate policy and are thus generally referred to as "hot air") under the Emissions Trading that is part of the Protocol; see Chapter 15.

38 CAN-CEE/CNE 1995.

reductions, future emission levels could be held down and even reduced. If this opportunity is not grasped, future emission levels will be affected for some time because of the long-term nature of these investments.

In Russia, parts of the political elite perceive the country as a potential winner from climate change. They argue in particular, that the large land areas permanently frozen ("permafrost") would shrink (although this will become relevant only in the very long-term) and thus could become suitable for agriculture. Furthermore, larger areas in the Arctic rim could become amenable for tourism. However, the retreating permafrost also is likely to damage infrastructure, including buildings, fossil fuel production facilities, pipelines, roads, railroads and communication networks. Adaptation will require considerable resources.[39] In addition, the melting areas would enhance the greenhouse effect by releasing additional CH_4 and CO_2.[40] Despite these dangers, Russian negotiators have tended to emphasise the potential benefits of global warming for their country and tended to downplay the hazards.

Russia's position on climate change has also been influenced by its large fossil fuel reserves. Despite the decline of the energy sector in the 1990s, available reserves are vast and expected to support a growing production and export of oil and gas. This scenario also applies to some of the Asian successor republics of the Soviet Union, most importantly Azerbaijan, Kazakstan and Turkmenistan who have significant reserves of oil and gas currently being prepared for intensive exploitation. As a result of the availability of low-price supplies, Russia has felt little need to save energy, even though this could improve opportunities for increased export earnings. Rather, Russia has had a vested interest in continued global demand for its fossil fuels.[41]

The CEITs have not been able to fully take advantage of the opportunity provided by the collapse of the Soviet Union to rebuild an efficient energy sector because of the limited capacity of their political and institutional systems. Modern administrations needed to be built, while the political capacity to steer societal and economic developments was lost. Successive governments have only been able to influence the transition process to a limited extent. More importantly, economic recovery has assumed the highest priority. Some progress towards climate protection has been inadvertently achieved, however, in that the opening of energy markets drove up energy prices and thus became an incentive for conserving energy. Climate change has generally had a low profile on national political agendas. Environmental movements, that have made a difference in other countries have tended to be weak and of little influence in Russia and other CEITs.[42]

A special interest in climate protection has been evident in those CEITs that have applied for membership with the European Union, i.e. the five "accession

---

39 IPCC 1996b: pp. 241–265.

40 IPCC 1996b: pp. 241–265.

41 See generally IEA 1996, Chapter 4; 1998; WRI 1996.

42 See Jancar-Webster 1993; Vari/Tamas 1993; Botcheva 1996.

countries" Czech Republic, Estonia, Hungary, Poland and Slovenia (in addition to Cyprus). As these countries will have to comply with EU legislation upon their accession to the EU, they have begun to anticipate the implementation of a number of climate-relevant measures. Moreover, their negotiating position vis-à-vis accession will be strengthened by "good" behaviour. This group of countries therefore had a vested interest in coming as close as possible to, and supporting, the position of the EU in the Kyoto process.

2.4 The Developing World: A Crumbling Block

Developing countries have a long tradition of solidarity that has its roots in their common history of colonialism. Their fight for independence and against their structural dependence on the North has had a strong unifying effect. This solidarity has remained an important factor, despite the growing economic differentiation within the developing world in the 1980s and 1990s. Aware of their limited weight if acting in isolation, developing countries attempt to develop common positions in the framework of the "Group of 77" (G-77). This group comprises some 130 developing countries. In addition, China often associates itself with the positions of the G-77. Thus, the Chair of the G-77, which rotates among members annually, usually speaks in the international climate talks "on behalf of the G-77 and China".

Despite this apparent solidarity, rather disparate interests with respect to climate change are assembled under the roof of the G-77. As in other fields of international environmental policy, different groupings within the G-77 have, at times pursued their own special interests. Two groups of countries, in particular, depart from the majority of the developing countries: the low-lying and island developing countries co-ordinated in the Alliance of Small Island States, commonly referred to as AOSIS, and members of the Organization of Petroleum Exporting Countries (OPEC), most notably Saudi Arabia and Kuwait.

2.4.1 Organization of Petroleum Exporting Countries (OPEC)

The negotiating position of the OPEC countries, especially Saudi Arabia and Kuwait, is dominated by their interest in exporting oil and natural gas. Production of fossil fuels has resulted in per-capita CO_2 emissions from these countries that exceed by far the average emissions from developing countries. For example, Saudi Arabian CO_2 emissions of nearly 11 t per capita in 1990 were more than two and a half times the world average (Figure 2.3). In absolute terms, however, Saudi Arabia accounted for less than 1% of global emissions.

Saudi Arabia and Kuwait have at their disposal more than one-third of proven global oil reserves, which under constant production would take over 100 years to deplete. Almost all foreign earnings for these countries come from oil exports, which have accounted, in the case of Saudi Arabia, for more than a third of the gross national product (Figure 2.4). Even efforts in these countries directed at product diversification tend to remain heavily dependent on fossil fuels, such as investments in fertiliser production and other branches of the chemical industry.

These countries' dependence on earnings from fossil fuel exports was particularly evident after the 1990/91 Gulf War that resulted in huge military expenditures and, at least in the case of Saudi Arabia, a large public deficit. Moreover, since the 1980s, the economic base of OPEC members has deteriorated due to declining oil prices. These have resulted in part from increased production by non-members like Norway and the United Kingdom, but they have also been caused by OPEC's failure to enforce its own production limits. For these countries, therefore, "climate change [policy] is a threat to their entire resource strategy and long-term prospects".[43]

Figure 2.4: Vulnerability of Energy Exporters

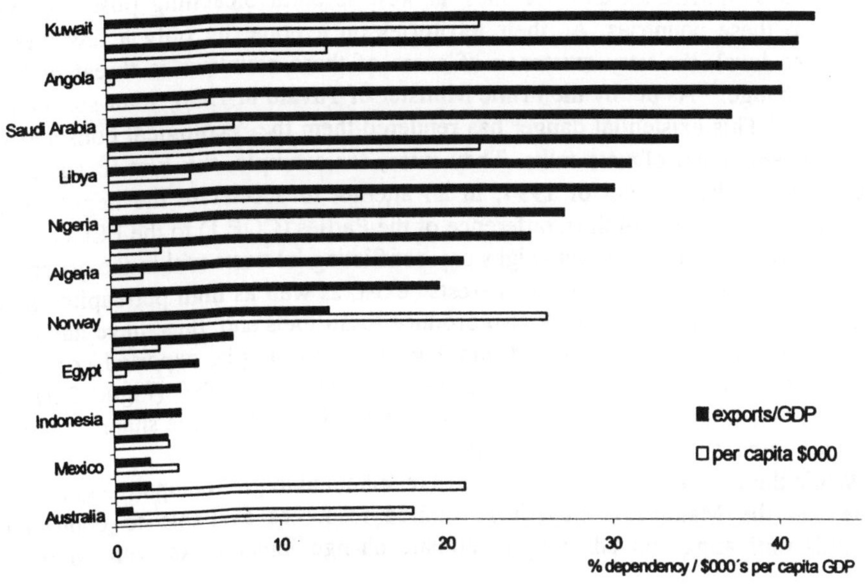

Source: Kassler/Paterson 1997, p. 48.

Given this situation, even potential advantages of international measures to combat global warming are not convincing enough to warrant support from OPEC

43 Grubb 1996, p. 152; for data on energy production and reserves as well as developments in Arabic OPEC members see IEA 1996; 1998; WRI 1996.

members.[44] Besides, the prevailing weather conditions (little rainfall, high temperatures) in the Middle East region render negative impacts of climate change rather insignificant. Export earnings from fossil fuels might be used to provide the means for required adaptation measures. But the monarchist political structure in most Arab countries further curbs the expression of public concern about the issue. Under these circumstances, OPEC members, and Saudi Arabia and Kuwait in particular, lobbied heavily by business NGOs, such as the Global Climate Coalition and The Climate Council (see below), missed few opportunities to slow down progress towards taking common action in the international negotiations.

2.4.2 Alliance of Small Island States (AOSIS)

The AOSIS group consists of 42 low-lying and island developing countries.[45] This group of countries share one common interest, namely extreme concern about the impacts of climate change. The likely sea-level rise, increased frequencies and intensities of hurricanes and cyclones as well as coral bleaching pose concrete threats to these countries. As their territories on average lie only a few meters above sea level, the very existence of some of these countries is threatened by climate change.[46] As put by the Prime Minister of Tuvalu in 1990, "Our survival is at stake".[47] This existential danger has rendered them the "ecological conscience" of the international climate talks. They were, for example, the first to submit a draft protocol by the end of 1994, in an attempt to accelerate the negotiations before the first session of the Conference of the Parties (COP 1) to the FCCC.

This overriding concern outweighs any conflicting interests within the group. In some AOSIS countries, "polluter interests" exist, as well as high per-capita emissions. Some members of AOSIS also produce fossil fuels and, thus, have had conflicting internal interests. Trinidad and Tobago, for example, exported crude oil and emitted more CO_2 per capita than the EU average in 1990 (Figure 2.3). In absolute terms, however, AOSIS members account for a negligible share of global CO_2 emissions due to their rather small size.

While the economic strength of AOSIS members diverges (compare e.g. Singapore and the Maldives), most have limited resources available to contribute towards mitigating and adapting to climate change impacts. As with all other

---

44 See Kassler/Paterson 1997.

45 American Samoa, Antigua and Barbuda, Bahamas, Barbados, Belize, Cap Verde, Comores Cook Islands, Cuba, Cyprus, Dominica, Federated States of Micronesia, Fiji, Grenada, Guam, Guinea-Bissau, Guyana, Jamaica, Kiribati, Maldives, Malta, Marshall Islands, Mauritius, Nauru, Netherlands Antilles, Niue, Palau, Papua New Guinea, Samoa, Sao Tome and Principe, Seychelles, Singapore, Solomon Islands, Surinam, St. Kitts and Nevis, St. Lucia, St. Vincent and the Grenadines, Tonga, Trinidad and Tobago, Tuvalu, U.S.Virgin Islands, Vanuatu.

46 IPCC 1997.

47 Address by the Prime Minister of Tuvalu, Rt Hon. Bikenibeu Paeniu to the Second World Climate Conference, Geneva, 6 November 1990 (on file with authors).

developing countries, their willingness to bear the costs of climate change has also been limited for reasons of equity. In the international climate policy arena, AOSIS has generally called for strong commitments from industrialised countries to reduce GHG emissions and to transfer sufficient resources to developing countries to adapt to any unavoidable climate change. In Kyoto and subsequently, many of them have acknowledged the need for future action by developing countries and supported the concept of "voluntary commitments" (see Chapter 17).

2.4.3 *Other Developing Countries*

Even without the two groupings mentioned above, developing countries are very heterogeneous because of their varying cultural, political and economic conditions. Such distinct countries as China, India, South Korea, Mali, and Costa Rica belong to this group. Obviously, China and India, along with Brazil, carry special weight when it comes to formulating positions in the G-77 and China group. Nevertheless, many other developing countries, such as Malaysia, the Philippines, Tanzania, Indonesia, Mexico participated actively in the Kyoto process.

The central concern of this group in the climate change negotiations has been equity. Its main objective has been to ensure that their economic and social development would not be hampered in order to tackle a problem caused mainly by industrialised countries. Per-capita emissions of CO_2 and other GHGs in developing countries are well below the global average (Figure 2.3). At the beginning of the 1990s, three-fourths of the world population living in developing countries accounted for less than one-third of global CO_2 emissions. Against this backdrop, some have contrasted the "survival emissions" of developing countries with the "luxury emissions" of industrialised countries.[48] Taking a historical perspective, this differentiated responsibility becomes even more pronounced. Of the total industrial CO_2 emissions since 1800, less than 20% stem from developing countries. Developing countries' historically cumulated share will not reach half of the global total before the middle of the 21st century.[49]

Nevertheless, the absolute contribution of some countries, such as China, India and a number of the newly industrialising countries, was already considerable in 1990 (Figure 2.2) and show a dynamic and increasing tendency. Chinese CO_2 emissions are expected to double by 2010. Around 2020, developing countries' CO_2 emissions could be higher than those of industrialised countries, with China superseding the US as the world's number-one emitter.[50] Taking co-operative steps to mitigate climate change in developing countries will be critical in the future. Since their capabilities are limited, however, their negotiating strategy was to link

48 Global Commons Institute 1991.

49 IPCC 1996c, p. 94; WRI 1996, p. 319; Kelly 1990; Global Commons Institute 1991; Agarwal/Nahrain 1991.

50 IEA 1996; 1998; WRI 1996.

costly measures to limit GHG emissions to the transfer of financial resources and technology by industrialised countries.

This has, however, not prevented them from contributing to the fight against global warming.[51] Although India and China are among the world's largest producers of coal and have large coal reserves at their disposal, the majority of the G-77 has very limited energy resources available. As most developing countries, including China and India, are net energy importers, they have a vested interest to use energy in an efficient manner. A number of countries, including China, India and Brazil have thus enacted climate-related measures even before Kyoto, such as cutting fossil fuel subsidies. These have resulted in reduced energy intensity and reduced growth of their GHG emissions.[52]

Current unfavourable natural conditions are likely to worsen in most developing countries as a result of global warming. These countries are particularly vulnerable to climate change because of high dependence on natural resources. As the IPCC puts it, "developing economies are much more dependent on agriculture and other aspects of natural resource flows".[53] But their capacity to adapt to changing climatic conditions, rising sea levels and increased frequency of extreme weather events is very limited due to a lack of economic and administrative resources. The helplessness of East African countries in the face of extreme rainfall and flooding resulting from the El Niño phenomenon in 1997/98 provided a tragic illustration.[54] Despite this vulnerability, public concern about global warming in these countries has remained limited. Although awareness of global environmental issues has risen in recent years,[55] environmental concern has been directed towards "classical" problems of local and regional air and water pollution, which has increasingly dominated the environmental agenda of several developing countries.[56]

Most developing countries also lack sufficient capacity to develop national and foreign environmental policies. Even policy-makers of large developing countries, such as India, do not usually command adequate scientific and other informational capabilities. Moreover, many of these countries lack administrative capacity (personnel and equipment), which limits their ability to contribute positively to policy development at the international level. This has resulted in feelings of uncertainty about own interests and a general scepticism about new policy developments.[57]

Economic and social development remain the overriding policy aim of developing countries. Rising energy consumption and corresponding CO_2 emissions are traditionally seen as an inevitable by-product of this development. While devel-

---

51 See Reid/Goldemberg 1997.

52 See FCCC/AGBM/1997/CRP.5; see also WRI 1996; IEA 1996; 1998; Baumert et al. 1999; Goldemberg 1999.

53 IPCC 1996c, p. 97.

54 Center for Health and the Global Environment 1998.

55 Gallup 1992.

56 See UNEP 1997; WRI 1996; Taylor et al. 1993; Jänicke/Weidner 1997.

57 Gupta 1997; 1998.

oping countries have generally been supportive of action taken by the industrialised countries, emphasis has been put on securing access to affordable energy supplies. They also share a concern that industrialised countries' commitments could lead to inequitable obligations for themselves in the future.[58] Calls from industrialised countries for strengthening action on climate change in developing countries have triggered strong reactions because such requests are perceived as unfair and inequitable. This foreseeable reaction has led some laggard industrialised countries to purposely create an uproar in order to slow down negotiations (see Chapter 17).

2.5 Non-Governmental and International Organisations

Although governments have been the principal players in the Kyoto process, they are not the only influential ones at the international level.[59] Even though they did not directly participate in the intergovernmental negotiations, non-governmental and public international organisations influence government actions by providing information and advice, making policy recommendations and sometimes by direct lobbying. In total, "an enormous number of organisations of myriad type and size" participate in the international policy-making on climate change (see Figure 2.5).[60]

International bodies such as the UN Environment Programme (UNEP), the UN Development Programme (UNDP), the UN Conference on Trade and Development (UNCTAD), the International Civil Aviation Organization (ICAO), the World Bank, the Global Environment Facility (GEF), the OECD and the International Energy Agency (IEA), the UN Industrial Development Organization (UNIDO), the UN Children's Fund (UNICEF), the World Meteorological Organization (WMO) and others, attended the negotiating sessions and influenced the discussions by providing background information and expertise in their fields of competence. UNEP and WMO, the parent organisations of the IPCC, are extensively involved in the discussions. Others like ICAO have been involved to the extent that their special concerns (e.g. emissions from international aviation) are the subject of deliberations.

Most importantly, however, nearly 250 non-governmental Organisations (NGOs) participated as observers in the conference at Kyoto.[61] Formally, three different NGO constituencies are officially recognised in the framework of the international climate talks. The two largest groups are the environmental and business NGOs, which themselves embrace a diversity of interests and positions. In addition, many local authorities have also been represented by the International Coun-

---

58 See in general Gupta et al. 1995; Gupta 1997; Moltke/Rahman 1996.

59 On the growing importance of non-governmental players in international environmental politics see Hurrel/Kingsbury 1992; Kamieniecki 1993; Princen/Finger 1994; Chatterjee/Finger 1994; Wapner 1996; Schmidt/Take 1997; Take 1997; Chartier/Deléage 1998.

60 Sprinz/Luterbacher 1996, p. 58.

61 For a general account and a compilation of positions see Talaab 1998. On the role of NGOs in the FCCC process see Take 1998; Schmidt/Take 1997; Janett 1996; Altvater et al. 1998.

cil of Local Environmental Initiatives (ICLEI).[62] In the Kyoto process, ICLEI represented about 240 cities with 100 million inhabitants.

Table 2.1: Participation of Observer Organisations in the Kyoto Process

	COP 1	COP 2	COP 3
NGOs	165	116	236
Inter-governmental Organisations	12	10	15
UN Secretariats and Specialised Agencies	19	20	27

Source: COP reports.

The "Climate Action Network" (CAN), established in 1989 in the run-up to the Second World Climate Conference of 1990, consists of eight regional subnetworks and more than 260 member organisations (as of March 1998). Modern communication technologies and e-mail systems in particular have allowed CAN members to exchange information and co-ordinate closely during the intersessional periods. Regular gatherings are held at negotiating meetings to discuss matters of strategy and exchange "intelligence" on latest developments. As a result, environmental NGOs have thus considerably contributed to the development of the climate regime and the adoption of the Kyoto Protocol.

United by their common concern for the global climate, CAN members act in a number of different roles in the climate process. Several are engaged in active lobbying of government representatives. International networks like the World-Wide Fund for Nature (WWF), Greenpeace and Friends of the Earth belong to these environmental activists, as do organisations with a national base, like Germanwatch (Germany) or Ozone Action (US, Canada). Some "green" NGOs are single-issue organisations that work only or mainly on the issue of climate change, others (e.g. WWF and Greenpeace) are multi-issue oriented. Organisations like the World

62 In addition, representatives of "pure" research institutions like universities have also attended negotiating sessions. Other NGOs taking part in the process, without a strong predetermination of position, have been Trade Unions and other societal groups that increasingly perceive the climate issue as the central environmental problem and paradigm of the turn of the millennium. The Global Commons Institute has played a special role by teaching conference participants about equity issues and advocating equal per-capita emission rights. It has thus had a special impact on the Indian position.

Council of Churches are also actively involved in the climate change process as part of their broader agenda, which stretches beyond the environmental field.

A special group of CAN members are not engaged in direct lobbying, but provide technical or scientific advice and policy recommendations (although they have been "activist" to varying degrees as well). Ecologic, the London-based Foundation for International Environmental Law and Development (FIELD), the Stockholm Environment Institute, VERTIC, the Woods Hole Research Center, the World Resources Institute, the Worldwatch Institute and the Wuppertal Institute for Climate, Environment, and Energy belong to this category.

In giving advice and lobbying, advocates for the environment make use of different channels and mechanisms. One of the most important instruments for lobbying is "ECO", the NGO newsletter regularly published at international environmental conferences since 1972. It serves as an influential common organ that publicises information and raises new issues. Delegates of most governments usually read it first-thing in the morning, in order to remain up to date in the negotiations – and for a good laugh, because the editors are well aware of the general lack of humour in the negotiations and include entertaining items. Some NGOs have been invited to become members of national delegations (e.g. on the Canadian and Danish delegations) and thus got direct access to the "inner circle" of international policy-making on climate change. For example, FIELD's international lawyers participated in the Kyoto process as members of AOSIS delegations. Apart from that, NGOs roam the corridors of the conference centres in their search of government representatives to influence.[63]

At the other end of the spectrum, business NGOs also play an influential role in the international process. Commanding more resources, representation of these groups at meetings has at times been twice that of "green" NGOs. The International Chamber of Commerce co-ordinates business NGOs. During the Kyoto process, hard line "grey" business NGOs, which were mainly concerned about the economic impact on their businesses of measures to combat climate change, were distinguishable from the "light-green" forces in the business community.

Fossil fuel interests are especially influential, with the US-based Global Climate Coalition being the most significant "grey" NGO. Like their major sources of funding, e.g. from Exxon, Mobil and others, their orientation has been transnational. Don Pearlman of The Climate Council, a New York-based lawyer advising the OPEC countries, was probably the most experienced representative in the Kyoto process of the "Carbon Club", as it is known informally. Several OPEC interventions in the Kyoto negotiations were indeed drafted by this group.

Other industry organisations have, however, taken a more careful stance. Business NGOs like the World Business Council for Sustainable Development (WBCSD) have called for prudent action that does not threaten to disrupt business development. Going even further, the insurance industry (e.g. Munich Re) has

63 On the functions of NGOs see Sprinz/Luterbacher 1996, pp. 59–62; Altvater et al. 1998.

become increasingly aware that it might actually lose in a changed climate (see Figure 6.1) and has become allied with the progressive industry forces in the negotiations on the Kyoto Protocol.[64]

Most importantly, however, an increasing part of the industry sector has begun to perceive climate change as a challenge that ought to be met with creativity.[65] As such, during the 1990s, they built and strengthened ties with environmental groups. For example, the Business Council for a Sustainable Energy Future, which assembles the likely winners of decisive action on climate change (i.e. industries offering energy-efficient appliances or renewable energy technologies), supported stringent action during the Kyoto process. For a growing part of the business community, climate change is an "enormous opportunity to make money", as was pointed out by the Chief Executive Officer of Trigen Energy Corporation. Multinational enterprises such as BP – whose Chairman John Browne enthusiastically applauded the adoption of the Kyoto Protocol and committed his company to a 10% emission cut by 2010 – actively take part in bringing green ideas into the business associations.

64 See Salt 1998, UNEP 1997.

65 Loske 1996.

3. The Framework Convention on Climate Change: The Legal Basis of International Action

The legal basis for the Kyoto process was established in the United Nations Framework Convention on Climate Change (FCCC), adopted on 9 May 1992 in New York. Negotiated over five sessions within 15 months, the Convention was opened for signature at the Earth Summit in Rio de Janeiro in June 1992.[1] Within one year, it was signed by 165 countries and the European Community. Three months after the 50th ratification had been submitted, the Convention entered into force on 21 March 1994. By the time of the Kyoto conference in December 1997, 167 States and the European Community were Parties to the international regime on climate change established by the Convention.

The FCCC is thus universally accepted as the basis of international politics on climate change. The price for this universal acceptance was high, however, because "hard", legally binding reduction targets had to be abandoned in favour of a "soft" approach. The FCCC was a result of the conflicts and the constellation of actors and interests that characterised the negotiations and which remained rather similar during the Kyoto process (see Chapter 2). The major importance of the Convention for the Kyoto process lies in the fact that it provided an objective, as well as basic principles and obligations. Equally important, the Convention established procedures and institutions, which provided the framework for political and diplomatic activities. These provisions of the FCCC delineated the future "playing field" and defined the basic rules of the future game. In the next section, the FCCC's objective, principles and obligations are sketched out. Afterwards, the chapter provides an overview of the most important institutional and procedural provisions. Finally, the chapter concludes with a brief account of the developments after the adoption of the FCCC in 1992.

3.1 Objective, Principles and Obligations

The Convention's "ultimate objective" is stated in Article 2, as the:

> "stabilisation of greenhouse gas concentrations in the atmosphere at a level that would prevent dangerous anthropogenic interference with the climate system".

This objective has been heralded as an achievement even by the normally sceptical environmentalists. While the "level" is not specified, the second sentence of Article 2 stipulates that:

---

1 The Convention is reprinted in: International Legal Materials (ILM), 13 (1992), pp. 851 et seq. Its provisions and the negotiating process have been the subject of numerous analyses; see e.g. Bodansky 1993; Oberthür 1993; Ott 1996a; Rowlands 1995; Sands 1992; 1994; Breitmeier 1996; Mintzer/Leonhard 1994; Sprinz/Luterbacher 1996; Rowbotham 1996.

"(S)uch a level should be achieved within a time-frame sufficient to allow ecosystems to adapt naturally to climate change, to ensure that food production is not threatened and to enable economic development to proceed in a sustainable manner".

The non-specific nature of the objective allows for different interpretations, while acknowledging the need for adaptation to and mitigation of climate change.

This ultimate objective is accompanied by the principles contained in Article 3 of the Convention. This provision establishes the principles of equity and precaution as the fundamental basis of all policies under the Convention, taking into account the:

"common but differentiated responsibilities and respective capabilities. Accordingly, the developed country Parties should take the lead in combating climate change and the adverse effects thereof".

Policies and measures are to be comprehensive, covering all sources, sinks and reservoirs of GHGs and means of adapting to climate change. Special consideration should be given to countries most affected by climate change or by measures to combat global warming. Finally, Parties should promote sustainable development as well as an open international economic system without disguised restrictions on international trade.

The negotiators in 1992 did not translate these principles into any legally binding targets for industrialised countries to cut GHG emissions ("targets and timetables"), mainly due to US resistance. The agreed compromise is framed as a reporting obligation of industrialised countries. The core sentence that contains 117 words refers to "the aim of returning individually or jointly to their 1990 levels" the anthropogenic emissions of CO_2 and other GHGs not controlled under the Montreal Protocol (Article 4.2(b) FCCC). It has been characterised as "the most impenetrable treaty language ever drafted".[2] Even the most far-reaching interpretations do not derive from this a legally-binding obligation of industrialised countries to limit GHG emissions.[3] The provision has also not been effective politically, in that only few OECD countries will reach this "aim".

The Convention includes, however, other substantial obligations that provide the basis for its subsequent development. All Parties shall:

- develop national inventories of anthropogenic GHG emissions and removals by sinks,
- elaborate and implement national and regional programmes containing measures to mitigate and facilitate adaptation to climate change,
- promote sustainable management of sinks and reservoirs,
- co-operate in preparing for adaptation,

2 Sands 1992, p. 273.

3 E.g. Bodansky 1993; Sands 1992; Ott 1996a.

- promote and co-operate in the integration of climate policy considerations into other policy areas and international co-operation in related fields (science, technology, education etc.), and
- report on inventories and relevant policies and measures (Articles 4.1 and 12 FCCC).

Further obligations apply only to industrialised countries of which two broad groups are distinguished according to different Annexes of the Convention. Annex I contains the EC and all industrialised countries existing in 1992, i.e. the 24 OECD countries of 1992 and the then existing European CEITs (with the exception of Yugoslavia).[4] In Kyoto, this list was adapted to include the new nation states in Central and Eastern Europe as well as some smaller countries that had been left out of Annex I in 1992: Czechoslovakia was deleted from the list, while the Czech Republic, Slovakia, Croatia, Slovenia, Liechtenstein and Monaco were added.[5] Annex II, which serves primarily as the reference list for financial obligations under the FCCC, contains all OECD countries as of 1992 and the EC. Thus, different provisions of the Convention apply to (1) all Parties (including developing countries), (2) all industrialised countries, (3) OECD countries or (4) CEITs.

Only industrialised Parties listed in Annex I are under the soft obligation to "aim" at reducing their GHG emissions to 1990 levels by the year 2000. They are also subject to special reporting obligations. Their communications under the Convention must include detailed descriptions of policies and measures, as well as their probable effects on future emissions (Article 12 FCCC). Moreover, industrialised countries have to report earlier than developing and least developed countries. Over time, the reporting guidelines for industrialised and developing countries have been further elaborated. For example, the guidelines for GHG inventories have drawn from the work of the IPCC.[6] In the Convention, a specific review of country reports was not foreseen, but it was envisaged that the Conference of the Parties would "assess (...) the implementation of the Convention by the Parties" (Article 7.2(e) FCCC).

Due to the economic decline in the CEITs and the concomitant decrease of CO_2 emissions, these countries were granted the flexibility of selecting a base year different from 1990 (Article 4.6 FCCC). They were thus allowed to select the year

---

4 "Countries that are undergoing the process of transition to a market economy" according to the footnote in Annex I of the Convention. The OECD membership was subsequently expanded with the accession of Poland, Hungary, the Czech Republic, Mexico, and South Korea. If not otherwise indicated, the term OECD in the following is used to refer to the former 24 member countries. Conflict has arisen from the listing of Turkey which is one of the original 24 OECD members, but at the same time a developing country, for which reason Turkey has yet refused to become a Party to the Convention.

5 Decision 4/CP.3 in FCCC/CP/1997/7/Add.1.

6 IPCC reporting guidelines; Decision 4/CP.1 in FCCC/CP/1995/7/Add.1; Decision 9/CP.2 in FCCC/CP/1996/15/Add.1; Decision 2/CP.3 in FCCC/CP/1997/7/Add.1.

with their highest emissions as the reference for any obligations under the Convention. Bulgaria, Hungary, Poland, Romania and Slovenia have invoked this provision when they submitted their first national communication, which was accepted by the second and fourth Conferences of the Parties (see Table 3.1). Additional requests will be decided upon by the Convention's Subsidiary Body for Implementation (SBI, see below).[7] Apart from these exceptions, the Convention has established 1990 as the base year for industrialised countries' commitments and the reference point for subsequent negotiations.

OECD countries and the EC, which are listed in Annex II of the Convention, have special reporting obligations with respect to their activities on financial assistance and technology transfer. Only for these Parties does the Convention establish substantive obligations to provide new and additional financial resources to developing countries. Specifically, industrialised countries have to bear the "agreed *full* costs" (i.e. all costs) of reporting by developing countries according to Article 12 FCCC and the "agreed full *incremental* costs" of other implementing measures including emission mitigation projects (Article 4.3, 4.4 and 4.5 FCCC). The incremental costs of a project are those resources that enable an additional reduction of GHG emissions as compared to a realisation of the project under business-as-usual conditions. AOSIS' influence furthermore ensured that support for adaptation to adverse effects of climate change was explicitly included. Financial assistance is channelled through a financial mechanism that was first operated by the GEF on an interim basis (Article 21 FCCC).

Table 3.1: Base Years of Countries with Economies in Transition

Country	Base Year	Country	Base Year
Bulgaria	1988	Poland	1988
Croatia	(1990)	Romania	1989
Czech Republic	1990	Russian Federation	1990
Estonia	1990	Slovakia	1990
Hungary	1985–87	Slovenia	1986
Latvia	1990	Ukraine	1990
Lithuania	1990		

Source: FCCC/CP/1998/11/Add.2 and Decision 11/CP.4 in FCCC/CP/1998/16/Add.1; Croatia had not submitted its first national communication.

7 Decision 9/CP.2 in FCCC/CP/1996/15/Add.1; Decision 3/CP.4 in FCCC/CP/1998/16/Add.1.

Two other issues addressed in the Convention have proven to be important for the development of international policy on climate change: the so-called "sinks" and Joint Implementation. Whereas the term "sink" in general refers to any process, activity or mechanism which removes GHGs from the atmosphere for an unspecified time, the major issue has been the absorption of atmospheric carbon by growing trees. The IPCC has estimated the regrowth of forests in the Northern hemisphere to account for an absorption of between 0.0 and 1.0 Gt carbon per year (equivalent to between 0 and 15–20% of industrialised countries' CO_2 emissions).[8] While the Convention determines that calculations of GHG inventories include emissions by sources and removals by sinks (Article 12.1 FCCC), large uncertainties exist with regard to the size of the carbon sink function for varying kinds of forests and the effectiveness of including sinks in the regime has remained disputed and doubtful (see Section 11.3)

The concept of Joint Implementation (JI), as referred to in Article 4.2 of the Convention, is an economic one. Mitigation costs differ among countries mainly due to differences in energy efficiency. For this reason, emission reductions are thought to be cheaper in developing countries and CEITs than in OECD countries. JI allows industrialised countries to save resources and maximise emission reductions, as well as get credit, for carrying out climate protection projects abroad. Despite JI's intellectual appeal, practical and political difficulties exist.[9] Nevertheless, JI has, upon the insistence by Norway and the support by other industrialised countries, found its way into the Convention in Article 4.2.a, which determines that Annex I Parties may implement policies and measures "jointly".[10] This has been the root of one of the most interesting and time-consuming discussions in the international negotiations on the way to Kyoto (see Chapter 4).

3.2 Institutional and Procedural Predeterminations

The international regime on climate change is process-oriented in that its institutions and procedures are designed to support a constant development and implementation of the regime.[11] In this respect, the FCCC has been a milestone because it established the institutional framework and some, although rudimentary, procedures that can be used by Parties to further elaborate the provisions of the Convention

> "in the light of the best available scientific information and assessment on climate change and its impacts, as well as relevant technical, social and economic information" (Article 4.2(d) FCCC).

8 IPCC 1996c.
9 See below and Loske/Oberthür 1994; Luhmann et al. 1997.
10 Bodansky 1993, pp. 520–523.
11 On the process dimension of international regimes see Ott 1998a; Gehring 1994.

The most important institution established by the Convention is the Conference of the Parties (COP), which is the supreme decision-making body of the regime (Article 7 FCCC). Among the powers and responsibilities of the COP are:

- reviewing the implementation of the Convention and the adequacy of commitments,
- promoting of the development and refinement of methodologies for GHG inventories,
- assessing the overall effectiveness of the Convention, and
- fulfilling any other function required for the achievement of the objective of the Convention.

The COP may establish subsidiary bodies (ad hoc or standing), make recommendations, take decisions, adopt Protocols to the Convention or amend the treaty itself (see Figure 3.1 and Chapter 18 on Institutions).

The Convention also establishes two standing bodies, the Subsidiary Body for Scientific and Technological Advice (SBSTA) and the Subsidiary Body for Implementation (SBI). They are open to participation by all Parties as well and operate under the guidance of the Conference of the Parties (Articles 9 and 10 FCCC). As the mandates of the subsidiary bodies are rather broad and general, specific tasks have subsequently been assigned to them by the COP, which has at times caused confusion because of overlapping competences.[12] Further complicating the picture, scientific and technological advice is provided not only by the SBSTA, but also by the IPCC (Article 21.2 FCCC). In communicating the IPCC results to the Parties, the SBSTA thus serves as another political filter for the IPCC's output.

Administrative support for the functioning of the Convention and its bodies is provided by a Secretariat (Article 8 FCCC) that was initially located on a preliminary basis in Geneva. Following a decision at the first meeting of the COP (COP 1), the Secretariat moved to Bonn, Germany, in 1996. In line with the importance of the climate change issue and the many tasks the Secretariat has to perform, it has a particularly large staff. As of January 1999, about 100 people worked at the Secretariat, including administrative staff and short-term consultants. In comparison, the Secretariat serving both the Vienna Convention for the Protection of the Ozone Layer and its Montreal Protocol has been run by four or five professional staff. The FCCC Secretariat's total annual expenditure is set to exceed US $ 10 million in 1999.[13] The high number of personnel at the FCCC Secretariat reflects the complexity of the issue and the huge amount of work resulting therefrom. It also indicates that industrialised Parties attach exceptional

---

12 Decision 6/CP.1 in FCCC/CP/1995/7/Add.1; Decision 2/CP.2 in FCCC/CP/1996/15/Add.1; Decision 13/CP.3 in FCCC/CP/1997/7/Add.1.

13 See Decision 15/CP.3 in FCCC/CP/1997/7/Add.1 and UNEP/OzL.Pro.9/12, Annex IX for the ozone regime.

importance to climate change – despite their apparent reluctance to take on substantive obligations.

Figure 3.1: The System of Institutions of the FCCC Prior to Kyoto

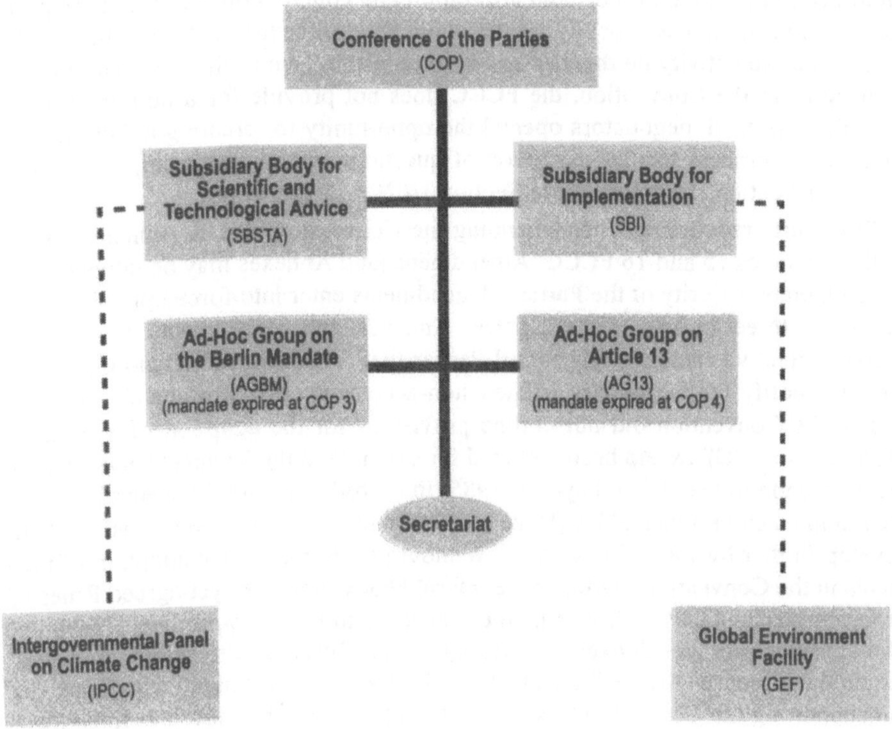

The Financial Mechanism "defined" by Article 11 of the Convention is another important element of its institutional structure. Providing evidence of a strong North-South divide, its establishment and design had been among the most controversial issues in the negotiations on the Convention. The mechanism functions under the guidance of the COP and is accountable to it. Because developing countries sought to establish an independent fund as the financial mechanism, in principle any international entity could have been entrusted with its operation. However, upon the insistence of industrialised countries, the GEF was designated as the operating entity on an interim basis. The Convention states, however, that the GEF should be restructured in order for its representation to become fair and equitable, its participation universal, and the mechanism transparent (Articles 11 and 21.3 FCCC). The arrangements between the COP and the GEF as well as the applicable policies, programme priorities and eligibility criteria were to be decided later (see below).

Furthermore, the Convention contains a mechanism for the resolution of conflicts and problems that might arise in its implementation (Article 14 FCCC). Disputes between two or more Parties may be settled by negotiations or other peaceful means, including resort to the International Court of Justice or arbitration if all the Parties involved have agreed to such a procedure. Otherwise, the dispute will be submitted to conciliation. For both arbitration and conciliation, the FCCC requires the adoption of annexes specifying the applicable procedures (Article 14.2 and 7 FCCC), but no activity on this has so far taken place. Due to the soft nature of the obligations in the Convention, the FCCC does not provide for a non-compliance procedure. Instead, negotiators opened the opportunity for creating a "multilateral consultative process for the resolution of questions regarding the implementation of the Convention" (Article 13, see Section 16.2).

Procedural requirements for amending the Convention and its Annexes are set forth in Articles 15 and 16 FCCC. Amendments and Annexes may be adopted by a three-quarters majority of the Parties. Amendments enter into force upon the ratification by three-fourths of the Parties. Annexes and amendments to Annexes become effective automatically for all Parties that do not explicitly choose to "opt out", i.e. notify the Depositary of their non-acceptance.[14] Unfortunately, negotiators of the Convention did not outline provisions for the adoption of a protocol (Article 17 FCCC), as had been included for example in the Vienna Convention on the Protection of the Ozone Layer of 1985. In accordance with the general trend in international environmental law,[15] the regime on climate change was most likely to develop further by way of a protocol. Without provisions for the adoption of protocols in the Convention, Parties were referred back to the not-yet-agreed Rules of Procedure of the COP to determine the procedure to be followed for the adoption of a protocol to the Convention. Adoption of Rules of Procedure, however, required consensus (Article 7.2(k) FCCC). This was an incentive for governments not supporting the Convention's objective to actually join because it enabled them to block consensus on the Rules of Procedure. And without rules of procedure, full consensus would also be required for the adoption of a protocol – this meant that they would also be in a position to veto such a protocol.[16]

In exchange for the "soft" language on substantive commitments, negotiators of the Convention agreed to review the adequacy of commitments at COP 1 "and periodically thereafter" (Article 4.2.b). The provisions for amending the Convention or adopting a Protocol became thus first relevant at COP 1, which was to take place within one year after the entry into force of the Convention.

---

14 On the opting-out procedure in international environmental law see especially Sand 1990.

15 See Ott 1998a.

16 This proved to be one of the major obstacles on the road to Kyoto; see Chapters 4, 7 and Section 19.3.

3.3 Development of the Convention

Several outstanding issues were placed before the first session of the COP for consideration, including the multilateral consultative process, the Rules of Procedure of the COP, the continuation of the interim arrangements for the Financial Mechanism, and criteria for Joint Implementation. Most importantly, COP 1 had to review the adequacy of commitments included in the Convention.

Before its entry into force, which required ratification by 50 countries, the Convention's institutions could not carry out the preparatory work. For the interim period, negotiators of the Convention thus agreed to a "prompt start" by mandating the Intergovernmental Negotiating Committee, the body which had negotiated the Convention, to prepare the decisions to be taken by COP 1.[17] Thereby, the process-orientation of the Convention became further pronounced and a proper transition to the operational phase of the Convention was ensured.

Although public interest has focused primarily on targets and timetables for the reduction of GHGs, a number of other important issues were dealt with and advanced by the Parties since the adoption of the Convention. For example, the development of the Financial Mechanism has been particularly important. Upon adoption of the Convention, debate continued on whether the GEF should be designated as the operating entity on a permanent basis. These discussions were deadlocked as long as the negotiations on the restructuring of the GEF remained unresolved.[18] The agreement finally reached on restructuring of the GEF in 1994 (to make its governance transparent and provide an equitable and balanced representation of all Parties),[19] was deemed insufficient by developing countries. Consequently, COP 1 prolonged the interim status of the GEF as the operating entity of the Financial Mechanism under the Convention.[20]

Before COP 1, the Parties also agreed on a three-stage process to determine funding for adaptation measures and on further policies, programme priorities and funding eligibility criteria. Concrete arrangements between the COP and the GEF were agreed after protracted discussions. These had to deal especially with the difficulty of finding a balance between keeping the nature of the COP and the GEF as formally independent institutions, while providing for close COP guidance to the GEF in matters related to the Convention. As a result, a Memorandum of Understanding between the COP and the GEF, and an annex thereto on the determination of funding necessary and available for the implementation of the Convention, were

---

17 See Resolution INC/FCCC/1992/1 in A/AC.237/18 (Part II)/Add.1; for the concept of the "prompt start" see Chayes/Skolnikoff 1992.

18 Oberthür 1996b, p. 200.

19 Ramakrishna 1994.

20 See Decision 9/CP.1 in FCCC/CP/1995/7/Add.1; on the GEF and the restructuring process see Fairman 1996.

agreed upon. Finally, COP 4 designated the GEF as the Financial Mechanism of the Convention on a permanent basis, but under regular review.[21]

A number of other issues were on the climate change agenda after Rio. The most important of these included technology transfer, the elaboration of a system for reviewing national communications and the multilateral consultative process in accordance with Article 13 of the Convention. With respect to the multilateral consultative process, COP 4 decided to approve the text elaborated by the Ad Hoc Group on Article 13, with the exception of the provisions on the composition of the Multilateral Consultative Committee envisaged therein and the applicable voting rules (see Section 16.2). As regards the system of implementation review, guidelines for reporting by industrialised countries were passed and an in-depth review procedure agreed on. Special arrangements have been agreed for different groups of countries (CEITs, developing countries, least developed countries), that are constantly being further developed (see Section 16.1). These issues are addressed in more detail in the Chapters dealing with the related provisions of the Kyoto Protocol.

---

21 Decision 10/CP.1 in FCCC/CP/1995/7/Add.1; Decision 12/CP.2 in FCCC/CP/1996/15/Add.1; Decision 13/CP.2 in FCCC/CP/1996/15/Add.1; Decision 1/SBI 4 in FCCC/SBI/1996/14, Annex I; Decision 12/CP.3 in FCCC/CP/1997/7/Add.1; Decision 3/CP.4 in FCCC/CP/1998/16/Add.1.

4. The Berlin Mandate and the AGBM Process

The political process that started immediately after the Rio Conference and included the negotiations leading to the Kyoto gathering is the subject of this Chapter. The FCCC established the playing field and some fundamental rules for the game of international climate policy. More specific rules of the game were developed later, in the ensuing process. The first session of the Conference of the Parties (COP 1), then set the stage for the Kyoto process with the adoption of the Berlin Mandate (Sections 4.1 and 4.2). The special negotiating body established to carry out the Berlin Mandate, the Ad hoc Group on the Berlin Mandate (AGBM), took further important strategic decisions (Sections 4.3–4.6). The process leading to Kyoto can be divided in two phases before and after COP 2, with the adoption of the Geneva Declaration at COP 2 being a watershed (Section 4.5). Up until Kyoto, governments used the AGBM process to define and narrow down the available options (Section 4.6). However, the most contentious issues remained unresolved before the gathering in Japan.

4.1 The Run-up to Berlin

Already at the Earth Summit in Rio in 1992, Germany offered to host COP 1.[1] This offer was accepted by the UN General Assembly and COP 1 was scheduled to take place in Berlin in spring 1995 (28 March–7 April). Six sessions of the Intergovernmental Negotiating Committee for a Framework Convention on Climate Change (INC) – the same body that was in charge of negotiating the Convention – were held in preparation for COP 1 (Table 4.1).

The negotiations started slowly, in 1992 and early 1993, primarily because the US was undergoing a change of government and it took time for the new administration of President Clinton to flesh out a strategy for the international talks. As such, the first INC sessions focused on procedural issues. One of the most important decisions was a formal one: the election of Argentine Ambassador Raúl Estrada-Oyuela as the chairman of the INC.[2] He replaced retiring Chairman Jean Ripert of France who had led the negotiations on the Convention.[3]

Many had hoped that the re-assessment of the situation by the new US administration would lead to a major shift in the US position. The new Vice President Al Gore had presented himself during the election campaign as an environmentalist especially concerned with global environmental problems, and had publicly

1 See Umwelt No. 12/1992, p. 460. "Umwelt" is a publication of the German Environment Ministry.

2 See Section 7.4 for an assessment of Estrada's role in the process.

3 See reports of the INC on the work of its sixth (UN Doc. A/AC.237/24, 4 January 1993) and seventh (UN Doc. A/AC.237/31, 22 April 1993) session.

denounced the preceding administration for its "abdication of leadership" on climate change (see Chapter 2). Other changes in the US administration, e.g. changes in personnel, further nourished the hope that the US would join the EU in a Transatlantic leadership coalition on climate change.[4] These hopes were in vain, however. Faced with substantial domestic opposition to substantial movement on climate change issues, the US delegation changed its rhetoric but not its fundamental position. The situation worsened in November 1994, when the Republican Party won a landslide victory in both houses of the US Congress, thereby virtually burying any hope of support for stringent climate policy in the US legislative branch.

Table 4.1: International Negotiating Sessions in the Run-up to Berlin

Session	Date and Venue	Main Issues Discussed
INC 6	7–10 December 1992 Geneva	Procedural questions
INC 7	15–20 March 1993 New York	Procedural questions
INC 8	16–27 August 1993 Geneva	Financial Mechanism and JI
INC 9	7–18 February 1994 Geneva	Adequacy of commitments, JI, Financial Mechanism
INC 10	22 August – 2 Sept. 1994 Geneva	Adequacy of commitments, review of country reports, JI, Financial Mechanism
INC 11	6–17 February 1995 New York	Adequacy of commitments

At the same time, prospects for effective climate policy darkened in Europe. The EU Member States were unable to agree on a proposed CO_2/energy tax, the perceived centrepiece of EU climate policy that was finally shelved in November 1994.[5] The combination of economic recession in the early 1990s and the unwillingness of several European governments to support stringent action on climate change did not bode well for EU leadership in the international process.

As a result, the process leading up to the Berlin Conference lacked leadership. With no major industrialised player pushing for new commitments, discussions proceeded at a slow pace. It was left to AOSIS to keep the discussion on substantive commitments alive by submitting a proposal for a protocol to the Convention

[4] See, e.g., Moltke 1993.
[5] Haigh 1996, pp. 164–166; Liberatore 1995, pp. 55–72.

six months prior to COP 1, which formally could have been adopted at Berlin (cf. Article 17.2 FCCC). The proposal called for a 20% reduction of CO_2 emissions of industrialised countries by 2005. It aimed at raising the expectations around the first review of the adequacy of commitments, which COP 1 was to undertake under Article 4.2.d of the Convention. This proposal was followed by one from Germany, which submitted "elements for a protocol" including a proposal for Annex I Parties to reduce emissions by "the year (x) ... by (y)%".[6]

None of these submissions were explored in any depth by the negotiating body. Instead, the INC was preoccupied with exploring the new concept of Joint Implementation (see below) and getting the institutional structure to work. Discussions on the Financial Mechanism (see Chapter 3) continued. Furthermore, as the US was unwilling to enter into new commitments, it triggered an intense debate on the design of an effective system of reporting and implementation review as a precondition for new commitments.[7] Significant progress was indeed achieved on this issue.

Finally, the Secretariat's compilation and synthesis of the first 15 "national communications" of industrialised countries injected some momentum into the discussion on the adequacy of commitments. The synthesis proved that in most of the cases reviewed, the action taken by the countries was insufficient to reduce GHG emissions to 1990 levels by 2000, the target envisaged by the Convention.[8] A special IPCC report pointed out that even if current CO_2 emissions were stabilised at the global level, atmospheric concentrations would continue to rise for at least two centuries.[9] Despite all the evidence, the OPEC countries, well-briefed by business lobbyists from the Global Climate Coalition and The Climate Council, refused to accept that the Convention's commitments be called "inadequate".

To make matters worse, the OPEC countries insisted that a protocol had to be adopted by consensus and that they should be granted a seat on the Bureau of the COP. This demand prevented agreement on the Rules of Procedure of the Conference of the Parties. Industrialised countries also contributed to stalling the talks by requesting consensus on decisions bearing financial implications. Approval of the Rules of Procedure was, until recently, a pure formality at international environmental conferences. But as governments have attached special importance to the issue of climate change, they were reluctant to give up this "veto power" without extracting any concessions.

Under these circumstances, the OPEC countries, with the tacit consent of the United States and other JUSSCANNZ delegations, were able to prevent substantive progress. Had the industrialised countries been united in their will to make progress, the OPEC countries would not have been able to pursue their obstructive course with such success. Given its general resistance to progress, the US was,

---

6 For the AOSIS protocol and the German elements paper see A/AC.237/L.23 and Add.1.

7 See Victor/Salt 1994a; p. 280; Victor/Salt 1994b.

8 See UN Doc. A/AC.237/81.

9 See IPCC 1994.

however, rather pleased to see the status quo continue without having to take the blame directly. The result was a deadlock in the negotiations prior to Berlin.[10]

4.2 The Berlin Mandate

In light of the inconclusive discussions during the preparatory process, a protocol – let alone the one proposed by AOSIS – clearly stood no chance of being adopted in Berlin. A clear mandate for negotiating, including a concrete target date for adopting strengthened commitments in 1996 or 1997, was the best-possible outcome. The fact that COP 1 in Berlin achieved this by passing the Berlin Mandate was the result of the special political dynamics at that conference. Given the lack of agreement on the Rules of Procedure, the draft rules were "applied" (not "adopted") with the exception of the voting requirements – a procedure that had also been followed in the framework of the Convention on Biological Diversity.[11] All decisions thus had to be taken by consensus.

In the end, a number of developments during the second half of the two-week COP helped to bring about the adoption of the Berlin Mandate. As press coverage and public pressure on delegations were mounting, environmental NGOs invested an enormous effort to bridge the gap between the more progressive forces of the G-77 and the EU. The host country, Germany, together with other EU Member States, cleared the way for progress by clarifying that they would not require developing country commitments to be on the agenda of protocol negotiations. On that basis, it was possible to forge a so-called "Green Group" of developing countries led by India (G-77 minus OPEC) which lent their support to a Berlin Mandate and entered into a coalition with the EU. The "green paper" on which the Green Group's position was founded was drafted, in part, by environmental NGOs. In the words of the Indian chief negotiator, it "was a rare example of co-operation between government representatives and non-governmental organizations".[12]

The heat was thus on the US and OPEC, with environmental NGOs using their excellent media contacts to make public who was to be blamed for the stalemate. Intensive negotiations between the two blocks, the EU plus the Green Group as against JUSSCANNZ plus OPEC, continued until the final hours of the Conference. During the last night, developing and industrialised country ministers involved in the negotiations were separated with the German President of COP 1, Environment Minister Merkel, engaging in "shuttle diplomacy" between the rooms.

In the face of mounting public pressure, the US delegation finally agreed to a compromise. Having lost their most powerful ally, neither Canada nor Australia

---

10 See Lanchbery 1997, pp. 7–8; on the INC process between 1992 and 1995 see in particular Oberthür 1993; Ott 1994; Oberthür/Ott 1995; Earth Negotiations Bulletin on the internet at <http://www.iisd.ca/linkages>.

11 Earth Negotiations Bulletin Vol. 9, No. 28, 28–29 December 1994.

12 FCCC/CP/1995/7, p. 23.

nor OPEC countries dared to take the blame for blocking consensus. However, Saudi Arabia, Kuwait and Venezuela voiced reservations.[13] The result was the Berlin Mandate, which was the beginning of the Kyoto process.

The Berlin Mandate (see Box 4.1) was built on the recognition that the commitments stipulated in the Convention were not adequate. To ensure support of developing countries, the mandate expressly stated that no new obligations for developing countries would be introduced in the next round of talks. This round was to be conducted by an "open-ended ad hoc group of Parties", to become known as the Ad hoc Group on the Berlin Mandate (AGBM). A clear target was set for the deliberations of the AGBM: to work out a protocol or another legal instrument that included targets and timetables for GHG emission limitations and reductions of industrialised countries. This work was to be completed in time for the adoption of the result at COP 3 in 1997, which Japan offered to host in Kyoto.[14]

Virtually no participant or observer had expected such a far-reaching result to be achieved in Berlin.[15] Concessions to the laggards were mainly of a rhetoric nature. Due to US resistance towards "negotiations," this word was replaced in the Berlin Mandate by "process". The envisaged obligations could, however, only be the result of negotiations. Furthermore, instead of "targets and timetables", the expression "quantified (emission) limitation and reduction objectives within specified time-frames" was used. Furthermore, the US managed to prevent mentioning the year 1990 as the base year of future obligations – a move that kept open the possibility of a higher baseline for potential obligations. Finally, the "process" was to be based on an "analysis and assessment" to identify appropriate policies and measures – a request introduced by the US mainly to slow down the Kyoto process – and could be contained in a protocol "or another legal instrument". In sum, however, the Berlin Mandate set the stage for a strengthening of industrialised countries' commitments to protect the global climate.[16]

In addition to the Berlin Mandate, the COP 1 decision on JI was potentially relevant to the future protocol negotiations. After extensive discussions, it was finally agreed to establish a pilot phase of "activities implemented jointly" (AIJ) to be reviewed before the turn of the millennium.[17] It thus seemed as if JI would not be on the agenda of the AGBM and would not be part of the deal to be struck in Kyoto in 1997. But given the high degree of political support for the instrument,

13 On COP 1 see FCCC/CP/1995/7 and Add.1; on the negotiations and decisions at COP 1 see especially Krägenow 1995; Oberthür/Ott 1995; Werksman/Yamin 1995; Grubb 1995a.

14 Oberthür/Ott 1995; Krägenow 1995.

15 See Lanchbery 1997, p. 8.

16 Oberthür/Ott 1995; Lanchbery 1997.

17 Decision 5/CP.1 in FCCC/1995/7/Add.1; Oberthür/Ott 1995: pp. 146–147 and Chapter 13.

and its notoriety, the concept itself nevertheless found its way onto the agenda of the protocol negotiations.[18]

Box 4.1: The Berlin Mandate (Decision 1/CP.1)

The Conference of the Parties, at its first session, (...)

Having concluded that these subparagraphs [Article 4.2.a and b] are not adequate,

Agrees to begin a process to enable it to take appropriate action for the period beyond 2000, including the strengthening of the commitments of the Parties included in Annex I to the Convention (Annex I Parties) in Article 4, paragraph 2(a) and (b), through the adoption of a protocol or another legal instrument: (...)

2. The process will, inter alia:

(a) Aim, as the priority in the process of strengthening the commitments in Article 4.2(a) and (b) of the Convention, for developed country/other Parties included in Annex I, both

– to elaborate policies and measures, as well as

– to set quantified limitation and reduction objectives within specified time-frames, such as 2005, 2010 and 2020, for their anthropogenic emissions by sources and removals by sinks of greenhouse gases not controlled by the Montreal Protocol, (...)

(b) Not introduce any new commitments for Parties not included in Annex I, but reaffirm existing commitments in Article 4.1 and continue to advance the implementation of these commitments (...); (...)

(f) Provide for a review mechanism.

3. The process will be carried out in the light of the best available scientific information and assessment on climate change and its impacts, as well as relevant technical, social and economic information (...). (...)

5. The protocol proposal of the Alliance of Small Island States (AOSIS) (...), along with other proposals and pertinent documents, should be included for consideration in the process.

6. The process should begin without delay and be conducted as a matter of urgency, in an open-ended ad hoc group of Parties hereby established, which will report to the second session of the Conference of the Parties on the status of this process. The sessions of this group should be scheduled to ensure completion of the work as early as possible in 1997, with a view to adopting the results at the third session of the Conference of the Parties.

---

[18] On the relationship between the Berlin Mandate process and JI see Werksman/Yamin 1995; Yamin 1995.

4.3 The Issues

Between COP 1 in Berlin and COP 3 in Kyoto, negotiators met eight times in the framework of the AGBM (see Table 4.2), which was chaired by former INC Chairman Estrada. COP 2 in July 1996 divided the sequence of negotiations into two parts. Three AGBM sessions preceded COP 2, which was held in conjunction with AGBM 4. The COP, as the supreme decision-making body of the Convention, could thus review progress and provide further directions for the second half of the game, namely AGBM sessions 5 to 8. From AGBM 6, negotiating sessions were held in Bonn, the new location of the Secretariat.[19]

Initially, it was unclear whether a protocol or another legal instrument, such as an amendment to the Convention, would be the result of the Kyoto process. While most countries favoured a protocol, some also considered that an amendment might be advantageous since it could be adopted by a majority of Parties (i.e. it did not require consensus). Although there was little doubt that the instrument would be legally-binding, it was uncertain whether the "quantified limitation and reduction objectives" would also be legally-binding (or phrased in "soft" language as in the Convention). Not surprisingly, the strength and timing of any targets and timetables was very controversial. Other issues concerning the character of targets and timetables also remained unresolved:

- Would the same targets be set for all industrialised countries (flat rate) or would there be differentiation?
- Would all GHGs be covered and any emission ceilings be "net" of sinks (comprehensive approach)?
- Would any target apply to a "basket" of GHGs or to single gases (gas-by-gas approach) (see Chapter 11)?

Regarding industrialised countries' obligations, the other hotly debated issue was whether the policies and measures (PAMs) to be included in the agreement would be legally-binding or not (see Chapter 10). Furthermore, the extent to which "flexibility mechanisms" would be part of the deal was unclear. These mechanisms, including Joint Implementation and international emissions trading, were demanded especially by JUSSCANNZ. The US proposed multi-year budgets instead of a single target year and intended to allow "banking" of "unused" emissions as well as "borrowing" from future emission budgets. The flexibility debate concerned also the EU's request to be able to fulfil any targets jointly, i.e. to merge Member States' individual targets under an "EU bubble" (Chapters 12–15).

Although the issue of additional developing country commitments had apparently been settled by the Berlin Mandate, some industrialised countries, and the US in particular, repeatedly called for such commitments during the delibera-

---

19 See Decision 16/CP.1 in FCCC/1995/7/Add.1.

tions.[20] Further negotiating issues concerned possible impacts of the implementation of any new commitments of industrialised countries on developing countries and institutional issues, such as the establishment of bodies of the protocol (or other legal instrument), decision-making procedures, the review of implementation and a non-compliance procedure.

Table 4.2: Chronology of the AGBM Process

Session	Dates and Venue	Characterisation/Main Developments
AGBM 1	21–15 August 1995 Geneva	Analysis and assessment
AGBM 2	30 October–3 November 1995 Geneva	Analysis and assessment, EU protocol structure
AGBM 3	5–8 March 1996 Geneva	Analysis and assessment, dispute over science/IPCC
AGBM 4 and COP 2	8–19 July 1996 Geneva	Dispute over science, US support for binding targets, Geneva Declaration
AGBM 5	9–13 December 1996 Geneva	Framework compilation of proposals for further consideration
AGBM 6	3–7 March 1997 Bonn	EU target prop., US Draft Protocol Framework, negotiating text adopted
AGBM 7	31 July–7 August 1997 Bonn	Consolidation of negotiating text
AGBM 8	22–31 October 1997 Bonn (and 30 November 1997 Kyoto)	Proposals by Japan, G-77, US, revised text under negotiation by Chairman

To resolve these issues, creative negotiating techniques were employed alongside the usual inter-governmental negotiations and a lively exchange of information was fostered in the AGBM process. From AGBM 4 to 6, for example, Chairman Estrada organised seven round-tables on various relevant topics, such as policies and measures (PAMs), quantified emission limitation and reduction objectives (QELROs) and possible impacts on developing countries of new commitments of industrialised countries (AGBM 4), on proposals from Parties (AGBM 5 and 6),

20 On the damaging effect of this demand see Chapter 17.

AIJ (AGBM 5) and differentiation (AGBM 6).[21] Without diplomatic constraints, these round-tables facilitated an open exchange of views and input by non-governmental experts. In addition, negotiating sessions increasingly served a clearinghouse function. Various side-events on relevant activities were held by international organisations, NGOs, governments and research institutes, which provided the opportunity to share information and establish contacts.

4.4 The AGBM Process: The First Year

During its first two sessions in 1995, the AGBM made little headway on substantive issues. The US and other laggard countries called for an in-depth analysis and assessment of PAMs before negotiations on any targets and timetables, launching what was described as a "largely diversionary discussion".[22] OPEC countries pushed for more studies on the possible negative economic impact of climate protection measures in industrialised countries on developing countries. The EU was stymied in internal discussions about concrete proposals on targets and timetables as well as PAMs. Some Member States intended to derive targets from PAMs, while others favoured the opposite. At AGBM 2, however, the EU was able to propose a structure for a protocol that included three annexes containing mandatory, highly recommended and voluntary PAMs.[23]

Laggard countries argued to wait for the IPCC Second Assessment Report before deciding on future commitments. The IPCC finally agreed on this report in December 1995. Against OPEC opposition, the report stated that "the balance of evidence suggests that there is a discernible human influence on global climate".[24] The IPCC also found that "10–30% energy-efficiency gains above present levels are feasible at little or no net cost in many parts of the world ... over the next two to three decades". Using most advanced technology, "efficiency gains of 50–65% would be technically feasible in many countries over the same time period".[25]

The US administration accepted and publicly supported the findings of the IPCC, which had drawn heavily on US research results. In a significant change of position at AGBM 3 in the spring of 1996, the US stated that it would now be time for "negotiations" – a term that was excluded from the Berlin Mandate less than one year ago because of US resistance. To start negotiations, the US delegation announced that it would table a position on targets at COP 2.[26] Concrete targets, however, were not submitted by the US until the autumn of 1997.

---

21 See reports of the sessions and Earth Negotiation Bulletin Vol. 12 No. 38, 22 July 1996; Vol. 12, No. 39, 23 December 1996; Vol. 12, No. 45, 10 March 1997.

22 Lanchbery 1997, p. 8.

23 For the reports of AGBM 1 and 2 see FCCC/AGBM/1995/2 and FCCC/AGBM/1995/7.

24 IPCC 1995, p. 22.

25 IPCC 1995, pp. 36–37.

26 For the report of AGBM 3 see FCCC/AGBM/1996/5; see also Earth Negotiations Bulletin Vol. 12, No. 27, 11 March 1996; Oberthür 1996a.

A split in the business community provided further evidence of the slowly changing balance of influence in the negotiations. Due to increasing pressure by "green" industry groups, business NGOs, for the first time, were unable to agree on a common position. In an effort to increase the momentum, Germany, supported by some other European delegations, proposed that industrialised countries should reduce 1990 levels of CO_2 emissions by 10% until 2005 and by 15–20% until 2010.[27] Although the German initiative indicated what the progressive EU Member States were pushing for, the EU, at this point, was still unable to resolve its internal discussion over targets versus PAMs (see above).

While forces aiming at progressive development of the regime engaged in discussions on safe maximum limits to atmospheric concentrations of CO_2 and GHGs (with ranges between 450 and 550 ppm CO_2), OPEC, Australia and Russia stood firm. None of them was prepared to accept the new IPCC conclusions as a basis of the AGBM discussions. As a result, discussions within the SBSTA were at a stalemate by the time AGBM 3 began.

4.5 Increasing the Pressure: The Geneva Declaration

COP 2 (and the concurrent meetings of the AGBM and the subsidiary bodies of the FCCC), which took place at about mid-way on the road to Kyoto, provided the opportunity to determine the future direction of the process.[28] OPEC members insisted that the prevailing uncertainties did not justify using the IPCC report as the basis for elaborating a protocol, as requested by the majority of the countries. Lobbied by the Global Climate Coalition and The Climate Council, the Russian delegation joined forces with OPEC. Australia voiced reservations as well.

As the Rules of Procedure were still not adopted and a decision based on a vote was thus impossible, the majority looked for other ways of guiding the Kyoto process. They soon coalesced around the idea of a Declaration that would be adopted by ministers at the end of COP 2. These efforts were supported by the US delegation, which followed up on its announcement made at AGBM 3. In the Ministerial Segment of COP 2, Under-Secretary Timothy Wirth called for intensified international negotiations on a "realistic, verifiable and binding medium-term emission target"[29] – language that had not even crossed the lips of US negotiators one year earlier. Taking advantage of the improved US economic situation and in an attempt to strengthen its environmental profile, the Clinton administration finally appeared to change its position on the issue of targets – thereby decisively altering the distribution of power as between progressive and laggard forces.

---

27 Ibid.

28 See FCCC/CP/1995/7, p. 41; Earth Negotiations Bulletin Vol. 12, No. 21, Monday, 10 April 1995, p. 9.

29 Speech by Timothy Wirth, Under Secretary of State for global affairs, reprinted in: Global Issues – Confronting Climate Change (Electronic Journal of the U.S. Information Agency), April 1997, Vol. 2, No. 2, 6–8.

Box 4.2: The Geneva Ministerial Declaration

The Ministers and other heads of delegations present at the second session of the Conference of the Parties to the United Nations Framework Convention on Climate Change, (...)

2. *Recognize and endorse the Second Assessment Report of the Intergovernmental Panel on Climate Change (IPCC) as currently the most comprehensive and authoritative assessment of the science of climate change (...). Ministers believe that the Second Assessment Report should provide a scientific basis for urgently strengthening action at the global, regional and national levels, particularly action by Parties included in Annex I to the Convention (Annex I Parties) to limit and reduce emissions of greenhouse gases, and for all Parties to support the development of a Protocol or another legal instrument; (...)*

8. *Instruct their representatives to accelerate negotiations on the text of a legally-binding protocol or another legal instrument to be completed in due time for adoption at the third session of the Conference of the Parties. The outcome should fully encompass the remit of the Berlin Mandate, in particular:*
   - *commitments for Annex I Parties regarding:*
     * *policies and measures including, as appropriate, regarding energy, transport, industry, agriculture, forestry, waste management, economic instruments, institutions and mechanisms;*
     * *quantified legally-binding objectives for emission limitations and significant overall reductions within specified time-frames, such as 2005, 2010, 2020, with respect to their anthropogenic emissions by sources and removals by sinks of greenhouse gases not controlled by the Montreal Protocol;*
   - *commitments for all Parties on continuing to advance the implementation of existing commitments in Article 4.1;*
   - *a mechanism to allow the regular review and strengthening of the commitments embodied in a protocol or other legal instrument;*
   - *commitments to a global effort to speed up the development, application, diffusion and transfer of climate-friendly technologies, practices and processes (...).*

9. *Welcome the efforts of developing country Parties to implement the Convention (...); and call on the GEF to provide expeditious and timely support to these Parties (...);*

10. *Recognize that the continuing advancement of existing commitments by developing country Parties (...) requires determined and timely action, in particular by Annex II Parties. Access to financial resources and to environmentally-sound technologies (...) will be most critical; (...)*

The US shift left OPEC (with Russia) isolated. This eventually enabled agreement on the Geneva Ministerial Declaration, in which ministers and other heads of delegations acknowledged the IPCC report as a scientific basis for urgently strengthening action, and instructed the AGBM to accelerate the negotiations. The Geneva Declaration extended the agreement reached at COP 1 by clarifying not only that the instrument resulting from the AGBM process would be legally-binding, but so too would be targets and timetables therein (Box 4.2).

The procedure for adopting the Geneva Declaration proved to be as important as its content. The Ministers who signed on to the Declaration wanted COP 2 to "endorse" or "take note of" it, so as to enhance its importance for the future negotiations. It was obvious that OPEC members and Russia (supported by Australia) would try to block any such move in the COP.[30] The overwhelming majority of Parties, however, decided that this small minority would not be allowed to hinder progress. With no agreed rule on voting in the COP, another procedure was found that satisfied the formal consensus requirement, despite the opposition of a small number of Parties.

The solution was to adopt the Declaration by "consensus minus x". Skilfully supported by the Secretariat, the President of the Conference, Chen Chimutengwende of Zimbabwe, proposed to "take note" of the Declaration in the final plenary meeting. Attempts to object to the procedure were swept away by the sustained applause of the great majority of delegations (and observers) supporting the President. The opponents could only file reservations (Australia), make qualifications (New Zealand) or object to the Declaration (OPEC and Russia), but they could not prevent its formal recognition by COP 2. Proving its ability to act against a minority, even in the absence of formal voting rules, the majority thus sent a strong signal to OPEC and Russia that it was unwilling to let these countries block progress on formal grounds.[31]

4.6 Approaching Kyoto

COP 2 took the important decision to accept Japan's offer to host COP 3 in Kyoto at the beginning of December 1997.[32] This improved the prospect of arriving at a reasonable compromise at Kyoto, because the host country was likely to face increased international as well as domestic pressure to secure a successful outcome. One consequence of this was that the Japanese delegation started immediately to hold meetings with the international environmental NGOs to learn more about world public opinion.

---

30 These countries were not represented at ministerial level and were not part of "other heads of delegations" (note the missing "the") mentioned in the first line of the Geneva Declaration.

31 For a more detailed account of COP 2 see FCCC/CP/1996/15 and Add.1; Oberthür 1996b; Earth Negotiations Bulletin, Vol. 12, No. 38, 22 July 1996.

32 See Decision 1/CP.2 in FCCC/CP/1996/15/Add.1.

COP 2 did not lead to an immediate acceleration of negotiations. At AGBM 5, in December 1996, neither the US nor the EU put forward any concrete proposals for targets and timetables. Instead, submissions of Parties focused on the general approach to be taken in the Protocol (or other legal instrument). After their defeat at COP 2, OPEC countries modified their strategy; instead of attempting to obstruct the protocol negotiations, they tried to broaden support for their position among G-77 by arguing costs to developing countries that might arise from action by industrialised countries had to be taken into account. AGBM 5 concluded by requesting the Chairman to prepare a framework compilation of all existing proposals so as to provide a uniform basis for the negotiations.[33]

Preceded by a one-day informal meeting on differentiated targets (1 March 1997), AGBM 6 intensified discussions by establishing two "non-groups" on commitments and on institutions, and adopted a negotiating text in March 1997.[34] EU environment ministers, co-ordinated by the Netherlands holding the EU Presidency, then played their first trump card. Following a last minute decision by the EU Environment Council on 3 March 1997, they proposed to cut the combined emissions ("basket") of CO_2, CH_4 and N_2O in industrialised countries by 15% by 2010 and backed this proposal with an internal burden-sharing agreement.[35] By contrast, JUSSCANNZ and the US in particular, were engaged in examining how to provide for maximum flexibility to Parties in implementing future obligations. The US introduced a "Draft Protocol Framework", that it had submitted in January 1997, including proposals for emission budgets, JI, emissions trading, banking and borrowing.[36] As any new proposals on substance (except pure numbers) had to be submitted by 1 June 1997 in order to meet the six-months deadline pursuant to the Convention, pressure on JUSSCANNZ was mounting. After the deadline, Chairman Estrada attempted to keep up the pressure by holding informal consultations with key delegations on industrialised countries' commitments in Bonn 16 and 17 June 1997.

One of the major negotiating issues pushed by a number of industrialised countries at AGBM 6 and subsequently was developing country commitments, even though these had effectively been excluded from the agenda by the Berlin Mandate. Although Chairman Estrada decisively ruled that proposals on "graduation" and "evolution" fell outside the terms of the Berlin Mandate on several occasions, the US, in particular, demanded that the "advancement of existing commitments in Article 4.1" pursuant to the Berlin Mandate be interpreted to allow for new com-

---

33 For the report of AGBM 5 see FCCC/AGBM/1996/11; for the framework compilation see FCCC/AGBM/1997/2 and Add.1.

34 For the report of ABGM 6 see FCCC/AGBM/1997/3; Ehrmann/Oberthür 1997; for the negotiating text see FCCC/AGBM/1997/3 and Add.1.

35 European Union, General Secretariat of the Council, Meeting Document CONS/ENV/97/1 Rev.1 (SN/11/97 Rev.1), Brussels, 3 March 1997; EWWE, Vol. 6, No. 5, 7 March 1997, pp. 7–9; see below Chapter 6.

36 FCCC/AGBM/1997/MISC.1.

mitments from developing countries. It proposed an Annex B (of the envisaged Protocol) for countries without legally-binding targets that would voluntarily adopt emission limitations. The US further proposed that commitments for all Parties should be agreed by 2005. The EU had introduced an Annex X containing industrialised countries plus new OECD members.[37] Prior to AGBM 7, Chairman Estrada invited key delegations to Geneva for informal consultations on the issue, which, however, did not result in a reconciliation of the diverging views. At that point, the industrialised countries' demands ran the risk of wrecking the entire process.

Still under the Presidency of the Netherlands, the EU Environment Council complemented its position in June 1997 by proposing an intermediate cut of GHG emissions of industrialised countries by 7.5% by 2005 compared to 1990 levels.[38] This proposal was submitted to AGBM 7 in August 1997, when JUSSCANNZ members still had not tabled any numbers. Deliberations then proceeded at an accelerated pace within four informal "non-groups" on: targets and timetables, PAMs, advancement of commitments under Article 4.1 of the Convention, and institutions and mechanisms. These groups drafted text and reported back to the Chairman. However, given the lack of numbers, the discussions remained inconclusive. It was thus left to the Chairman to prepare a consolidated negotiating text for the last pre-Kyoto negotiating session, on the basis of available material (see Table 4.3).

With only one session of the AGBM left before the Kyoto conference, time for elaborating an agreement was running out. Although JUSSCANNZ were playing for time, they needed to put forward their own proposals prior to AGBM 8, if they wanted their views to be reflected in a Kyoto Protocol (or another legal instrument). Japan, wishing to secure a successful outcome at the Kyoto conference, proposed a differentiated formula for reducing emissions of a basket of CO_2, CH_4 and N_2O by up to 5% until 2010.[39] Although insufficient in ecological terms, this proposal paved the way for an eventual differentiation of commitments and was thus an important contribution.

On the opening day of AGBM 8, the G-77 and China tabled another proposal. This proposal called for industrialised country gas-by-gas reduction targets relating to CO_2, CH_4 and N_2O with reductions of 7.5% by 2005, 15% by 2010 and 35% by 2020. OPEC's concerns were balanced by requests for minimising adverse impacts on developing countries and establishing a Compensation Fund (for countries negatively affected by industrialised countries' action on climate change).[40] Many observers had felt that, up to this point, too little recognition had been given to the concerns of developing countries, which were not to take on new commitments but nevertheless had to be part of the final deal. However, given the differing interests

37 FCCC/AGBM/1997/MISC.1.

38 EWWE, 20 June 1997 (special issue), p. 2.

39 FCCC/AGBM/1997/MISC.1/Add.6 and 10; see also Chapter 11.

40 FCCC/AGBM/MISC.1/Add.6, 8, 10; see also Chapter 11.

of developing countries, most industrialised countries were caught by surprise by the G-77 initiative.

Table 4.3: The Sequence of Negotiating Texts on the Way to Kyoto

Date	Text	Document No.
19 November 1996	Synthesis of proposals by Parties	FCCC/AGBM/1996/10
February 1997	Framework compilation of proposals from Parties	FCCC/AGBM/1997/2 and Add.1
April 1997	Negotiating text	FCCC/AGBM/1997/3/Add. 1
October 1997	Consolidated negotiating text	FCCC/AGBM/1997/7
November 1997	Revised text under negotiation	FCCC/CP/1997/2

The stage was thus perfectly set for the US – the one major player whose input was still missing. Shortly after the announcement of the G-77 and China's position, the AGBM meeting room became almost completely deserted as most participants rushed to television sets to watch the expected disclosure of the US position by President Clinton on CNN. In his speech, President Clinton proposed to return emissions of a basket of CO_2, CH_4, N_2O, HFCs, PFCs and SF_6 to 1990 levels in the budget period 2008–2012, and to reduce them afterwards. In addition, he demanded the inclusion of JI and emissions trading in the Protocol and, although falling outside the Berlin Mandate, stated that "(T)he United States will not assume binding obligations unless key developing nations meaningfully participate in this effort".[41] The next day, the US delegation officially submitted this position to the AGBM.[42]

While proposals from all major players (except Russia) were thus on the table, the positions differed substantially. AGBM 8 largely failed to show the way to close the gap, but Parties did narrow down the options.[43] Agreement had been reached on the Preamble and some institutional questions, including the Secretariat, the Subsidiary Bodies, dispute settlement, amendments to the instrument and the Multilateral Consultative Process (see the respective Chapters of Part II). On some other subjects, e.g. methodologies, reporting and the review of implementation, Parties had come close to agreement. A bargain on the most contentious

41 Remarks by the President on Global Climate Change, National Geographic Society, Washington, D.C., 22 October 1997, reprinted in: Environmental Policy and Law, Vol. 27, No. 6, pp. 503–504; available at <http://www.whitehouse.gov/Initiatives/Climate/199710226127.html> as of 26 May 1998.

42 See FCCC/AGBM/1997/8; Earth Negotiations Bulletin, Vol. 12, No. 66, 3 November 1997.

43 Ibid.

issues, however, was still not in sight. On this basis, the Chairman was instructed to produce a draft Protocol as a "revised text under negotiation" for COP 3 (see Table 4.3). Thus, most of the crucial issues needed to be resolved at Kyoto (see Chapter 7) or during the coming inter-sessional period. Many influential delegations also asked Estrada to "produce additional text", indicating their approval of the Chairman's talks behind-the-scenes.

5. Outside the Climate Arena: Multilateral and Bilateral Diplomacy

Mitigating climate change requires changing the dominant pattern of economic and social development. As such, it is one of the most important issues on the sustainable development agenda, cutting across all areas of society. Consequently, the Kyoto process received a degree of attention of world leaders that has been unprecedented in international environmental affairs. Unlike other environmental issues, "foreign environmental policy"[1] on climate change has involved foreign ministers and Heads of State, elevating climate change to "high politics".

Climate diplomacy before Kyoto extended far beyond the AGBM process. In fact, there was hardly any high-level international meeting in the end-phase of negotiations where heads of state, foreign ministers, economic ministers, environment ministers and others did not "talk climate". In addition, many high-level bilateral meetings in the last months of negotiations focused solely on reaching an agreement in Kyoto. The following sections sketch out relevant developments within international organisations, deals with climate-related events of the UN General Assembly Special Session in June 1997 and tracks some of the most important regional and bilateral activities.

5.1 International Organisations

Governments used various international organisations as negotiating fora on climate change, and several international organisations also provided substantive input into the Kyoto process. Of these organisations, the OECD and the GEF were of special importance and are dealt with below. The World Bank's role in furthering JI by establishing a Carbon Investment Fund (CIF) within the Bank's Global Carbon Initiative and UNCTAD's effort to develop an emissions trading scheme also deserve to be mentioned (see Chapters 14 and 15).

The OECD had already begun to serve as an important forum for negotiations in respect of the FCCC in 1992. During an OECD meeting in April 1992, the US declared that it would not accept any binding targets and timetables in the Convention.[2] During the Kyoto process, the OECD provided the forum for the "Annex I Expert Group", i.e. a group of experts from industrialised countries. The Expert Group held workshops on specific issues and regular meetings, approximately twice a year, to discuss policy issues of special concern to industrialised countries. Several working papers were prepared to assist the Expert Group.[3] Its work and internal conflicts very much mirrored the international process, because

---

1 Prittwitz 1984.

2 Oberthür 1993, pp. 52–53; Bodansky 1993, pp. 489–491.

3 See <http://www.oecd.org/env/cc/freedocs.htm> as of December 1998.

the overall political dynamics were such that the main contentious issues had to be resolved first within this group of countries prior to settlement at the global level. Thus, the Expert Group focused on various PAMs until COP 2 when JUSSCANNZ pressed for a change of focus. Subsequently, it provided input to the discussion on emissions trading, which also remained controversial in the Expert Group. In sum, the OECD proved to be a useful forum for discussion among industrialised countries, but could not heal the divides within OECD.

The GEF provides grants and concessional funds to developing countries and CEITs for projects that aim to protect the global environment. Launched in 1991 as a pilot programme implemented jointly by the World Bank, UNDP and UNEP, the GEF focuses on four global environmental issues, namely climate change, biodiversity, depletion of the ozone layer, and international waters. As the entity operating the Financial Mechanism of the FCCC, the GEF is the main multilateral financial institution for financing GHG mitigation and adaptation measures in developing countries. About US $ 2 billion had been pledged by industrialised countries for the budget period 1994–1997. In 1995, 35% of the available US $ 766 million were used for climate protection measures. In 1995–1996, 45 out of 86 GEF projects were related to climate change.[4]

In the GEF Council, discussions on the second replenishment of the GEF Trust Fund had started in early 1997 and thus overlapped with the Kyoto process. As the GEF was operating as the Financial Mechanism of the Convention, the replenishment would be decisive for the level of funding available over the next years for mitigating, and adapting to, climate change in developing countries. However, agreement on the replenishment could not be reached before Kyoto.[5] This did not bode particularly well for the Kyoto process. It was bound to sharpen developing country scepticism about the intentions of industrialised countries, who were asking them to take on enhanced commitments under the FCCC pro-cess while turning down requests for more funding under the GEF.[6]

5.2 Rio + 5: UNGASS

The schedule of diplomatic events leading up to Kyoto provided a unique opportunity for world leaders to push along the climate talks: the UN General Assembly Special Session (UNGASS, 23–27 June 1997), which was to review the implementation of the decisions taken at the Earth Summit in 1992. However, the major players were not willing to show their cards before the endgame approached and the formal results at UNGASS did not further the climate negotiations. The

4 Fairman 1996, p. 85; Ehrmann 1997, p. 594; GEF Doc. GEF/C.9/Inf.7, 31 March 1997.

5 GEF Doc. GEF/C.10/Inf.4, 3 October 1997.

6 In February 1998, a replenishment of US $ 2.75 billion was agreed for the GEF for 1998–2002; see World Bank Press Release No. 98/1637S available at <http://www.worldbank.org/html/extme/1637.htm> as of December 1998.

part of the final UNGASS declaration that dealt with climate change merely summarised the current state of negotiations by stating that

> "there is already widespread but not universal agreement that it will be necessary to consider legally-binding, meaningful, realistic and equitable targets for Annex I countries that will result in significant reductions in greenhouse gas emissions...".[7]

Nevertheless, important developments with respect to climate change did occur at UNGASS. First, the Heads of State of Germany, South Africa, Brazil and Singapore demonstrated their ambitions by launching an initiative to establish a Global Environmental Umbrella Organisation and an agreement on GHG emission reductions in Kyoto.[8] Thus, important developing countries from different regions signalled their support for a strong agreement in Kyoto.

Second, and probably more importantly, US President Clinton took a firm stance on climate change in his statement. He not only commended European leadership on climate change, but clarified that his own inability to exert international leadership was due to domestic conditions, in particular the reluctant US Congress. However, he committed himself to work to "convince the American people and the Congress that the climate change problem is real and imminent".[9] He thus sent a strong political signal to his opponents as well as to the negotiating partners that the US administration was determined to strike a deal in Kyoto.

Finally, the performance of Japan at UNGASS was subject to special scrutiny by observers because of its important role in hammering out a deal in Kyoto. Japan had a prominent role by co-chairing a ministerial group on climate change at UNGASS (along with Argentina). However, it hardly mediated between the EU and the US and objected to a stronger wording in the final declaration. This raised doubts about the ability and willingness of the host of COP 3 to press for a strong agreement in Kyoto. However, being criticised by environmental NGOs and "world public opinion" for its performance at New York gave the Japanese government a strong taste of what was expected from them in Kyoto.[10]

5.3 Regional Developments

Many of the key governments met in regional frameworks and other smaller groupings. In a G-7 Expert Group on Environment and Natural Resources Management,[11] the world's richest countries exchanged views on climate change.

---

7 See UN Doc. A/S-19/AC.1/L.1/Add.1.

8 See "Sonderteil: Ergebnisse der VN-Sondergeneralversammlung zur Überprüfung der Umsetzung der Rio-Ergebnisse im Juni 1997 in New York," in: Umwelt, No. 9/1997.

9 Remarks by the President in Address to the United Nations Special Session on Environment and Development, 26 June 1997, available at <http://www.whitehouse.gov/Initiatives/ Climate/ 19970909-3216.html> as of 26 May 1998.

10 On UNGASS see Beyerlin/Ehrmann 1997; Earth Negotiations Bulletin Vol. 5, No. 88, 30 June 1997.

11 G-7: Canada, France, Germany, Italy, Japan, UK, USA.

The global environment, and especially climate change, had been on the agenda of the G-7 summits since the late 1980s. Due to the high profile of climate change before Kyoto, it figured prominently on the agenda of the 1997 G-7 Summit held in Denver just prior to UNGASS. Russian President Boris Yeltsin participated officially in such a summit for the first time, making it the G-8 (also sometimes referred to as the G-7+1). Like their negotiators in the AGBM process, European Heads of State and Government canvassed support from other members of the G-8 for the EU proposal of a 15% cut in GHG emissions by 2010. With three JUSSCANNZ members participating, the final communiqué fell short of that goal when participants expressed their intention "to commit to meaningful, realistic and equitable target that will result in reductions of greenhouse gas emissions by 2010".[12] The Denver Summit, however, marked another turning point of the US stance in the Kyoto process. President Clinton reportedly was impressed by the vigour of the concerted European request and realised the seriousness of the issue. Consequently, he became personally engaged and his strong speech delivered at UNGASS was the first result of this change.

The South Pacific Forum in September 1997 assembled two groups of countries with very different interests and views: Australia and New Zealand, as two industrialised laggards, and 14 Pacific AOSIS members. When these island states demanded to see their concerns reflected in the Forum's official communiqué, a diplomatic stand-off with Australian Prime Minister Howard occurred. He insisted instead that the reference to GHG emission reductions be deleted from the communiqué. Being economically dependent on Australia and New Zealand (their biggest donors of foreign aid), the attending small island states were, as Greenpeace put it, "bullied into submission".[13]

Australia was more comfortable in the framework of the Asia-Pacific Economic Cooperation (APEC) whose leaders met for three days at the end of November 1997 in Vancouver, Canada. Japan, Canada, Australia, New Zealand and the US met with several Asian developing and newly industrialising countries, including China, South Korea, Indonesia, Malaysia and the Philippines. On the eve of the Kyoto Summit, the APEC meeting provided JUSSCANNZ members with the opportunity to co-ordinate their strategies and to enlist support from important developing countries. Not surprisingly, no mention was made of limiting, let alone reducing, GHG emissions in the final statement. Instead, the APEC summit simply recognised "the importance of accelerating action on a global level to deal with emissions of greenhouse gases".[14]

---

12 Denver Summit of the Eight, Communiqué, Denver, 22 June 1997, available at <http//www.g7.com/comm.htm> as of 17 March 1998.

13 "Australia in Hot Water over Global Warming Stance," 20 September 1997, available at <http://CNN.com/WORLD/9709/20/pacific.forum/index.html>.

14 APEC Economic Leaders Declaration: Connecting the APEC Community (Vancouver, Canada), 25 November 1997, available at <http://www.apecsec.org.sg/econlead/vancouver.html> as of March 1998.

Other relevant regional gatherings in 1997 included the Commonwealth Heads of Government Meeting held in Edinburgh in November 1997, which supported "realistic and achievable goals, significant reductions in greenhouse gas emissions and recognition that we all need to play a role".[15] Several conferences dealing with various aspects of climate change, such as JI and emissions trading were convened prior to Kyoto in all regions (including some cross-cutting fora). As said above, there was hardly a high-level meeting in 1997 that did not deal with global warming, at least in part.[16]

5.4 Bilateral Diplomacy

Laggards and leaders in climate change also engaged in extensive bilateral diplomacy on climate change, in order to explore how much room to manoeuvre existed and to enlist the support of other countries. Due to the importance of the US in climate politics, and its new willingness to come to a meaningful agreement, European policy-makers put special emphasis on improving transatlantic communication. Among numerous other meetings and activities, a high-level European delegation met with chief US negotiators Stuart Eizenstat and Timothy Wirth after the last AGBM session in November 1997, to explore the degree of common ground and to raise the US's proposed stabilisation target. The delegation included EU Environment Commissioner Ritt Bjerregaard and environment ministers from the EU "Troika", i.e. the current EU Presidency Luxembourg, as well as its predecessor (the Netherlands) and successor (the UK). The UK environment minister John Prescott subsequently travelled to Australia and New Zealand to build on the UK's traditionally close relations with the Commonwealth countries.[17]

While these European efforts did not yield immediate results, bilateral US climate diplomacy was more effective. When touring the Latin American countries in mid-October, where US influence has traditionally been strong, President Clinton solicited wide support for including the JI concept in a Kyoto agreement. In Argentina, President Menem even issued a joint statement with the US President, which acknowledged that any meaningful solution to the problem of global warming had to address the question of developing countries emission limits. For the first time, a large and influential developing country indicated publicly that concrete emission limits for developing countries might be acceptable. Not surprisingly, President Menem was the subject of serious criticism from other developing countries for this move, since it undermined the agreed G-77 position. The US

15 Commonwealth Secretariat, Promoting Shared Prosperity: Edinburgh Commonwealth Economic Declaration, Commonwealth Heads of Government Meeting, Edinburgh 1997, para. 10 (on file with authors).

16 See, for example, the descriptions of environmental meetings provided in /linkages/journal/, 2 (1997) 2–4 available at <http://www.iisd.ca/linkages/journal> as of December 1998.

17 EWWE, Vol. 6, No. 21, 7 November 1997, p. 15.

administration also engaged in consultations with other developing countries. Little concrete results were reported.

The diplomatic efforts of the US were balanced and complemented by similar acts by EU Member States. For example, German environment minister Angela Merkel visited China and Japan in August 1997 to try to narrow the differences in approach towards the Kyoto conference. Other German high-level environmental officials met with counterparts from Russia and other CEITs.[18] With respect to developing countries, the UK and France consulted their former colonies. The UK also revived its ties with the US, thereby causing fears that they might tacitly break ranks with the EU and instead broker a compromise solution, as they did in the FCCC negotiations in 1992.[19]

OPEC, and Saudi Arabia in particular, were the target of a number of diplomatic activities. German officials visited Saudi Arabia in mid-1997. More influential might have been talks in Riyadh in autumn 1997 between the US administration and Saudi Arabia. Highest diplomatic levels became involved when Japanese Prime Minister Hashimoto visited Saudi Arabia at the beginning of November 1997. Whereas the US leverage on Saudi Arabia mainly flows from the security situation in the Gulf region, Japan's impact is based on strong economic ties. While promising to deepen economic relations between the two countries further, Hashimoto asked for constructive co-operation from Saudi Arabia in Kyoto. Crown Prince Abdullah indeed indicated support for the Japanese position at Kyoto, while urging Japan to buy more crude oil from Saudi Arabia. In response, Hashimoto promised to look into the situation of oil imports carefully and agreed to establish a bilateral commission, hold regular talks at the level of vice-foreign ministers, and launch negotiations on an investment protection agreement to boost Japanese investment in Saudi Arabia.[20]

---

18 See Umwelt No. 1/1997, p. 6; Umwelt No. 10/1997, p. 399; Umwelt No. 11/1997, p. 447.

19 Bodansky 1993, pp. 489–491.

20 "Fahd appears to favor renewing oil concessions in talks with Hashimoto", Daily Yomiuri On-Line, November 1997 (on file with authors).

6. Changing the Balance: Governmental and Non-governmental Developments

The complexity of the Kyoto process was not only due to the many layers of diplomacy. Governments also had to play domestic policy games. Indeed, it was primarily the major players' domestic options and constraints that influenced the prospects for a successful outcome at Kyoto. Given that the Kyoto Protocol was designed to introduce targets and timetables for *industrialised countries*, the following section will elaborate some developments in the European Union, the US, and the other JUSSCANNZ members. As developments within the non-governmental world also had an important impact in the run-up to Kyoto, the chapter will conclude with a description of the diverse developments in the business community and the activities of environmental NGOs.

6.1 Strengthening EU Leadership

The game before Kyoto was particularly complex for the EU Member States, because they had to co-ordinate their behaviour at the European level in addition to designing their national positions. After UNCED, the activities of a few countries within the Union, including Germany, the Netherlands, Denmark and Austria, had largely driven the EU's stance. However, EU leadership aspirations in the Kyoto process had suffered from the fact that its own emission levels benefited from the effects of German unification in 1990. The massive closure of inefficient East German industry and the subsequent enormous investments in infrastructure and housing resulted in a fall of German CO_2 emissions by about 12% by the mid-1990s. Alluding to the fall of the Berlin wall in November 1989, this drop was termed "wall-fall profits" by environmental NGOs.[1] As these reductions had been achieved without any special effort aimed at climate protection, the EU's negotiating partners were able to point out that special circumstances allowed Germany and the EU to demand emission reductions.[2]

On the way to Kyoto, the EU's credibility was increasingly enhanced because it worked out a more coherent climate strategy. Several factors contributed to this achievement. At the EU level, Member States had intensified their co-ordinating efforts after COP 1 in Berlin by establishing a special EU Ad hoc Group on Cli-

---

1 By contrast, CO_2 emissions in the "old" West German Länder increased slightly during the first half of the 1990s, see Germany 1997; see also Ziesing et al. 1998; Beuermann/Jäger 1996.

2 Attempting to counter the argument that the "wall-fall profits" had been acquired "for free", German officials pointed to the high costs of unification, since more than DM 750 billion (US $ 450 billion) had been transferred to the "new" Länder between 1991 and 1996 alone. Informally, they stated their willingness to share the emission reductions with anyone willing to shoulder a corresponding part of these costs.

mate Change, served by an Expert Group on Common and Co-ordinated Policies and Measures. These groups met regularly (every month) to review existing policies and explore future options. This internal institutional arrangement improved the basis for developing a common understanding and for co-ordinating the EU position prior to Kyoto.

Table 6.1: The EU Burden-sharing Agreement of March 1997

Member State	Commitment by 2010
Austria	-25.0%
Belgium	-10.0%
Denmark	-25.0%
Finland	0.0%
France	0.0%
Germany	-25.0%
Greece	+30.0%
Ireland	+15.0%
Italy	-7.0%
Luxembourg	-30.0%
Netherlands	-10.0%
Portugal	+40.0%
Spain	+17.0%
Sweden	+5.0%
United Kingdom	-10.0%
EU-Total	9.2%

Encouraged by the positive experiences resulting from its role in other environmental negotiations, the European Commission requested a negotiating mandate to co-ordinate the EU position. The Environment Council declined to grant it, however, because many governments were opposed to the idea of transferring even more competence to the EU level.[3] EU policy-making on climate change thus remained dependent to a large degree on the country holding the rotating Presidency of the Council. It is up to the Presidency to determine the agenda of the Council, chair sessions of preparatory groups and to co-ordinate and represent the EU and its Member States at international negotiations. Limited progress was

3 See European Commission 1996d. This refusal has severely restricted the EU's capacity to play a strong role in the final negotiations.

made in 1996 when Italy and Ireland held the Presidency, as neither country was known for a progressive stance on climate change. When the Netherlands took over at the beginning of 1997, however, climate was high on the agenda of the two meetings of the Environment Council under their Presidency, both of which were scheduled to take place prior to AGBM 6 and 7.

Driven by the Dutch commitment, decisive progress was made at both meetings of the Environment Council. The EU Expert Group, under the leadership of Dutch experts,[4] concluded in early 1997 that emissions of CO_2, CH_4 and N_2O in the Union could be reduced significantly by 2010, if existing and planned policies were implemented. As a result, the Environment Council agreed to propose an emission reduction of these three GHGs by 15% by 2010. Surprisingly, even though the Presidency was one of the strongest supporters of PAMs, the conflict within the EU over PAMs versus targets was actually resolved in favour of targets. The Council also put forward a scheme for internal burden-sharing among Member States (see Table 6.1).[5] Although the combined commitments only added up to a reduction of 9.2%, this burden-sharing agreement enhanced the credibility of the EU proposal by showing that it was backed by arrangements for its implementation. In June 1997, the Council was able to agree to an intermediate 7.5% target by 2005.[6]

The European Commission further complemented the progress of the Environment Council. Soon after AGBM 6 in March 1997, the Commission requested an increase of resources for its ALTENER programme to promote renewable energy. Concurrently, the Commission tabled new or revised legislative initiatives to introduce and increase minimum tax levels on energy products (oil, coal, natural gas, kerosene and electricity) and to require application of demand-side management techniques in the electricity and gas sectors.[7] In addition, the Commission outlined strategies and prospects for emission reductions in the last year of the Kyoto process. This included

- a strategy paper on options for reducing CH_4 emissions by 40% by 2010,
- a communication specifying action in the energy sector,
- the publication of the EU's approach for Kyoto identifying options to reduce the emissions of the three main GHGs, and
- a White Paper outlining a strategy and an ambitious action plan for promoting the use of renewables.[8]

---

4 See Blok et al. 1997.

5 "EU Calls for 15% Climate Emissions Cut By 2010", EWWE, 7 March 1997, 7–9; European Union (General Secretariat of the Council), Meeting Doc. CONS/ENV/97/1 Rev.1 (SN/11/97 Rev.1), Brussels, 3 March 1997.

6 EWWE, 20 June 1997 (special issue), p. 2.

7 European Commission 1997d; 1997e; 1997f.

8 European Commission 1996b; 1997b; 1997c; 1997d.

EU leadership and coherence also benefited from governmental changes in 1997 in two major Member States: the UK and France. While the former conservative UK government of Prime Minister Major had been sceptical of a stringent EU policy on climate change, the new Labour government under Tony Blair was more inclined to play a pro-active role in European environmental policy-making. To implement an election campaign promise, the new government set a far-reaching goal of reducing CO_2 emissions by 20% by 2010. At the EU level, the new UK government became more co-operative and increasingly joined the leading Member States, thereby changing the balance in EU climate policy-making (although the UK remained hesitant about passing new European legislation on climate change).[9]

The change in France (May 1997) pointed to the same direction, although the French Socialists led by Lionel Jospin had demonstrated scepticism about European integration in their election campaign. While being concerned about the social implications of the Euro – the common European currency agreed to by the previous government – the socialists were thought to give more weight to the environment. The change in the position on climate change took time, however, as the new government first had to reassess existing French climate policies.

The Environment Council, the European Commission and the Member States, thus, increasingly pulled together in the direction of strengthened action on climate change. The enhanced credibility of the Union at the international level that resulted was again undercut, however, by the EU's request for the right to merge the obligations of its Member States' under an "EU bubble", allowing for differentiation of commitments within the Union. While from the European point of view the bubble was the logical consequence of the special division of competence between the EU and its Member States, the proposal rendered the EU position inconsistent. While the EU asked all other industrialised countries to commit to flat-rate emission reductions, it requested internal differentiation, including the right to increase emissions for some Member States. Furthermore, the EU claimed the special right to change the internal burden-sharing agreement until five years before the expiry of the commitment period. Hardly surprising, JUSSCANNZ countries immediately attacked this vulnerable point in the EU position, which provided a welcome opportunity to undermine EU leadership at the climate talks and thus ease the pressure on themselves (see Chapter 12).

6.2 Developments in the US

The political scenario in the US was complex, in that President Clinton and the US administration in general were prepared to enter into a reasonably strong international agreement, but faced fierce opposition by industry and domestic political forces. After raising its environmental profile in the re-election campaign

---

9 See EWWE, Vol. 6, No. 10, 16 May 1997, p. 13 and Vol. 6, No. 14, 18 July 1997, p. 9.

in 1996, the Clinton-Gore administration stepped up its domestic efforts on climate change in 1997. In February 1997, the White House Climate Change Task Force was established under the US Council on Environmental Quality and the National Economic Council to work with the public on the administration's position on climate change mitigation. In April 1997, the President's Council on Sustainable Development created a task force on climate change to advise the President on efficient domestic policy options and activities that could reduce GHG emissions.[10]

However, the main effort followed President Clinton's speech and commitment at UNGASS in June 1997 (see Chapter 5 above). The US administration designed a campaign, the White House Initiative on Global Climate Change, to raise public awareness about the issue and generate support for action. The starting point was a meeting of President Clinton and Vice President Gore with seven scientists, including three Nobel Laureates, on 24 July 1997 at the White House. The campaign included regional conferences held across the United States on the probable impacts of climate change, which generated wide media coverage of the issue. In October 1997, the President himself toured the US to explain the need to act. The campaign culminated in a White House Conference on climate change in Washington, D.C., on 6 October 1997. It was attended by some 200 politicians and representatives of civil society, and broadcast live over the internet and via satellite to 32 locations across the country.[11]

The US administration's efforts built upon strong support from some parts of society. Environmental NGOs worked to raise the profile of the issue and consistently challenged the arguments of the small minority of climate sceptics. As a result, a New York Times Poll shortly before the Kyoto gathering showed that 65% of US citizens supported steps to cut GHG emissions "regardless of what other countries do". In the same poll, 49% attributed global warming to the release of GHGs, but only 16% to normal climate fluctuations (as argued by climate sceptics).[12] In addition, 2,400 scientists lent their support to cost-effective measures to reduce US GHG emissions in June 1997 and similar statements from experts followed.[13]

But the opponents of action on climate change also proved their ability to play the cards. The fossil fuel lobby sponsored a US $ 13-million media campaign, which not only denounced the scientific basis of climate change but also painted a bleak picture of alleged economic hardship resulting from reduced use of fossil fuels. The excuse for not taking action: dramatically rising emissions in developing

---

10 <http://www.whitehouse.gov/CEQ/ctask.html; http://www.whitehouse.gov/PCSD/charter> as of March 1998; on the internal US politics before the Kyoto conference see Harris 1998.

11 Information about the White House Initiative on Global Climate Change can be accessed at <http://www.whitehouse.gov/Initiatives/Climate/main.html>.

12 "Public Backs Tough Steps for a Treaty on Warming", The New York Times, 28 November 1997, A 35.

13 See Harris 1998, pp. 56–57.

countries would render any emission reductions by industrialised countries meaningless. Emission limitations by developing countries were presented as a *sine qua non* – which the "Carbon Club" knew very well would not be acceptable to the G-77. Given the historic responsibility of industrialised countries (the US in particular) and their failure to implement emission reductions and the Berlin Mandate, developing countries would certainly not agree to emission limitations for themselves. The argument was therefore strategically aimed at sinking the international process.

However, the actual physical rise in emissions of developing countries was not debatable and provided fertile ground for the strategy of the Carbon Club. The major outcome was a bipartisan resolution of the US Senate. Named after its sponsors, the Byrd-Hagel Resolution of 12 June 1997 made "meaningful participation" of developing countries a precondition for US ratification of the Kyoto protocol. Accordingly, the Senate would not accept an emission reductions agreement,

> "unless the protocol or other agreement also mandates new specific scheduled commitments to limit or reduce greenhouse gas emissions for Developing Country Parties within the same compliance period".[14]

Passed by consensus (95-0), the Resolution of the Senate marked a dark day for climate protection, leaving little hope for flexibility in the US position in the short- to medium-term.[15]

The Byrd-Hagel Resolution did not miss its target. President Clinton and his advisors decided to adapt their negotiating position to conform to the Senate's request, although they violated the consensus of the Berlin Mandate. As one of its key demands, US participation in a Kyoto Protocol was made dependant on the "meaningful participation" of key developing countries (see Chapter 5). On the other hand, by calling for a stabilisation of GHG emissions on 1990 levels by 2008–2012 the President was opting for the most progressive proposal on targets among those under consideration by the US.[16]

6.3 Entrenching Positions: Other JUSSCANNZ Countries

In contrast to the US administration, the Australian government was not keen on a stringent international agreement. The situation became even more complicated in early 1996 when the Australian conservatives won the national elections. While Australia had certainly not been a climate activist under the previous Labour

14 105th Congress, 1st Session, Senate Resolution 98 of 12 June 1997.

15 Consensus on this resolution in the Senate came at a price for its sponsors, however, since those senators supporting stringent action successfully demanded an affirmation of the IPCC's findings. The scientific basis of the Kyoto Process was therefore accepted even by the hardliners in the Senate.

16 Handelsblatt, 20 October 1997, p. 10.

government, the attitude of incoming Prime Minister Howard towards climate policy was almost hostile.

The Australian government's sceptical position was bolstered by the economic modelling analysis by a government department, the Australian Bureau of Agricultural and Resource Economics (ABARE). Their results purported to demonstrate that Australia would be overly disadvantaged by international action to curb GHG emissions. If emissions were reduced by 10% below 1990 levels by 2020, wages would allegedly be some 20% lower than under business-as-usual conditions, tens of thousands of jobs lost and each Australian would have to shoulder a burden 22 times as heavy as the average European.[17]

The results of the modelling were soon revealed to be biased as it was largely funded by the fossil fuel industry and no expert review was conducted. The fee for sitting on the steering committee overseeing the development of the modelling work was put at AUS $ 50,000 (ca. US $ 35,000). As a result, the Steering Committee was composed of the coal, oil and other resource industries.[18] It was only after the Kyoto conference that the Commonwealth Ombudsman upheld a complaint filed by the Australian Conservation Foundation. He ruled that the modelling exercise was designed to exclude environmental groups' input, that it did not comply with public administration standards and failed to get a peer review.[19]

Before Kyoto, the government did not accept this criticism of the modelling exercise. It also failed to respond to the remarkable public support for action on climate change that prevailed, even in large parts of industry, or to the increased activities by environmental NGOs. On the contrary, Prime Minister Howard (followed by other leading officials including the Foreign Minister), threatened to withdraw from the FCCC if staying in meant accepting reduction targets – destroying any hope that Australia would play a constructive role in the Kyoto process.[20]

The situation in the host country of COP 3 was different. Japanese policymaking on climate change was characterised by severe interagency conflict between the Ministry for International Trade and Industry (MITI) and the Environment Agency. MITI had already opposed hosting the conference, but Prime Minister Hashimoto overturned its vote in Cabinet and took the final decision. In the ensuing discussion about the Japanese position in the Kyoto process, the dominant MITI was put under pressure because Japan was expected to act so as to secure a successful outcome of the conference.

---

17 DFAT/ABARE 1995; ABARE 1997; Hamilton 1997.

18 Hamilton 1997.

19 "Government Must Accept Umpires Decision" and "ACF Complaint about Government Greenhouse Research Upheld" at <http.//www.pag.apc.org/~acfenv/mr0398.htm> as of 3 February 1998.

20 "The Australian Government's greenhouse journey" at <http://www.peg.apc.org/~acfenv/ghfact8.htm> as of 13 March 1998.

In addition, a growing environmental NGO community supported the Environment Agency. Based on the example of the KlimaForum, which co-ordinated German environmental NGOs at COP 1 in Berlin, the KikoForum was established in 1995.[21] Supporting stringent climate protection, KikoForum was a hybrid, bringing together environmental groups as well as representatives of business and public administration (in their personal capacity). In addition, international environmental NGOs, such as Greenpeace and Friends of the Earth, stepped up their efforts in Japan. This build-up of non-governmental capacities benefited from the general opening up of the Japanese political system to non-governmental input that took place in the 1990s.[22] Furthermore, a poll of June 1997 showed that 86% of the Japanese population were concerned about global warming and 84% supported emission reductions, even if they entailed some economic costs.[23] Consequently, the Japanese stance at the climate talks was increasingly pushed towards mediating between the US and the EU (see Chapter 7).

Political developments in Canada and Norway also appeared to improve the prospects of the Kyoto process. In both countries, elections in 1997 were won by parties that were presumed to be more inclined towards climate protection than their opponents. In the international talks, this was hardly felt, but it was clear that each of these governments would face very high political costs at home if it was blamed for failure at Kyoto.

6.4 Business: Some Going Dirty...

Not only governments, but also the business community was preparing for Kyoto. In their attempt to obstruct the international talks, the "Carbon Club" focused on the US and developing countries. After the victory won with the passing of the Byrd-Hagel Resolution in the US Senate, the "grey" industry launched a 13-million dollar campaign across the US. Advertisements in newspapers, radio and television alleged that US citizens would have to pay an enormous bill as a result of a protocol and that the UN global climate treaty was "not global". Pressure was thus mounting on the US administration to insist on commitments for developing countries.[24]

At the same time, the Carbon Club disseminated a somewhat different message to the developing world, and Asian developing countries in particular. This major

---

21 Both the German and Japanese names mean "Climate Forum".

22 Schreurs 1996; 1997.

23 Asahi Shinbun, 21 June 1997.

24 There are indications, however, that the campaign was only moderately successful, since it was widely perceived as partisan by the public. On the 13 million dollar campaign see Greenpeace Press Release, "Oiling the Machine: Fossil Fuel Dollars funneled into the US Political Process", 20 October 1997, available at <http://www.greenpeace.org/~climate/kdates/october20b.html> as of 18 June 1998; Greenpeace Magazin 6/1997, 23–25; Die Zeit, No. 43, 17 October 1997, 23; see also the Global Climate Coalition's own webpage at <http.//climatefacts/org>; Greenpeace 1998.

emerging market for the fossil fuel industry was entering economic crisis during the second half of 1997. The fossil fuel industry's "advice" was that the poor countries of the world could afford neither emission limitations for themselves nor emission reductions of the industrialised countries, because both would limit their economic development. Thus, the Chairman and Chief Executive Officer of the Exxon Corporation, Lee R. Raymond, at the World Petroleum Congress in October 1997 in Beijing, China, called upon developing nations to "work with us to resist policies that could strangle economic growth". Defying the scientific evidence and neglecting the vast no-regret potential of reducing GHG emissions, Raymond declared that the most pressing environmental problems of developing countries were related to poverty. Addressing this would "require economic growth, and that will necessitate increasing, not curtailing, the use of fossil fuels".[25]

These activities on both sides of the Pacific were part of a strategy of playing both sides against each other. If the US insisted on developing country commitments and China and others were sceptical even of emission limitations in the industrialised world, the Kyoto process would eventually lead nowhere – the preferred option for this part of industry co-operating in and working through the Global Climate Coalition and The Climate Council.

6.5 ...Others Going Green

In contrast to the "Carbon Club", an increasing number of industry associations went "green" by supporting, or at least accepting, a reasonably strong agreement in Kyoto. The FCCC negotiations in the early 1990s had seen little involvement of the "green" industries, but the potential winners of strong climate policy made themselves increasingly visible in the Kyoto process, including renewable energy industries, energy efficient businesses and others. 1996 saw the establishment of the European Business Council for a Sustainable Energy Future (e^5), which includes the influential cogeneration industry (that is engaged in combined production of heat and power). The insurance industry, alarmed by rising damages caused by natural catastrophes, also openly supported a reduction agreement (see Figure 6.1).[26] The insulation industry, which could benefit from energy conserving efforts in the building sector, called for urgent action to reduce CO_2 emissions and pointed to the Lisbon Declaration of June 1997.[27]

---

25 "Energy – Key to Growth and a Better Environment for Asia-Pacific Nations", Remarks by Lee R. Raymond, Chairman and Chief Executive Officer, Exxon Corporation, World Petroleum Congress, Beijing, 13 October 1997 (on file with authors).

26 Salt 1998; UNEP 1997.

27 "The Lisbon Declaration on CO_2 Reductions, Submission of the Insulation Industry to the Conference of Contracting Parties of the UN International Framework Convention on Climate Change", Kyoto, December 1997, 13 June 1997.

Figure 6.1: Natural Disasters and Economic Damages

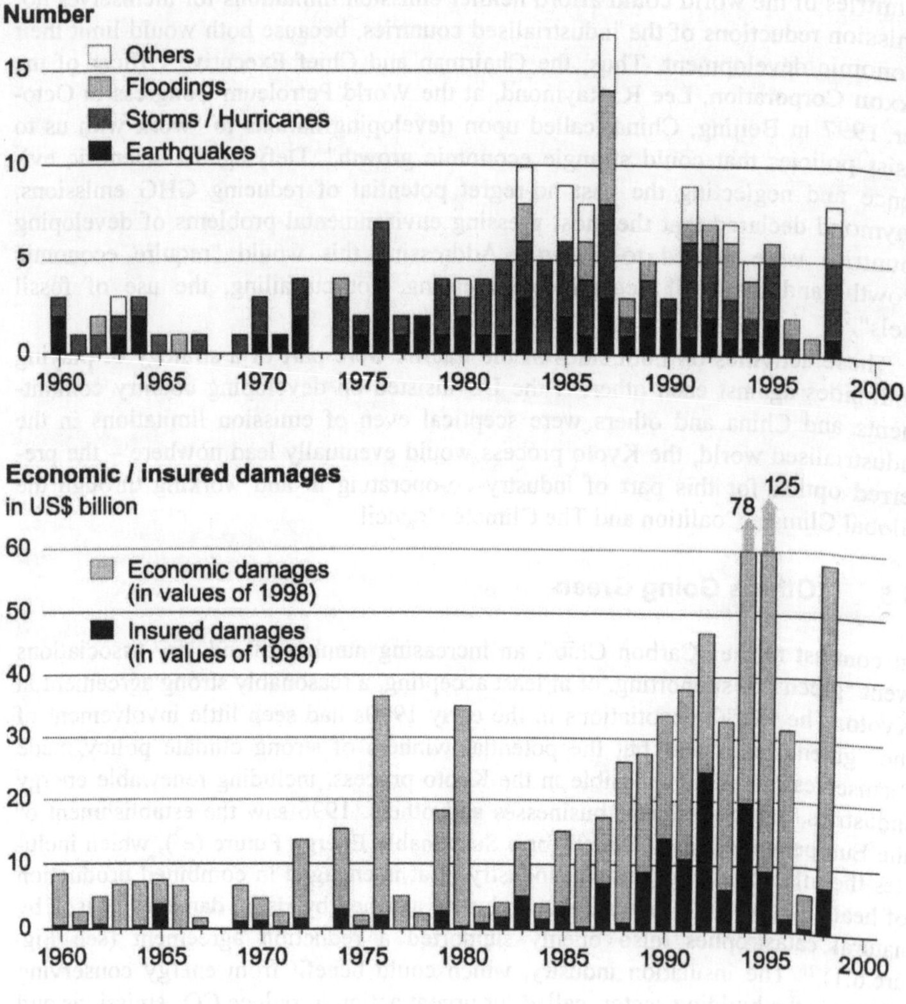

Source: Munich Re 1998

Although not pushing for stringent action, the World Business Council for Sustainable Development, the International Chamber of Commerce and others (like the German industry association BDI) were increasingly disassociating themselves from the more radical Global Climate Coalition. In essence, the business community became increasingly divided, leaving the above mentioned "moderates" to bridge the gap between the "green" and the "grey" industry[28] – an increasingly impossible task. At AGBM 3 in spring 1996, the business community, for the first time, could not agree on a common statement (see Section 4.4).

The growing support for action led to a reassessment by some oil companies with headquarters in Europe of their position on climate change. On the occasion of a speech at Stanford University in May 1997, British Petroleum Company's Group Chief Executive John Browne became the first oil industry official to acknowledge that "it would be unwise and potentially dangerous to ignore the mounting concern" about climate change. At the same time, Browne declared that BP would expand its investment in solar power, in which it already held a 10% share of the world market, "to position the business to reach US $ 1 billion in sales over the next decade".[29] Although cautious, other industry officials followed suit. The Chairman of Sun Oil, for example, supported "initiation of prudent mitigation measures now" in a letter to President Clinton in autumn 1997.[30] Finally, Shell gave a boost to renewable energy in October 1997 by establishing Shell International Renewables. The new company was to triple Shell's spending on renewables within five years by investing US $ 500 million.[31]

6.6 Environmental NGOs

Environmental NGOs, co-ordinated under the framework of the Climate Action Network (CAN), tried to counter the activities of the "grey" industry. They formally supported the AOSIS 1994 draft protocol by demanding a 20% reduction of CO_2 emissions in industrialised countries by 2005. As Kyoto approached, however, their focus increasingly shifted towards fighting the "loopholes" that they felt could be used by industrialised countries to escape domestic action while still formally agreeing to a numerical reduction target. In particular, the inclusion of sinks in GHG inventories, "banking" and "borrowing" of emissions, JI, emissions trading and a rumoured shift in the baseline for calculating emission reductions, from 1990 to 1995, were seen as glaring loopholes.

Environmental NGOs intensified their campaign at both governmental and general public levels prior to Kyoto. Although commanding fewer resources than the Carbon Club, they engaged in active lobbying not only during international negotiations, but also in their home capitals. This was complemented by public relations activities made possible through close contacts with the news media. Their efforts focused on the three big industrialised players in the game, the US, the EU and Japan.

---

28 See Taalab 1998.

29 Browne 1997, pp. 4 and 13.

30 Martha M. Hamilton, "Oil Executives are Shifting Their Stance", Washington Post, 3 March 1998.

31 "Shell Makes a Push for Renewable Energy", ENDS Daily, 16 October 1997.

In Japan, the KikoForum co-ordinated numerous conferences and events. In the US, environmental NGOs organised town-hall meetings across the country to generate public support for climate change policies. Limited resources, however, prevented successful lobbying of the US Congress, which was heavily influenced by the Carbon Club (see Section 6.2). European environmental NGOs also engaged in public awareness campaigns to generate the necessary momentum for the Kyoto process. A key event was a two-day conference on "targeting Kyoto and Beyond" held in Bonn on 16 and 17 October 1997, organised by CNE and GLOBE Europe, a forum of European Parliamentarians. There were more than 300 participants including many negotiators and ministers. The event also received wide press coverage.[32]

Although CAN members spoke with one voice in public, the CAN membership of more than 250 NGOs from more than 70 countries, of course, held different views on various subjects. Cultural differences between the New and the Old World existed. For example, US NGOs, operating in a society that highly values its free-market ideology, were much more positively inclined towards market-based concepts, such as emissions trading and JI, than their European partners, who rather tended to stress the advantages of regulatory systems. During the negotiations, differences also became apparent on other issues, such as a possible shift of baselines and the size of the basket of controlled substances by the inclusion of fluorinated GHGs. However, environmental NGOs were quite aware that their only chance of influencing the process was to act coherently and in unison. Therefore, they successfully co-ordinated their actions and strategies before and at Kyoto, at both regional and global levels.[33] This increased their weight. In fact, observers of international environmental negotiations tend to agree that those NGOs active in climate change are particularly effective because of this close co-ordination.

---

32 CNE 1997; on the pre-Kyoto activities of environmental NGOs in the US see also Harris 1998: pp. 54–55.

33 In addition to the global CAN strategy sessions, European environmental NGOs, for example, discussed their position in an internal workshop held at the end of September 1997 in Berlin.

7. Kyoto: The Endgame

When delegates flew to Japan at the end of November 1997, all the big players were holding their cards firmly to their chests. None was sure of the outcome, but they were acutely aware that world public opinion expected an agreement at Kyoto. Japan, in particular, was keen to host a successful conference. On that basis, the Kyoto Summit became one of the most extraordinary and remarkable events in international environmental diplomacy, bringing together more than 2,200 delegates from 158 Parties to the Convention and six observer states, nearly 4,000 observers from NGOs and international organisations and more than 3,700 media representatives.[1]

This last chapter of the negotiating history highlights the special dynamics of the Kyoto conference. To begin with, the role of the host Japan in securing agreement will be explored (Section 7.1). Next, the slow progress during the first week of the conference will be examined (Section 7.2). Several key factors that decisively influenced the outcome will be the subject of further analysis, most importantly the role of the media and NGOs, the availability of modern communication technologies (Section 7.3) and Chairman Estrada's leadership (Section 7.4). As usual at large international conferences, the deliberations accelerated with breathtaking speed as the conference drew to the end (Section 7.5). As negotiators hammered out the ambiguous compromise package during the final night, sheer exhaustion became another decisive factor in bringing about the final compromise (Section 7.6).

7.1 Japan

The decision at COP 2 (July 1996) that Japan would host COP 3 raised hopes, since Japan was expected to show strong leadership and a willingness to compromise in order to preside over a successful conference. However, it was not in Kyoto but during the months prior to COP 3 that Japan made its main contribution to the Kyoto process. According to the Japanese style of diplomacy, the careful balancing of interests, rather than strong leadership, is its main strength. This proved to be important in the run-up to Kyoto, but prevented the Japanese Government from exercising the kind of forceful shuttle diplomacy that the German Minister Merkel had showed at COP 1 in Berlin. Japan deserves considerable merit, for example, for the diplomatic effort to accommodate OPEC, and Saudi Arabia in particular (see Chapter 5). Furthermore, Japan invited key delegations for three informal meetings to Japan during 1997. The first two of these meetings focused on the commitments of industrialised countries. Representatives of 10 OECD countries (including representatives of JUSSCANNZ

---

1 FCCC/CP/1997/INF. 5, 9 December 1997.

and the EU) met in Tokyo on 23 and 24 April and again on 9 and 10 September to discuss the issues. However, these meetings served mainly to bring together key negotiators for an informal exchange of views and no major progress was reported from these meetings.

The third unofficial two-day meeting, held at Ministerial level in Tokyo on 8 and 9 November 1998, proved more successful. The Japanese government had invited Environment Ministers from the US, Canada, Australia, New Zealand, the UK, Germany, the Netherlands, Luxembourg, Norway, France, Italy, Russia, the European Commission, South Korea, Mexico, China, India, Indonesia, Brazil, Argentina, Tanzania, Samoa, Saudi Arabia, as well as Chairman Estrada and Executive Secretary Michael Zammit Cutajar. The meeting was divided into two parts: industrialised countries discussed the issues among themselves at the first day and developing countries joined on the second day. As a sign of the poor relations between North and South at the time, both India and China declined to come. As a further signal, most of the industrialised countries were represented by ministers but most developing countries only sent lower-level officials.

Among themselves, industrialised countries discussed targets and flexibility mechanisms, PAMs, and developing country involvement. Japan tried to achieve some common ground between the US and EU proposals on targets and proposed a package of financial and technical assistance to developing countries, including a replenishment of the GEF. The EU indicated its willingness to compromise. Italy and France reportedly admitted to Japan that the 15% target proposed by the EU might be impossible to reach. After the meeting, all industrialised countries supported (legally binding) targets expressed in budget periods of up to five years. Otherwise, existing disagreements prevailed (see Chapter 4).[2]

The second day of the meeting was devoted to the involvement of developing countries in the Protocol. The US indicated that developing countries could get involved in various ways, for example by an agreement on talks about developing countries' commitments. Discussions reportedly made some progress. While developing countries insisted that industrialised countries should take the lead by reducing their GHG emissions, Brazil and the US developed a common understanding of a mechanism for some kind of JI between developing countries and industrialised countries. Building on a Brazilian proposal to generate funds for climate change mitigation in developing countries, both countries elaborated what was to become the Clean Development Mechanism (see Chapter 14).

In conclusion, the Tokyo meeting helped the ministers of key countries to get involved and better understand each other. In addition, the talks were perceived as paving the way for involving developing countries in a long-term emission-cutting effort. According to the Japanese hosts, "the first concrete step was made towards

2 Informal Ministerial Consultations Between Annex I Countries: Tokyo, 8 November 1997, Chairman's Report (on file with authors).

compromise".[3] The non-attendance of the two most important developing countries, China and India, however, severely handicapped the talks.

At the Kyoto conference itself, the Japanese role remained limited. The process was driven by Chairman Estrada, rather than the President of the Conference, Japanese Minister Ohki. It seems that after UNGASS in June 1997 (and partly as a result of the criticism of its role at UNGASS, see Chapter 5), the Japanese government decided to keep a relatively low profile by trying to find some middle ground between the US and the EU. As a major contribution to the Kyoto conference, Japan provided the excellent technical facilities and infrastructure at Kyoto that facilitated proceedings at the conference (see Section 7.3).

7.2 The Dynamics of the Kyoto Meeting

Many issues regarding a Protocol were still unresolved when delegates arrived at Kyoto. The Chairman, in close co-operation with the Secretariat, had prepared a "revised text under negotiation",[4] but only a few of its provisions were uncontested. The preamble and the provisions relating to the role of organs, dispute settlement, amendments, right to vote and the final provisions (except for the entry into force) were largely agreed. Furthermore, agreement was in sight for the definitions, methodological issues, reporting, review of implementation and some aspects of the design of targets (especially the choice of commitment periods instead of single year targets and the possibility of banking unused emissions for the next compliance period). Most of the substantive issues, however, remained controversial, including:

- the size, timing, design (flat rate versus differentiation, number of GHGs covered, "borrowing" of emissions from future commitment periods) and the legal character of quantified emission limitation and reduction objectives for industrialised countries,
- the role of PAMs in the final agreement (binding obligations versus general guidelines),
- whether and to what extent sinks could be used to offset any emission reduction obligations,
- how to involve developing countries, including the question of voluntary commitments and compensation payments for negative effects of industrialised countries' measures on their economies,
- whether and to what extent JI, emissions trading and the EU "bubble" would be allowed to be used to achieve emission obligations,
- how cases of non-compliance would be treated, and

---

3 On the Tokyo meeting see "Pre-Kyoto talks conclude, nations upbeat on summit", Yomiuri Shimbun (on-line) (on file with authors).

4 FCCC/CP/1997/2.

- a large part of the institutional arrangements, including the role of the bodies of the Convention for the instrument to be adopted, the provisions for entry into force and the review of commitments.

It was still not clear whether a Protocol or "another legal instrument" (i.e. an Amendment to the Convention in particular) would result from the process and what it might be called. To increase the pressure on the opponents of a Protocol, the EU had previously submitted a proposal for an Amendment of Article 17 of the Convention. It introduced the possibility of majority voting for the adoption of Protocols, thus circumventing the requirement for consensus that arose from the failure to adopt the Rules of Procedure (see Chapter 4). The proposal stipulated that prior to the amendment entering into force it was to be applied provisionally, thus allowing for the immediate adoption of a protocol. Kuwait countered this proposal by submitting a proposal to amend Article 4.3 of the Convention so that industrialised countries would be required to provide financial resources, including technology transfer as determined by the COP, to meet the "full incremental costs" of developing countries' obligations.[5]

To resolve all pending issues, the gathering in Kyoto took place in three parts. Before the start of the COP, the AGBM resumed its eighth session for one day to take stock of the state of discussions and put forward any new conclusions to the COP.[6] The COP itself was then divided into two phases. During the first week, Parties tried to establish as much common ground as possible, in order to allow the final deal to be struck during the three-day Ministerial Segment of the Conference that took place during the second week of the meeting.

Not surprisingly, the Sunday gathering of the AGBM on 30 November, which focused on the issue of sinks, proved inconclusive. Chairman Estrada reported that he "had planned to add conclusions to the AGBM report but no conclusions had been reached".[7] The next day, delegates assembled again, this time as the COP. Under the chairmanship of the chief of the Japanese Environment Agency, Minister Hiroshi Ohki, the plenary considered the more general business of the COP. Simultaneously, the Protocol negotiations proceeded in the Committee of the Whole (COW), chaired by Estrada in a smaller meeting room.

Because of the complexity of the matter and the many unresolved issues, the COW established several negotiating groups on the following issues: institutional questions, the involvement of developing countries and consideration of their concerns in the Protocol, and PAMs. Chairman Estrada himself presided over the negotiations on targets and timetables. Furthermore, the COW created numerous informal groups on sinks, emissions trading, JI, voluntary commitments of devel-

---

5 FCCC/SBI/1997/15.

6 The resumption of AGBM 8 immediately prior to COP 3 helped avert any claims that the work that had proceeded since AGBM 8 in October 1997 under the leadership of Chairman Estrada was beyond the mandate of the AGBM.

7 Earth Negotiations Bulletin, Vol. 12, No. 76, 13 December 1997, p. 3.

oping countries etc. Added to the already busy schedules of negotiators were the regular consultations of the country groupings (the EU, JUSSCANNZ, OECD, the G-77 and China, AOSIS etc.). Moreover, bilateral and multilateral contact groups, cutting across established coalitions, met to find common ground.[8]

Under these circumstances, it was difficult for both observers and government delegates to keep track of developments. New and existing proposals were explored and weighed against each other in different settings by government representatives meeting in smaller conference rooms closed to the public. In most cases, the proposals somehow found their way into the corridors where government negotiators, sometimes looking for expert advice, mingled with business, environmental NGOs and the media. Certain members of the NGO-community were usually better informed than many negotiators. Press conferences by delegations and NGOs alike had the potential of revealing new information.[9]

This information was not always correct – in the corridors of the conference centre rumours merged easily with the truth about progress (or the lack of it) in the informal groups. While negotiators met their delegations regularly to stay informed about discussions, NGOs held daily meetings to exchange "intelligence" and discuss strategies. As the Ministerial Segment drew nearer, the number of people and events and the intensity of negotiations increased. Negotiations slowed down over the weekend, after which ministers were scheduled to arrive.

A "non-paper" by Chairman Estrada summarised the state of affairs after the first week of negotiations. It provided evidence of a slowly growing convergence. Broad agreement now existed on most of the institutional issues, and the proposal to "borrow" emissions from future emission budgets had been abandoned. Agreement on the inclusion of sinks and the issue of compensation for developing countries was in the offing. Many of the substantive issues mentioned above, however, were still far from being resolved, although positions had converged to varying degrees.[10] At the very least, the Conference was now clearly aiming for the adoption of a Protocol, since both amendments to the FCCC proposed by the EU and Kuwait were withdrawn at the end of the first week.[11]

Activity heightened at the Kyoto conference centre during the ministerial meeting from 8 to 10 December. The final countdown was reached at the end of the day of 10 December 1997, when the conference was officially scheduled to finish its business. The final days and the final night of negotiations are dealt with below. Before, some distinct features of the proceedings in Kyoto are highlighted so as to shed some light on how the Kyoto Protocol emerged.

---

8 See, e.g., Earth Negotiations Bulletin, Vol. 12, No. 76, 13 December 1997.

9 As in previous negotiating sessions, various side-events hosted by research institutes, international organisations and interest groups contributed to the busy agenda of participants.

10 FCCC/CP/1997/CRP.2, 7 December 1997.

11 See FCCC/CP/1997/7, p. 26.

7.3 Modern Communication Technologies

The Kyoto Summit attracted an unprecedented degree of attention of the world's public for climate change. Especially, in industrialised countries, the public had high expectations for a deal in Kyoto. TV, radio and newspapers reported prominently on the conference with increasing intensity. Media reports were immediately fed back into the negotiating process, as newspaper articles were distributed at the conference itself and government officials received information directly from their home countries where public pressure was mounting.

This public pressure was partly the result of the interplay between media and environmental NGOs. In the run-up to Kyoto, NGOs had invested considerable efforts in raising public awareness, by organising conferences and by providing background material to the media. There was thus a high degree of awareness about the issues at stake and about the acceptable outcome in environmental terms (see also Section 6.6). Some NGOs established particularly close contacts with journalists, who transmitted the issues to the broader public.

This "symbiosis" between journalists and environmental NGOs proved instrumental at the Kyoto conference, where they were situated together in one hall. On the one hand, NGOs were able to amplify their messages and expose the laggards to public pressure by providing relevant information to journalists. The advocates of the environment were not only welcome as sources of information, but also frequently appeared as interviewees in the media. Journalists, on the other hand, profited from receiving inside information on the negotiations. Through their consistent awareness campaigns, NGOs and media thus ensured that public pressure was maintained and increased during the negotiations.

In its exchange with the external world, as well as in internal communication, the conference relied on modern communication technology to a degree hitherto unknown – which may have been due to the location of the conference in Japan, a country known for its interest in modern technology. The Japanese hosts and the Secretariat provided free and constant internet service to the attendees. As a result, news on the conference were spread instantaneously. Negotiators and observers had detailed reports on the previous day of the conference available before breakfast. Newspaper articles, comments on the state of the conference, stories about the latest developments were easily immediately transferred world-wide from Kyoto. News media made use of this new opportunity. CNN and BBC, for example, transmitted news on the state of negotiations virtually around the clock via internet, in addition to regular reports on their TV programmes. Similarly, news on the situation at home was easily received via these channels. Even official documents, most importantly the Kyoto Protocol itself, were available world-wide via internet within minutes of their release.

Internally, the internet had already played an unprecedented role in the intersessional period before Kyoto. Draft negotiating texts and documents had been extensively circulated and discussed via e-mail. In general, e-mail allowed all those involved in the preparation of the Kyoto conference to liase closely with each other. The Secretariat in Bonn and Chairman Estrada, then Ambassador to China, were in close touch, sending e-mails back and forth several times per day. Elements of a "virtual diplomacy" were thus emerging even before Kyoto. At Kyoto, the US and the UK delegations used online connections to computer centres in their home countries for an immediate evaluation of proposals.

The exposure of negotiators to the world's watching eye was enhanced decisively by the fact that, for the first time in the history of the international climate negotiations, the official sessions and selected events (like the press conference held by US Vice President Gore during his visit to Kyoto) were broadcast live. TV screens accessible everywhere in the conference centre enabled negotiators and observers to follow the official negotiations, while engaging in bilateral talks or going for a coffee. More importantly, however, the broadcasting far exceeded the conference building, as the official proceedings were also fed into the internet. The biggest live video broadcast on the internet to that date enabled anyone anywhere with internet access to follow the speeches and interventions of government representatives at the official meetings. TV crews were even allowed to be present at the last final session of the COW – unprecedented for a working meeting. However, this presence of TV cameras considerably hightened the sense of urgency and importance of this historical moment.

One of the most prominent features in Kyoto were the constantly ringing mobile phones. Arguably, the big number of formal, informal and unofficial meetings could hardly have been organised without these devices – that at the same time disturbed these meetings by their frequent ringing. Mobile phones allowed attendees of the conference to be contacted from their home countries directly, receive the latest information on the state of affairs, arrange for interviews etc. They also enabled, or at least facilitated, internal co-ordination of delegations that had to come to terms with several concurrent meetings.

Mobile phones furthermore changed the lobbying strategy of NGOs in the negotiating process, as they were not allowed to participate directly in informal negotiating sessions. Mobile phones ensured access to policy-makers during open meetings of the plenary, where NGOs were denied access to the floor and had to follow the negotiations from the balcony. This had been the procedure since INC 11, because of heavy interference by grey business NGOs (especially The Climate Council) with certain national delegations on the floor. By using mobile phones, NGO experts could now stay in touch with their contact points in key delegations to give their views, provide new information and make themselves heard. Interested delegates also called key NGO observers to spread new information and seek advice. These telephone contacts grew particularly intense during the final phases of negotiations, when time became increasingly scarce.

The Kyoto Summit thus marked the final victory of modern communication technology in the sphere of environmental diplomacy. As a result negotiators felt that the world was watching. Any attempt to block progress was therefore unlikely to go unnoticed and the accountability of negotiators was improved. This increased the willingness to compromise, since any country that would openly attempt to obstruct the success of the Kyoto meeting knew it would be blamed later. Moreover, the role of modern communication technology in Kyoto provided evidence that environmental diplomacy is quickly adapting to changing circumstances as it enters the 21st century. This notion has been enhanced post-Kyoto and modern communications have become a constant feature of the climate negotiations. However, as an unwanted side-effect, this development further threatens to marginalise least developed countries, because they frequently do not have access to such technologies.

7.4 The "Estrada Factor"

Throughout the negotiating history of the Kyoto Protocol, Chairman Estrada played an outstanding leadership role. The outcome of the Kyoto Process cannot be fully understood without paying tribute to the then Ambassador of Argentina to China, who served as Chairman of the INC after Rio, then chaired the AGBM and later the COW at the Kyoto Conference. During the five years of his chairmanship, he assumed a leadership role that created momentum and pushed negotiators towards meaningful solutions.

His chairmanship contrasted sharply with the conduct of his predecessor, Jean Ripert of France, who had led the negotiations on the FCCC in a more traditional diplomatic style and stayed neutral in discussions. Occasionally straying beyond diplomatic prudence, Estrada frequently reminded delegates of their mandate and intervened forcefully in debates that were mere repetitions of well-known statements. At times he argued with delegations, requiring them to check their arguments before presenting them. For example, when Saudi Arabia requested that the effects of measures by industrialised countries on developing countries should be investigated before any negotiations on a Protocol (at AGBM 3), Estrada noted that he "did not recall anyone talking about the effect of measures on developing countries when OPEC raised oil prices 20 years ago".[12] Furthermore, he left no doubt that he would not tolerate obstruction and requested OPEC to come forward with concrete proposals for how to deal with their concerns in a constructive manner. Similarly, he repeatedly ruled that requests by the US to discuss developing country commitments were outside the Berlin Mandate.

Estrada was determined to reach agreement in Kyoto and constantly sought for ways out of the problem of requiring consensus for adopting protocols (see Section 4.5). One of the possibilities was demonstrated at the end of the last AGBM before Kyoto, when he ruled that there was consensus on a draft text, except for three countries. When one of those countries raised a point of order, Estrada noted that delegates could challenge his ruling, but would need a two-thirds majority to revert it. A formal challenge was avoided on this occasion and Estrada thus demonstrated that a Protocol could be adopted "by consensus" against a minority of

12 Earth Negotiation Bulletin, Vol. 12, No. 27, 11 March 1996, p. 5; see also Oberthür 1996a.

opponents, even in the absence of a formal rule on voting.[13] At AGBM 8, he therefore laid the basis for the adoption of the Kyoto Protocol.

Estrada, in close co-operation with the Secretariat and its Executive Secretary Michael Zammit Cutajar, intimately knew the positions of the key players on important issues and closely followed their development. As a central focal point, he was also able to get a feeling for the importance of different issues for the various actors. One result of his intimate knowledge was his conviction already in July 1997 that differentiation of commitments would be the key to the solution, and therefore he steered the process in this direction.[14]

Assisted by the Secretariat, Estrada was able to keep track of all developments during the accelerating process at Kyoto itself. He had the best overview of the processes in the various negotiating groups, although even he was not directly involved in all cases (e.g. the trilateral talks). Consequently, he had no sleep for two nights from 9 to 11 December. Nevertheless, he remained master of the process in Kyoto – especially during the final session. Without Chairman Estrada's leadership and stewardship during the last night, events could have evolved quite differently. How Estrada virtually forced delegates to consensus, making him the "Hero of Kyoto"[15] is dealt with in the following sections.

7.5 Approaching the Final Show-down

The arrival of US Vice President Gore in Kyoto for the Ministerial Segment at the beginning of the second week marked the beginning of the final phase of the conference. Uncertain about whether to come until a few days before, his decision to address the conference demonstrated both his personal and the US administration's commitment to reach an agreement in Japan. His appearance at the opening plenary of the ministerial segment on 8 December 1997 was awaited with utmost interest and the large conference hall was thus full to capacity. Gore's address was preceded by speeches of Prime Minister Hashimoto of Japan, President Figueres Olsen of Costa Rica and President Clodumar of Nauru. As President of a small and threatened island state, President Clodumar further increased the pressure by recalling President Clinton's promised reductions and stating to Gore, "We await your announcement with bated breath".[16]

Vice-President Gore was in a delicate situation. On the one hand, the US wanted a meaningful result in Kyoto and Gore had to show willingness to com-

---

13 Earth Negotiations Bulletin, Vol. 12, No. 66, 3 November 1997, p. 6.

14 Statement by Raúl Estrada at the OECD Forum on Climate Change, see Raúl Estrada: Keynote Address on the Kyoto Protocol Negotiation; OECD Climate Change Forum, Paris, 12–13 March 1998 (transcript on file with the authors).

15 The Executive Secretary of the UNFCCC Secretariat Michael Zammit Cutajar at the meetings of the Subsidiary Bodies of the UNFCCC on 4 June 1998 (mimeo).

16 Speech to Plenary by His Excellency Kinza Clodumar, President and Minister for External Affairs, Republic of Nauru, 8 December 1997, Kyoto, Japan.

promise. On the other hand, like in poker, negotiators often hold out until the last possible minute, since early concessions might draw more requests from the negotiating partners. Furthermore, major concessions at this stage of the negotiations certainly would have provided fuel to the US "Carbon Club" in its attempt to mark Gore as betraying the American interest. Gore mastered this challenge with great candour: throughout his statement, he reiterated the well-known US position and continued to make US agreement dependent on the inclusion of market mechanisms, "realistic" targets and timetables, and the meaningful participation of key developing countries. At the end of his speech, however, he provided a sign of compromise, something which he had reportedly included against the wishes of his advisers, but with President Clinton's approval. Gore instructed the US delegation "to show increased negotiating flexibility if a comprehensive plan can be put into place".[17] This somewhat subtle statement was not what the media had hoped for, but it did not fail to make an impression on the negotiators.

After the opening ceremony for the Ministerial Segment, the COP plenary began a three-day exercise of listening to statements from 125 ministers, several international organisations and NGOs, which were made before a greatly reduced audience. Instead of listening to the re-statement of well-known positions, negotiators shifted their attention back towards the COW, which met in a room on the other side of the corridor, and the many negotiating groups that met virtually around the clock. Talks about the exact size of the targets intensified within a trilateral framework, involving the US, the EU and Japan. Obviously, these talks were closely intertwined with the negotiations on sinks, emissions trading, and joint implementation. Other delegations were merely left waiting for the outcome of the trilateral talks in order to position themselves.

As the speed and intensity of negotiations increased, the EU's leadership abilities were heavily constrained by its complex internal decision-making procedure. Given the limited capacity of Luxembourg as the EU Presidency, negotiations were conducted by a troika of the Netherlands, as Luxembourg's predecessor, and the UK, as its successor. Due to the diplomatic capacity and negotiating power of the UK, Deputy Prime Minister John Prescott quickly assumed leadership. Nevertheless, changes in position or concessions had to be approved formally in internal co-ordination meetings between Member States. Adding to the already full schedule of Member States' negotiators, internal EU co-ordination was a time-consuming process, as consensus was required. Nevertheless, many Member States not directly involved in the external negotiations felt excluded and were suspicious of the troika conceding too easily. Danish Environment Minister Svend Auken was most outspoken when he complained publicly about the negotiating behaviour of the troika, thereby attracting considerable media attention.

---

17 Remarks by Vice President Al Gore, The United Nations Committee on Climate Change (mistake in the original), Conference of the Parties, 8 December 1997, Kyoto, Japan.

This cumbersome decision-making procedure was burdensome and prevented the EU from reacting flexibly to initiatives and proposals by others in the negotiations. On at least one occasion, the EU group discussed issues that had already been decided in the intergovernmental negotiations (see Section 7.6 below). As a result of the intra-European focus, little time, effort and attention were available for EU officials to network with non-EU delegations, and developing countries in particular, so as to build international coalitions. The situation worsened as the Conference approached the final night. Even institutionalised consultations with non EU-members from Central and Eastern Europe broke down during the final days. Caught in internal dynamics and revolving around itself, the EU became known for its "bunker mentality" in Kyoto.[18]

Despite the limited number of players, the trilateral talks between the EU, Japan and the US were not completed by the evening of 9 December. With time running out, the Chairman tabled a new draft that for the first time included numbers that he felt mirrored the state of the debate. The draft indicated further convergence of government positions on the following:

- differentiated targets for the commitment period 2006–2010,
- no legally binding PAMs,
- no compensation fund,
- JI between industrialised countries, as well as the "joint fulfilment of commitments ("bubble"),
- review of commitments,
- a provision on non-compliance, and
- some categories of sinks.[19]

Before any discussion could take place on the new draft, the COW was suspended to continue the informal negotiations that were incessant. Observers waited for the COW to be convened to agree the final deal. Hours passed, but there was no sign of a final agreement. Instead, rumours spread. Prime Minister Hashimoto in Tokyo was reported to have announced a press conference, but kept postponing it. A White House spokesperson – in the middle of the night in Kyoto – was said to have announced that President Clinton was "hopeful but not confident" about the prospects of the ongoing negotiations.

In their trilateral talks on numbers, Japan, the US and the EU narrowed down the options and came close to agreement during that night. Other interrelated issues, however, were still outstanding (emissions trading, JI with developing countries, developing country involvement), as were the commitments of other industrialised countries. Due to the importance of the issues, world leaders kept in close touch with their delegations as well as with each other. As contacts between

18 See CNE "Hotspot" No.1, April 1998.

19 FCCC/CP/1997/CRP.4, 9 December 1997.

capitals intensified, the telephone increasingly became the medium for diplomacy. President Clinton, Prime Ministers Blair and Hashimoto and Chancellor Kohl were among the most active, reaching out to other OECD countries, CEITs and Russia as well as developing countries. Vice President Gore called Prime Minister Hashimoto the next day and asked him to accept the deal on numbers that was on the table.[20] British Prime Minister Blair reached out to the US and Australia, etc. These were all signs that global climate change had become part of "high" politics.

As the night progressed, diplomatic activity inside and outside the conference building continued without pause. Around 3 am, Chairman Estrada gave a report to the COW on the progress made. The agreement would cover CO_2, CH_4, N_2O as well as fluorinated GHGs and include some categories of sinks. The adoption of the Protocol was scheduled for next morning.

7.6 Negotiation by Exhaustion

The final session of the COW was postponed from 11 am to 1 pm and then again to 4 pm on 10 December. When the Chairman finally convened the COW at 7 pm, he suspended it immediately to allow for further informal discussions. It was already 1 am in the morning of 11 December 1997 when the Chairman resumed the session on one of the most exciting nights of international environmental diplomacy. The COW was convened for its final session in the smaller meeting room instead of the main conference hall. That room provided an atmosphere conducive for making the final deal, in that negotiations would have been possible without a microphone. The camera crews lined up in the back of the room added a strong element of urgency. As a result, delegates felt almost physically the pressure to arrive at a deal. It was clear to all that agreement had to be reached now in some way or else COP 3 would end in failure.

While a number of issues had been resolved during the conference (see above), several key issues were still contentious, such as emissions trading and participation of developing countries (including a Clean Development Mechanism for Joint Implementation with developing countries which had emerged during the previous days at the Conference; see Chapter 14). In addition, the targets of a number of industrialised countries still needed clarification, as did some aspects of the wording on sinks. Furthermore, solutions arrived at in informal groups with limited participation needed to pass the plenary, and it was the common understanding that no single issue was settled definitively until the whole package was agreeable at least in principle for each delegation.[21]

Observers and delegations had been awaiting this decisive session for over 24 hours and many of them had got little or no sleep during the last two days. Thus,

---

20 Press Briefing by Gene Sperling (Assistant to the President for Economic Policy), Jum Steinberg (Deputy Assistant to the President for National Security Affairs), and Leon Fuerth (National Security Advisor for the Vice President), 11 December 1997.

21 See the respective Chapters for an account of the discussions on the single issues.

when the Chairman finally "resumed" the 14th meeting of the COW, many participants were already exhausted – and this was set to worsen as the night progressed. The delegate of Tanzania, in an early-morning intervention hit the point by calling the procedure of the final night "negotiation by exhaustion". As rumour had it that another conference was scheduled to start in the conference building the next morning, the pressure was mounting. A new draft text of the Protocol was distributed at the start of the session.[22] The annex containing the exact numbers of industrialised countries' commitments was left blank, drawing immediate interventions by Russia and the Ukraine stating that they would not accept any emission reduction but were only prepared to stabilise emissions.

On that basis, an extraordinary night of international environmental policy took a historic course. The Chairman followed the route tested by him at AGBM 8. Whenever discussions did not lead to agreement on an issue, he took a decision on behalf of the assembly. Anyone not accepting a decision had to formally challenge the ruling and a two-thirds majority could then overturn it. In the event, nobody dared to raise a challenge, because even if a delegation had perceived a chance to win such a challenge, mounting it could have jeopardised the whole process. Others could easily have responded by challenging the Chairman's ruling on other issues, thus eroding the sense of compromise and threatening to throw the discussions into chaos.

Under these circumstances, Chairman Estrada's use of the gavel became the most prominent feature of policy making during the early hours of 11 December 1997. Whether a lack of consensus led to the deletion or retention of an article in its original form was dependent on the judgement of the Chairman. His sense of the overall feeling of the meeting and his knowledge of those issues fundamental to any of the key players meant that he could satisfy these demands before taking a final ruling (see Section 7.4). Estrada's experience and strategy finally bore fruit. It was part of the rules of this game that even the Chairman himself could not re-open an issue once he had taken a decision, as this would have seriously undermined his authority. If he was able to revert to an issue that had been closed, representatives of sovereign states would be entitled to do the same.

The negotiators proceeded first through the central Article 3 relatively smoothly. When the provisions on emissions trading were taken up, they triggered forceful opposition by India, China, and a number of other developing countries. Although the G-77 was not united – with several Latin American countries, South Korea, the Philippines and small island states advocating or at least accepting emissions trading – the vigorous opposition by the most powerful developing countries drove the session to the brink of total failure. It was clear to everyone in the room that, unless a solution was found for emissions trading, neither the US nor Japan would be part of the agreement. Japan feared especially that the US might buy up all emission allowances leaving nothing to be acquired by itself. It

22 FCCC/CP/1997/CRP.6, 10 December 1997.

thus insisted on specific rules ensuring transparency of the emissions trading market. Shortly after 4 am, Chairman Estrada decided to interrupt the session "for five minutes" to conduct last-minute informal exchanges between the major counterparts. After three-quarters of an hour, the Chairman was able to present a compromise; a slightly revised wording on emissions trading was put into a separate article (see Chapter 15) and the Protocol thus rescued.

With this achievement, the session had overcome the major hurdle. Despite a lot of heated and prolonged discussions, it never again reached the brink of total deadlock. Winning their point of principle, developing countries had shown their muscles, and industrialised countries had received the message. When discussion reached the draft Article on voluntary commitments of developing countries (see Chapter 17), these countries had little choice but to accept the fundamental opposition of Brazil, China and India to the provision. Although some developing countries did not agree with the most powerful members of their group, Estrada deleted the draft article to save the Protocol.

The long final night went on and the discussions proceeded article by article. Changes were proposed, rarely accepted unanimously, frequently discussed, sometimes adopted but in most cases dismissed in the usual procedure of those hours: by the Chairman's gavel. In parallel to the discussions in the COW, other meetings took place outside the plenary hall. Several groups discussed substantive issues and legal experts made a last minute effort to cast the emerging texts into proper legal language in a separate room. Reportedly, EU Ministers tried to coordinate their positions, but agreement among the Member States at times took longer than decision-making in the COW. On the Clean Development Mechanism for example (see Chapter 14), EU Member States informed the Chairman of their opposition against pre-budget crediting of emission reductions only when the decision was already taken and could not be reopened.

The longer the discussions lasted, the more decisive became the degree of fatigue. The US delegation was particularly well organised and rotated its chief negotiator every few hours. On a few issues, the advantage of being awake and willing to invest some energy made a decisive difference. On balance, however, tiredness and exhaustion slowly undermined the eagerness of all participants to engage in extensive discussions.

On the last major unresolved issue, Article 10 on commitments for all Parties (i.e. including developing countries), Estrada took the opportunity to compensate partially for the deletion of voluntary commitments of developing countries. Asked by the chairman of the G-77 Tanzania to use his gavel, Estrada opted for the compromise text elaborated by the Swedish Ambassador and opposed by Tanzania. Then, the COW dealt with the decisions of the COP that were to accompany the adoption of the Protocol. Although the exact targets of industrialised countries had still not been revealed, an increasing number of delegates and ministers left the conference to catch their scheduled flights. At 8.45 am, interpreters finished their work and negotiators had to use English only.

Contributing to the spreading feeling that something strange was happening, the President of the conference, Minister Ohki, suddenly appeared around 9 am announcing that he had important business to do in Tokyo and would thus need to resign and be replaced as the President of the conference for the final COP plenary that was still to come. When he left, attention of the somewhat puzzled assembly soon shifted to the final numbers for the targets of industrialised countries in Annex B of the Protocol. For tactical reasons, but also because some countries were waiting for clearance from their capitals, the numbers were only released at the end of the session, when no country dared to voice any opposition. Estrada's final gavel marked the recommendation of the COW to adopt the Kyoto Protocol by unanimity. The sense that something bizarre was happening was then briefly revived, when it was announced that Minister Ohki had retracted his resignation and would thus be available to chair the final plenary.[23]

When the COW completed its business at 10.17 am on 11 December 1997, the final plenary of the Conference of the Parties still had to formally adopt the Protocol and agree on accompanying decisions of the conference. By 1 pm, the tired staff of the Secretariat had produced the final documents including a preliminary final version of the Protocol, and Minister Ohki convened the final plenary. Chairman Estrada introduced the Protocol text, his slow speech giving a clear sign of his exhaustion. Nevertheless, the "Kyoto Protocol to the United Nations Framework Convention on Climate Change", as it was finally named, along with a number of decisions of the conference was adopted. Due to the haste of the last days and hours the adoption was dependent on the understanding that editorial changes would be made before opening the Protocol for signature on 16 March 1998. The third Conference of the Parties was thus able to complete its business, roughly one day late, shortly before 3 pm on 11 December 1997.

When observers and delegates eventually left the Kyoto Conference Hall, they had been part of a historic event. Nevertheless, exhaustion clearly outweighed the feeling of satisfaction. Moreover, hardly any of the delegates that had just agreed to the Kyoto Protocol knew exactly what had been achieved. It took several months for negotiators, NGOs and scientists to fully grasp the meaning of this Protocol. Participants were well aware, however, that Kyoto marked as much the end as the beginning of a negotiating process as the contents of the agreement had to be fleshed out and clarified.

---

23 Reportedly, he had received a phone call at the train station shortly before his departure informing him that the Parliament session he had to attend had been postponed. However, his resignation may also have been intended to open up the opportunity for a more capable diplomat to preside over the final plenary of the COP. After the unanimous adoption of the Protocol in the COW, there was no fear that special diplomatic skills would be needed in the concluding session of COP 3 to save agreement.

Part II
The Provisions of the Kyoto Protocol: A Commentary

8. Overview of Part II

The Kyoto Protocol to the United Nations Framework Convention on Climate Change, being one of the most ambitious treaties ever adopted, it is probably also one of the most ambiguous legal instruments. Much of its content represents "unfinished business" and requires further elaboration in the future. The Protocol features 28 Articles of varying length and detail and two Annexes (see Table 8.1). This chapter gives a brief overview of the main provisions that are dealt with in more detail in the subsequent chapters 9–20. The full text of the Kyoto Protocol can be found in the Annex to this book.

At the centre of the Kyoto Protocol is Article 3. It contains the joint commitment of industrialised countries to reduce their aggregate greenhouse gas (GHG) emissions by at least 5% below 1990 levels in the commitment period 2008–2012. Corresponding country-specific "quantified emission limitation and reduction commitments" (QELROs) are contained in Article 3.7 and Annex B of the Protocol. The Protocol thus contains differentiated binding targets and timetables for industrialised countries.

All GHGs that were on the agenda of the negotiations are subject to control under Article 3: CO_2, CH_4, N_2O, hydrofluorocarbons (HFCs), perfluorocarbons (PFCs), and sulphur hexafluoride (SF_6). These elements of the so-called "six-gas basket" are listed in Annex A of the Protocol together with a list of "sectors/source categories" for emissions of GHGs that are, however, not referred to in the Protocol. The base year for the reductions of CO_2, CH_4 and N_2O will be 1990 and Parties may choose the base year 1995 for the fluorinated gases.

To a limited extent, Article 3 also allows the inclusion of sinks in the calculation of "net" GHG emissions. The related provisions of the Protocol, however, leave room for expanding the definition of relevant activities (currently limited to afforestation, reforestation and deforestation). There is uncertainty about the scope and meaning of this particular provision of the Protocol. Article 3 is analysed and assessed in Chapter 11.

A full assessment of the quantified targets is hardly possible without taking into account the so-called "flexibility mechanisms" (as they provide "geographical flexibility" to Parties in implementing their commitments). These encompass the "joint fulfilment" of commitments by several Parties under Article 4 ("bubbling"), Joint Implementation (JI) of projects among industrialised countries under Article 6, Joint Implementation between industrialised and developing countries within the multilateral framework of the Clean Development Mechanism (CDM) defined in Article 12, and the establishment of a scheme for Emissions Trading among industrialised countries under Article 17 of the Protocol.

The flexibility mechanisms aim mainly at reducing the costs of complying with the Kyoto targets. Emissions Trading and bubbling should allow the transfer of emission allowances between industrialised country Parties. By financing specific

GHG mitigation projects in both Central and Eastern Europe (JI) and in developing countries (CDM), Northern OECD countries hope to tap the low-cost potential for emission reductions in these countries.

The provisions on bubbling are clearly defined, and allow a group of Parties (in particular the EU) to redistribute their assigned amounts of emissions among themselves, as long as their combined amounts specified in Annex B are not exceeded. The other three mechanisms (JI, CDM, Emissions Trading) are generally referred to as the "Kyoto Mechanisms". Concrete rules for these mechanisms were not elaborated and belong to the "unfinished business" of Kyoto. Depending on the eventual design of the Kyoto Mechanisms, they could become a loophole for evading domestic policies and measures in industrialised countries. If this can be avoided, the mechanisms might be considered a major innovation, as they introduce market instruments in international environmental policy to an unprecedented degree. A more detailed assessment is included in the Chapters dealing with the relevant articles of the Protocol (see Table 8.1) and in Chapter 22.

Furthermore, global climate protection will only be successful if the growth of emissions in developing countries is limited. Developing countries have an important role in the Kyoto Protocol (especially with respect to the CDM). Quantified targets for developing countries were not agreed upon in Kyoto, largely because that was not the objective of the process. The provisions in Article 10 applying to all Parties (and thus also to developing countries) remain general in nature, and Article 11 on the Financial Mechanism hardly introduces new elements that can be expected to lead to significant emission limitations for developing countries. The subject of how developing countries should participate in the global effort to curb GHG emissions is thus bound to be one of the major and most contentious issues of the future political process (see Chapter 17).

The effectiveness of the Kyoto Protocol may also hinge upon the implementation of other provisions. The policies and measures specified in Article 2 have a non-binding character. If these provisions are to have any significant effect, mechanisms for transparent implementation of concrete policies and measures need to be developed within the framework of the Protocol (see Chapter 10).

Successful implementation of the Kyoto Protocol will also depend on the design of an effective system for implementation review. The Kyoto Protocol contains related provisions on reporting (Articles 5 and 7), implementation review (Article 8), a multilateral consultative process on questions of implementation (Article 16), a non-compliance procedure (Article 18) and dispute settlement (Article 19). Especially the design of the non-compliance procedure will need to be elaborated in the future process (see Chapter 16).

The institutional structure of the Kyoto Protocol is rather special. The Parties' determined that the Conference of the Parties to the FCCC will also serve as the meeting of the Parties to the Kyoto Protocol. However, since membership is different and voting rights only accrue to Parties to the Protocol, this body is independent from the Convention, despite its name. The same approach was chosen for the Subsidiary Bodies for Scientific and Technological Advice and for Implementation. Furthermore, the Convention's Secretariat has been entrusted with the functions of the Secretariat. The Protocol furthermore makes use of the FCCC's Financial Mechanism (Articles 11, 13–15, see Chapter 18).

Table 8.1: The Provisions of the Kyoto Protocol and the Corresponding Chapters

Protocol Provision	Content[a]	Relevant Book Chapter
Article 1	Definitions	Chapter 9
Article 2	Policies and Measures (PAMs)	Chapter 10
Article 3	Quantified Targets (including Sinks)	Chapter 11
Article 4	Joint Fulfilment (Bubbling)	Chapter 12
Article 5	Methodologies for Emission Inventories	Chapter 16
Article 6	Joint Implementation (JI)	Chapter 13
Article 7	Reporting/National Communications	Chapter 16
Article 8	Implementation Review	Chapter 16
Article 9	Review of the Protocol	Chapter 19
Article 10	General Obligations	Chapter 17
Article 11	Transfer of Financial Resources	Chapter 17
Article 12	Clean Development Mechanism (CDM)	Chapter 14
Article 13	Conference of the Parties serving as the meeting of the Parties (COP/MOP)	Chapter 18
Article 14	Secretariat	Chapter 18
Article 15	Subsidiary Bodies	Chapter 18
Article 16	Multilateral Consultative Process	Chapter 16
Article 17	Emissions Trading	Chapter 15
Article 18	Non-compliance Procedure	Chapter 16
Article 19	Dispute Settlement	Chapter 16
Article 20	Amendments to the Protocol	Chapter 19
Article 21	Amendments to Annexes	Chapter 19
Article 22	Voting	Chapter 20
Article 23	Depositary	Chapter 20
Article 24	Signature and Ratification	Chapter 20
Article 25	Entry into Force	Chapter 20
Article 26	Reservations	Chapter 20
Article 27	Withdrawal	Chapter 20
Article 28	Languages	Chapter 20
Annex A	GHGs and Sectors/Source Categories	Chapter 11
Annex B	Quantified Emission Limitation or Reduction Commitments	Chapter 11

[a] The Articles of the Kyoto Protocol do not have titles; short description of content by authors.

The Kyoto Protocol also provides for a periodic review of its operation in the light of the best available scientific, technical, social and economic information (Article 9). As new information becomes available and political "windows of opportunity" open up, this will generally allow Parties to strengthen (or weaken) the Protocol in the future. In so doing, they can make use of the provisions for amending the treaty and its annexes. Substantial changes of the obligations, in any event, require the consent of a Party to enter into force (for it). Thus, a dynamic development of the Protocol is possible, but will require a high degree of agreement among Parties.

There is some uncertainty as to when and whether the Kyoto Protocol will enter into force. This would require the ratification by 55 Parties representing at least 55% of the total CO_2 emissions of industrialised countries in 1990 (Article 25). By mid-1999, all major industrialised countries were still waiting for the outcome of the negotiations on the Kyoto Mechanisms before deciding on their ratification. Decisions on these mechanisms are scheduled for the sixth session of the Conference of the Parties in late 2000 or early 2001. Even if the major issues can be resolved then, ratification, especially on the part of the US, will remain uncertain due to the domestic political circumstances. Other actors, the EU in particular, might thus take a lead in order to push international climate policy forward. These issues are elaborated with in the Chapters 24 and 25.

The Kyoto Protocol is a remarkable achievement because it introduces binding quantified targets for limiting and reducing GHG emissions of industrialised countries. However, it is unclear as to what extent the Protocol will be able to induce a behavioural change. Uncertainty about the concrete interpretation and the implications of agreed rules is not unusual in the context of international environmental agreements. However, the degree of uncertainty inherent in the Kyoto Protocol has hardly any precedent.

The following Chapters of Part II provide a detailed analysis of the provisions of the Kyoto Protocol. The analysis generally follows the structure of the treaty, as the articles reflect the substantial issues of the negotiations. By referring to Table 8.1, the reader can identify easily in which chapter the respective Articles are dealt with. Each Chapter provides the negotiating background of the respective provisions, analyses their content and concludes with an assessment and outlook.

9. Preamble and Definitions (Article 1)

International treaties usually begin with a preamble, often followed by definitions. So too does the Kyoto Protocol. Although these are standard elements of international law, neither the preamble nor the definitions in the Protocol create specific obligations and thus they will have limited impact on the development of the regime. Neither provision caused major disputes during the negotiations and the text was agreed early on in Kyoto. The following Chapter is confined to a brief analysis of the content of these introductory elements of the Kyoto Protocol.

9.1 Preamble

Preambles to international treaties traditionally provide an introduction, recall some of the motives for their conclusion and the diplomatic steps that led to their adoption, and recognise previous activities in other international fora.[1] They can also contain demands of states that were not incorporated in the substantive part of the treaty. As such, preambular paragraphs do not create legally binding obligations or rights that may be invoked, but they are certainly part of the treaty and thus more than simply material that might help in its interpretation.[2]

In contrast to the many paragraphs that constitute the preamble of the UNFCCC, the Kyoto Protocol is only a few lines long, which mainly recall some provisions of the Convention. This brevity is in part due to the fact that the Protocol is an offspring of the Convention and thus does not require a lengthy introduction. It was also considered prudent to avoid elaborating preambular paragraphs, since they might have aroused unnecessary disputes. One need only compare the different proposals submitted by the G-77 and China, Kuwait and AOSIS to grasp the potential for conflict that was contained in these formulations.[3]

As to the motives and background of the Kyoto Protocol, the Preamble states that the Protocol was agreed pursuant to the Berlin Mandate adopted at COP 1 and recalls the provisions of the Convention. In terms of substance, the Preamble states that the Protocol was agreed in pursuit of the ultimate objective of the Convention as stated in Article 2. This is merely a reminder of the fact that Article 2 of the Convention applies to all related legal instruments. Therefore, the Kyoto Protocol was adopted in order to stabilise GHG concentrations in the atmosphere at a level that would prevent dangerous anthropogenic interference with the climate system (see Chapter 3).

The Preamble further states that the Parties were guided by Article 3 of the Convention in the elaboration of the Protocol. The Principles enshrined in Arti-

1 See also Bodansky 1993, p. 497.
2 See Article 31.2 of the Vienna Convention on the Law of Treaties (1969).
3 FCCC/AGBM/1997/2, pp. 7 et seq.

cle 3 of the FCCC are one of its major elements and are often cited in the sessions of the COP or its subsidiary bodies.[4] Their legal status in the Convention is somewhat ambiguous, since they do not codify any concrete obligations. They nevertheless contain legal standards that may confer certain legal rights upon individual Parties.[5] The formulation of the Preamble in the Protocol, however, does not incorporate these principles into the Kyoto Protocol. Their significance will thus be mainly on the political level, but still with considerable force.

9.2 Definitions

In addition to the definitions in the FCCC (which apply also to the Protocol), Article 1 of the Kyoto Protocol contains some definitions that clarify the exact meaning of several terms and concepts. Some of these are rather obvious statements, such as the Convention meaning the UNFCCC. Other definitions are more remarkable for what they do not provide. For example, it is stated that "Conference of the Parties means the Conference of the Parties to the Convention". This definition, however, fails to define the status of the Conference of the Parties serving as the meeting of the Parties, which is the supreme decision-making body of the Protocol. This omission is due to the rather ambiguous institutional character of this body, which the Parties carefully avoided dealing with (see Chapter 18).

The definition of the Intergovernmental Panel on Climate Change (IPCC) is not, strictly speaking, necessary, but its presence in the definitions article is a political victory that indicates the growing acceptance of the IPCC as the principal scientific advisory body to the FCCC process. The term "Party" is defined as meaning "a Party to this Protocol", an inconspicuous but important clarification that further ensures that the Protocol is legally and politically distinct from the FCCC.

The appearance of a definition of "Parties present and voting" as meaning a Party present and casting an affirmative or negative vote is a novelty in a treaty text, since the usual place for this clarification is in the Rules of Procedure. However, the Parties to the Convention have been unable to agree on a voting procedure and the related provision of the draft Rules of Procedure is not applied in practice.[6] Article 13.5 of the Kyoto Protocol determines that the same rules applied under the Convention will also apply under the Protocol, thereby importing the stand-off under the Convention into the Kyoto Protocol. This definition in the Protocol text therefore is an important step in the process of voting to take place under the Protocol.

---

4 For a discussion of the principles see Bodansky 1993, pp. 501 et seq.; Ott 1996a, pp. 64 et seq.

5 Op. cit. and Dworkin 1972, pp. 24 et seq.; Alexy, Robert 1995, pp. 177 et seq., pp. 182 et seq.

6 See FCCC/CP/1996/2 for the draft rules. These Rules of Procedure have not been adopted but are applied in practice since COP 1, with the exception of Rule 42 on voting.

The final definition provides that a party included in Annex I means a Party included in Annex I to the Convention, as may be amended, or a Party which has made a notification under Article 4.2(g) FCCC. Under the latter provision of the Convention, a developing country Party may become an Annex I Party by way of a simple notification. It may thus profit from those provisions in the Kyoto Protocol that define eligible Parties as Annex I Parties, without having to accept any quantitative target for GHGs in the Protocol. In particular, such a Party will be allowed to participate in Joint Implementation, but not in Emissions Trading since Article 17 requires that any Party wishing to participate in trading must be listed in Annex B to the Protocol.

10. Policies and Measures (Article 2)

Common and co-ordinated policies and measures for combating climate change – informally referred to as PAMs – were a major issue during the pre-Kyoto process because of the high priority the EU attached to this subject. However, the provisions on PAMs finally agreed on in the Kyoto Protocol are hardly far-reaching. In particular, they lack specificity and do not legally require Parties to take any further action. This sharply contrasts with the great importance which the PAMs issue had possessed over long stretches of the negotiations.

Since the EU was the main proponent of common and co-ordinated PAMs at the international level, Article 2 was a defeat for the Union. At the same time, its negotiating history provides an excellent example of the interaction and interrelation between European and international policy making. This is dealt with in more detail in the next section. The Chapter concludes by investigating the future prospects for making progress on common and co-ordinated PAMs within the framework of the post-Kyoto process, where so far much less attention has been paid to this issue than in the run-up to Kyoto. However, Article 2 has the potential to be the basis for a renewed effort to co-ordinate PAMs and to enhance mutual learning from successful strategies to counter climate change.

10.1 Negotiating History

Like the Kyoto process as such, the negotiations on PAMs had roots in the Berlin Mandate. Mainly at the insistence of the EU, the negotiation mandate provided that both PAMs and quantified emission limitation and reduction objectives (QELROs) should be elaborated for Annex I Parties (see Chapter 4 above). Based on this provision, the EU in particular, which has a long history of internally developing and implementing common and co-ordinated PAMs,[1] initiated an effort to elaborate proposals for PAMs. Already at AGBM 2 in October/November 1995, the Union suggested that three types of PAMs be included in the Protocol. These PAMs would be distinguished as follows: the first would be mandatory, the second would not be mandatory but be given "high priority" and the third, also not mandatory, would be assigned high priority in line with national circumstances.[2]

Two factors in particular motivated EU leadership on PAMs: (1) internal EU politics and (2) early US opposition to targets and timetables. Internal EU politics may be illustrated by the history of the proposed European CO_2/energy tax. In that case, the smaller EU Member States, which are comparatively dependent on foreign trade (see Figure 10.1), pushed for common and co-ordinated action at the EU

---

1 On the issue of common and co-ordinated PAMs on climate change in the EU see, for example, Oberthür/Bär 1998; WWF 1996.

2 See the report of AGBM 2, FCCC/AGBM/1995/7; see also Section 4.4.

level by a proposed energy tax. Other Member States were more sceptical of this approach and won the battle in 1994 by pointing out the need to get the other major competitors on board. As a result, the tax proposal was shelved until the US and Japan would be willing to introduce equivalent measures – i.e. the European tax proposal was in effect abandoned.[3]

In subsequent years, similar arguments with respect to PAMs were brought up repeatedly in the EU Ad-hoc Group on Climate Change by the same countries. Essentially, these discussions, as well as those in the AGBM, dealt with the whole array of policy instruments available to policy makers, ranging from information and awareness building, voluntary agreements and economic instruments (like taxes and emission fees) to classic command-and-control instruments. Unable to agree on effective measures within the Union, the problem was transferred to the international level in hope of attracting support from the main competitors, i.e. Japan and the US.

Figure 10.1: Value of Exports of Selected Countries as Percentage of GDP (1997)

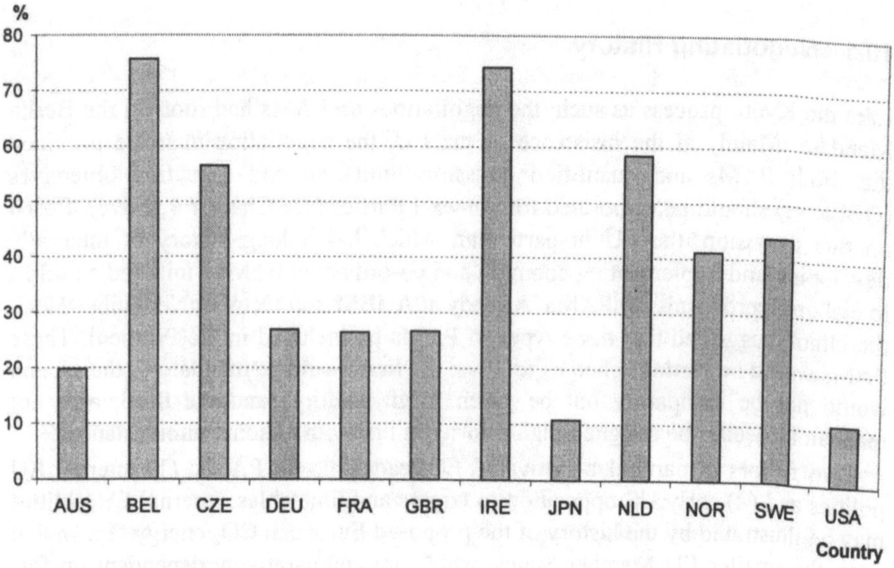

Note: AUS: Australia; BEL: Belgium; CZE: Czech Republic; DEU: Germany; FRA: France; GBR: Great Britain; IRE: Ireland; JPN: Japan; NLD: Netherlands; NOR: Norway; SWE: Sweden; USA: United States of America.

Source: <www.oecd.org/publications/figures/trade_a.pdf> as of July 1999

3 Haigh 1996, pp. 164–166; Liberatore 1995.

This internal driving force was further reinforced by the early US opposition to targets and timetables. Trying to avoid a discussion on QELROs, the US insisted on a process of "analysis and assessment to identify possible policies and measures for Annex I Parties" during the first year of the AGBM.[4] It even invested some effort in evaluating PAMs for reducing energy related GHG emissions.[5] As a consequence, some Member States of the EU were under the impression that the US might be willing to discuss common and co-ordinated PAMs rather than targets and timetables. In line with this, discussions in the OECD expert group during the first year of the AGBM process focused on PAMs. Under the heading "Policies and Measures for Common Action", a series of Working Papers was issued on, *inter alia*, subsidies, carbon taxation, and emissions from bunker and aviation fuels.[6]

Based on these two main factors, EU leadership soon acquired a self-reinforcing dynamic. During 1996 and 1997, the EU submitted long lists of PAMs to be included in the proposed Annexes.[7] Having reached that stage, the EU doggedly insisted on its approach and showed little willingness to compromise, even when it became clear that there was little support from others. While this might have been intended to maintain pressure on JUSSCANNZ regarding other issues (QELROs, flexible mechanisms), the ineffectiveness of such a strategy became increasingly apparent. Given the widespread opposition to mandatory PAMs, the EU proposal had virtually no prospect of being accepted, and was thus of very limited value as a bargaining chip.

Already at COP 2, the US strongly rejected any attempts to agree upon any international harmonisation of PAMs after it decided to support legally binding QELROs.[8] As the US chief negotiator at Kyoto put it, the pre-Kyoto US government position "firmly opposed mandatory, harmonised policies and measures that would be imposed upon us in order to reach our target".[9] Indeed, even its early analysis and assessment of PAMs (see above) was in no way directed at *co-ordination*. This was in line with the views of the great majority of Parties, including most of JUSSCANNZ, most developing countries, and OPEC. While the G-77 stressed that any PAMs implemented by industrialised countries should not

4 Decision 1/CP.1 in FCCC/CP/1995/7/Add.1, para. 4; see also the reports of the first three sessions of the AGBM in 1995 and 1996: FCCC/AGBM/1995/2 and 7 and FCCC/AGBM/1996/5.

5 US Department of Energy 1996.

6 See <http://www.oecd.org/env/cc/freedocs.htm> as of December 1998; see also Chapter 5.

7 See the respective parts in FCCC/AGBM/1996/10; FCCC/AGBM/1997/2 and Add.1; FCCC/AGBM/1997/3/Add.1. The last-mentioned document contains all the major proposals on PAMS submitted by Parties. See also Box 10.1.

8 Speech by Timothy Wirth, Under Secretary of State for global affairs, reprinted in: Global Issues – Confronting Climate Change, April 1997, Vol. 2, No. 2, pp. 6–8.

9 Stuart Eizenstat, Statement before the House International Relations Committee, Washington, D.C., 13 May 1998, as released by the Bureau of Oceans and International Environmental and Scientific Affairs (on file with the authors).

have a negative impact on developing countries, OPEC in particular requested the exclusion of taxation from the menu of available PAMs. In addition, these oil exporting countries requested a compensation mechanism for any negative effects on their economies due to the implementation of policies and measures under the Protocol by industrialised countries (see also Chapter 17).

Only AOSIS and some of the CEIT countries of Eastern Europe striving to become EU members showed sympathy for the EU approach – without, however, investing major efforts to push the issue. AOSIS proposed a procedure-oriented approach that might have provided a solution to the deadlock. A co-ordination mechanism was to be set up to provide advice on the full range of PAMs for which co-ordination might be beneficial. Since agreement on substance was not in sight and was perceived to require far more detailed discussions and evaluations, AOSIS's objective was to establish a post-Kyoto process under the Protocol to gradually develop a common understanding about common and co-ordinated PAMs in different policy domains.[10]

Japan also acknowledged the usefulness of co-ordinating PAMs and proposed that each industrialised country choose a certain number of them from a menu ("pick and choose"). In addition, Japan suggested that industrialised countries elaborate a national plan containing policies and measures. Progress made in implementing these policies and measures would then be evaluated using internationally-agreed performance indicators. This would have made it possible to ascertain whether and to what extent countries had reached their quantitative emission objectives through implementing PAMs. Other countries were only prepared to consider co-ordinated PAMs for specific areas (e.g. New Zealand in respect of bunker fuels).[11]

With the Kyoto conference approaching, it became increasingly evident that the EU approach to PAMs stood little chance of being adopted. A round-table discussion at AGBM 4/COP 2 in mid-1996 clearly revealed the extent of the general and widespread opposition.[12] As a result, PAMs moved increasingly lower down the agenda of the negotiators as Kyoto approached. In Kyoto, PAMs were among the issues that were resolved comparatively easily and the issue was settled already at the beginning of the second week of Kyoto. The text of Article 2 as contained in the Chairman's draft of 9 December 1997, which reflected virtually none of the EU's preferences, appeared almost unchanged in the Kyoto Protocol.[13] This indicates that the EU itself had eventually realised it had lost the battle. Consequently, Article 2 was adopted without a major discussion of details during the final night of Kyoto (see Chapter 7).

---

10 FCCC/AGBM/1997/2, pp. 16, 72; see also Yamin 1998a, p. 116.

11 For the different positions/proposals on PAMs and their development see FCCC/AGBM/1996/10; FCCC/AGBM/1997/2 and Add.1; FCCC/AGBM/1997/3/Add.1.

12 Earth Negotiation Bulletin, Vol. 12, No. 38, 22 July 1996.

13 Compare Article 2 of this draft, FCCC/CP/1997/CRP.4 with Article 2 of the Kyoto Protocol.

10.2 PAMs in the Kyoto Protocol

Given the long-standing efforts in support of PAMs, especially from the EU, the four paragraphs of Article 2 of the Kyoto Protocol containing soft legal language appear to be a rather meagre result. Contrary to the Union's goal, Article 2 does not require industrialised countries to pursue specific PAMs. On the positive side, and in contrast to the FCCC, Article 2 emphasises a number of priority areas for policy action. These may serve to support political arguments in the process of domestic implementation and may provide the basis for further work on internationally co-ordinated PAMs in the future.

Article 2.1(a) states that each Annex I Party shall "implement and/or further elaborate policies and measures in accordance with its national circumstances" in areas "such as" (replacing the stronger "in particular" proposed by the EU) energy efficiency, sinks and reservoirs, agriculture, new and renewable energy sources and carbon sequestration, market instruments[14] and market imperfections, transport, and waste management as well as other sectors.

Furthermore, Article 2.1(b), in conjunction with Article 2.4, provides the basis for engaging in a process of further discussions on international co-ordination of PAMs. Article 2.1(b) requests Parties to co-operate in order "to enhance the individual and combined effectiveness of their policies and measures". The COP/MOP is called upon to consider ways to facilitate such co-operation during its first session or as soon as practicable thereafter. The prospect for a future process is further enhanced by Article 2.4, which calls upon the COP/MOP to "consider ways and means to elaborate the co-ordination" of PAMs. However, this is contingent upon the COP/MOP finding such co-ordination to be beneficial. While both paragraphs leave open the option of future co-ordination, their wording, mainly due to the intransigence of the US and other JUSSCANNZ members, clearly falls short of placing a binding commitment on Parties.

Article 2.2 specifically addresses emissions from aviation and marine "bunker" fuels (i.e. fuel used in international air and maritime transport), thus acknowledging their particular importance in the context of co-ordinating PAMs. Emissions from bunker fuels accounted for approximately 3% of the CO_2 emissions reported by Annex I Parties in 1990, and these emissions have been steadily increasing.[15] The demand for policy co-ordination in this area is particularly high because international competition in the relevant businesses is intense. As a consequence, governments face particular political difficulties to regulate bunker fuel emissions unilaterally. Also, because of their international character, emissions from bunker fuels are reported separately under the FCCC and do not count towards the emissions of a country. The incentive to restrict those emissions is thus lacking. A decision on the allocation of bunker fuel emissions had proven impossible to reach in

14 On the prospect for harmonised ecological taxes after Kyoto see Weizsäcker/Ott 1998a.

15 See FCCC/SBI/1997/19 and Add.1; FCCC/CP/1998/11 and Add.1 and 2.

the pre-Kyoto process (see Section 10.3). Since there was no prospect of resolving the issue at Kyoto, bunker fuels had a low profile on the agenda of COP 3.

Under these circumstances, the inclusion of Article 2.2 into the Kyoto Protocol was mainly due to the continuing concern of a number of Parties (including the EU, New Zealand and Switzerland) about an area that is not, strictly speaking, under the full umbrella of the climate regime. However, the Protocol in effect transfers the issue to other international fora: Parties are called upon to work towards the limitation and reduction of emissions from bunker fuels, through the International Civil Aviation Organization (ICAO) and the International Maritime Organization (IMO), which are the leading international bodies that deal with air and maritime transport.

Finally, Article 2.3 requests Parties to "strive to implement policies and measures ... in such a way as to minimise adverse effects, including the adverse effects of climate change, effects on international trade, and social, environmental and economic impacts on other Parties". The paragraph reflects an amalgam of Parties' concerns. While reference to trade and economic impacts can be said to reflect concerns raised by G-77 and OPEC in particular, this is balanced by the references to environmental effects, the negative impacts of climate change and the principles contained in Article 3 of the Convention.

Nevertheless, Article 2.3, together with Article 3.14 and in conjunction with Decision 3/CP.3, is generally seen as the "sweetener" for OPEC that facilitated its acceptance of the Kyoto Protocol.[16] Decision 3/CP.3 adopted in Kyoto requested the SBI "to undertake a process to identify and determine actions necessary to meet the specific needs of developing country Parties ... arising from adverse effects of climate change and/or the impact of the implementation of response measures" (para. 1). These provisions provide a basis on which OPEC could build its case for compensation in the post-Kyoto period.

10.3 Assessment and Outlook

10.3.1 The Prospects for Co-ordinating PAMs Under the Kyoto Protocol

The international discussions on PAMs have lost momentum in the post-Kyoto period. Even the EU seems to have accepted that there will not be mandatory PAMs in the near future. Article 2 assigns the task of taking further action to the COP/MOP of the Protocol, but does not indicate a need for any preparatory work. More importantly, COP 3 did not request the Subsidiary Bodies to engage in preparing decisions of the (first) COP/MOP – as it did with respect to methodological issues and the Kyoto Mechanisms. Thus, few discussions were conducted on the issue of PAMs in the immediate post-Kyoto period.

---

16 See Yamin 1998a, p. 117.

However, based on the provision in Article 2.1(b) that the COP/MOP should become active "at its first session or as soon as practicable thereafter", PAMs were back on the international climate policy agenda at COP 4 in Buenos Aires. The Parties at COP 4 determined that SBSTA should conduct preparatory work for the "consideration of ways to facilitate co-operation to enhance the individual and combined effectiveness of policies and measures under Article 2.1(b)".[17] The Secretariat was requested to prepare a report on "best practices" by the end of 1999 as a basis for SBSTA's consideration of PAMs at SBSTA 11. Subsequently, a workshop on best practices with respect to PAMs is to be organised and the results will be reported to COP 6 in 2000.

These activities could lay the basis for a *learning approach* to further international discussion on PAMs. In the end, all countries will need to fulfil their commitments to a large extent by domestic action. Indicators could be developed to measure the effectiveness of various PAMs in domestic implementation by different countries and to increase their comparability and transparency, as called for in Article 2.1(b). Development of such performance indicators, that were envisaged in the Japanese proposals on PAMs in the Kyoto process (see above), could be linked up with other international processes concerned with indicator development that are ongoing in the framework of the CSD, OECD and others.[18] This might ultimately provide a basis for a "PAMs ranking" of countries that could provide a useful incentive for implementing effective PAMs.

Not only the Kyoto process, but the entire history of international environmental policy provides ample evidence of the difficulties in agreeing on *common* PAMs. This is because national governments and parliaments fear losing flexibility and sovereign control over political developments. However, *co-ordination of PAMs* does not necessarily face the same obstacles.[19] Based on the learning process envisaged above, the growing demand for internationally co-ordinated PAMs in conditions of economic globalisation might eventually be realised.[20] Finding agreement on co-ordinating PAMs or even taking common action (e.g. introduce international taxation) will nevertheless require a long-term effort in trust and consensus building, as envisaged in the proposals of AOSIS in the Kyoto process (see above). In addition to addressing the concerns about loosing sovereign control and democratic legitimacy, this effort will have to take diverging national interests

17 Decision 8/CP.4 in FCCC/CP/1998/16/Add.1, p. 38.

18 See UN 1996; 1999; OECD 1993a and b.

19 While "common" PAMs generally refers to the harmonisation of policies (e.g. the introduction of common technical standards for several countries), "co-ordination" of PAMs is less stringent as it requires the introduction of similar policy instruments without exact harmonisation (e.g. the co-ordinated introduction of a CO_2/energy tax without harmonisation of the tax level).

20 On the relationship between economic globalisation and environmental policy see OECD 1997.

rooted in political cultures (e.g. opposition to taxation in the US) and the export dependency of some national economies into account (see Figure 10.1).

Because of these difficulties, discussions on co-ordinated (and in some cases also common) PAMs might benefit from the establishment of an institutionalised process initiated by the COP (see Chapter 25). To develop trust and consensus, it might initially be best to focus on areas where the demand for (common and) co-ordinated action is most obvious (see also Box 10.1). These are, first, the areas where countries fear competitive disadvantages or interference with international free trade as a result of unilateral action. This applies to policies and measures such as energy taxation, removal of subsidies for climate-damaging activities (like burning coal) and bunker fuels. In some cases, second, co-ordinating efforts promises to create substantial synergies (for example, regarding energy efficiency standards of appliances).

Box 10.1: Selected Priority Policies and Measures as Proposed for Co-ordination by the EU Prior to Kyoto

- *Removal of barriers which prevent the penetration of potentially cost-effective renewable energy routes in the market*
- *Energy efficiency standards and labelling for household appliances and other equipment*
- *Minimum excise duties on fuels to be applied to all Annex I countries*
- *Target value(s) for average fuel efficiency/average CO_2 emissions of newly registered cars*
- *Reduction/removal of certain subsidies (e.g. for fossil fuels)*
- *A framework for the introduction of an environmental taxation scheme for all Annex I Parties*
- *Introduction of international voluntary agreements in internationally oriented industrial sectors*
- *International co-ordination on standards for energy efficiency and on the use of fiscal incentives*
- *International co-operation in PAMs development with industrial sector organisations*

Adapted from FCCC/AGBM/1997/3/Add.1.

The co-ordination of PAMs under the Kyoto Protocol could take the form of political agreement and legal commitment through decisions of the COP and the COP/MOP. Far-reaching agreements have been laid down in decisions by the supreme governing bodies in other international environmental agreements that have proven effective.[21] With respect to PAMs it is important not to force other Parties to take measures they do not like, but to enhance the mutual understanding, political will and consensus that co-ordination can be beneficial. Economic globalisation and growing economic and political interdependence is set to increase the demand for such co-ordination with time. By the same token, if policy co-ordination is pursued seriously, the need to consider possible conflicts with other international regulations (e.g. trade rules) might also increase (see also Chapter 23).

The issue of compensation is likely to be discussed separately from the PAMs co-ordination, although OPEC has in the past tried, and has to some extent been successful, to link both issues. This is evident by Article 2.3 of the Protocol, which calls for the minimisation of diverse negative impacts of PAMs on other Parties. COP 4, however, clearly separated the issues by including them in two separate decisions.[22] Compensation will encompass the broader issue of dealing with countries particularly affected by climate change and response measures, of which compensation requested by OPEC will only form a part. It is hard to imagine that Annex I Parties would agree to establish a compensation mechanism as envisaged by OPEC. While engaging in constructive discussions on the situation in countries heavily dependent on fossil fuel exports may serve the purpose of getting energy exporters involved, the above-mentioned provisions also provide the basis for OPEC requests for compensation, and as such have the potential to further disrupt the process in the future.

10.3.2 Marine and Aviation Bunker Fuels

Bunker fuel emissions have been rising dramatically for a number of years. In its special report on aviation, the IPCC provided new impetus to the discussions on international transport emissions.[23] The IPCC concluded that aviation was accountable for about 3.5% of the man-made climate change in 1992 and that CO_2 emissions from this source might increase up to tenfold by 2050.

The issue has been on the agenda of the international climate negotiations for several years. In December 1996, SBSTA identified eight options for resolving the pending issue of allocating emissions from marine bunker and aviation fuel. It rejected three: allocation in proportion to national emissions, according to the country of origin of the passenger/cargo, and according to the location of actual emission. The remaining five options were selected as the basis for "further work on the issue",[24] i.e. no agreement could be reached on their importance:

---

21 See Ott 1998a, pp. 166 et seq.

22 Decision 8/CP.4 in FCCC/CP/1998/16/Add.1 on PAMs (see above) and Decision 5/CP.4 in FCCC/CP/1998/16/Add.1 establishing a work programme for implementation of Article 4.8 and 4.9 of the FCCC and Articles 2.3 and 3.14 of the Protocol.

23 IPCC 1999.

24 FCCC/SBSTA/1996/20, p. 12.

- no allocation,
- allocation to the country where the bunker fuel is sold,
- allocation to the country of the transporting company, the country of registration of the aircraft/vessel, or the country of the operator,
- allocation to the country of departure or destination of the aircraft/vessel (including some kind of sharing of emissions between them), and
- allocation to the country of departure or destination of the passenger/cargo (including some kind of sharing of emissions between them).[25]

Parties did not make decisive progress on the allocation of these emissions for a combination of two reasons. First, given the rising trend of aviation and marine bunker fuel emissions, different countries would have been disadvantaged in meeting emission reduction obligations depending on which option (or combination) had been selected. As a consequence, for example, those countries that serve as bases for international air and sea transport, such as the Netherlands, with large ports such as Rotterdam, have rejected responsibility for emissions that occur from burning the fuels they sell at those transport platforms (see Figure 10.2). Second, the options available for these countries to unilaterally limit emissions from bunker fuels are rather limited because of the international character of the businesses and agreements concerned, and the roles of ICAO and the IMO. For example, guidelines of the ICAO, on which a number of bilateral air-traffic agreements are based, preclude unilateral taxation of aviation fuel. Norway tried to overcome the obstacles and introduced taxation of kerosene in spring 1999, but was forced to abandon the tax because of complaints of international airlines and their refusal to pay.[26] Competition among ports and the competence of the IMO over marine bunker fuel emissions further stifles unilateral action in this area.[27] These were also the main reasons why bunker fuel emissions were granted a special status under the FCCC (i.e. not allocated to national emissions).

Under these circumstances, there is a strong case in favour of common or co-ordinated action by Parties on aviation and bunker fuels. Article 2.3 acknowledges this by calling for action by ICAO and IMO (see above). Since no agreement on allocation could be reached, the Kyoto negotiators transferred the issue to the other international authorities mentioned. While this approach might help keep the issue on the international agenda and avoid conflict with ICAO and IMO, it also moved the issue away from the international climate arena. If ICAO/IMO do not act, several years could be lost in relation to this important area of international climate policy.

Deferring the regulatory authority for an important aspect of climate policy to international organisations in which generally economic and transport ministries

---

25 FCCC/SBSTA/1996/20 and FCCC/SBSTA/1996/9/Add.1.

26 See ENDS Daily, 22 January 1999; EWWE, Vol. 8, No. 3, 5 February 1999, p. 9.

27 On taxation of marine bunker and aviation fuel see, for example, Michaelis 1997a and b.

have the lead might look like setting the fox to keep the geese. Not surprisingly, the 32nd Session of the Assembly of ICAO in 1998 showed little willingness to take action but decided to study policy options to limit or reduce the GHG emissions from civil aviation until the next assembly in 2001.[28]

The IPCC special report on aviation mentioned above might lead Parties to the FCCC to increase their pressure on ICAO and IMO. Irrespective of action by ICAO/IMO, however, Parties to the FCCC/the Kyoto Protocol can decide to allocate emissions from aviation and bunker fuels on the basis of one (or a combination) of the criteria mentioned above. This would provide a much needed incentive to Parties to take domestic and/or co-ordinated action on this issue.

Figure 10.2: Bunker Fuel Emissions of Selected Countries as Percentage of their Total CO_2 Emissions in 1990

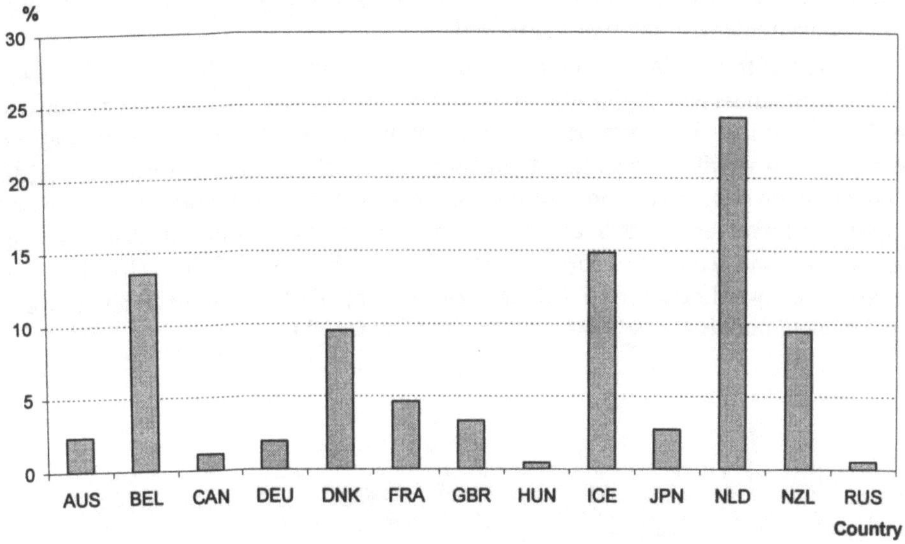

Note: Allocation to the country where the fuel has been sold. AUS: Australia; BEL: Belgium; CAN: Canada; DEU: Germany; DNK: Denmark; FRA: France; GBR: Great Britain; HUN: Hungary; ICE: Iceland; JPN: Japan; NLD: Netherlands; NZL: New Zealand; RUS: Russia.

Source: FCCC/CP/1998/11/Add.2

28 See Statement from the International Civil Aviation Organization (ICAO) to the Fourth Session of the Conference of the Parties to the United Nations Framework Convention on Climate Change (Buenos Aires, 2–13 November 1998) available at <http://www.icao.int/icao/en/env/cop-4.htm>.

10.3.3 Conclusion

Like various other provisions of the Kyoto Protocol, Article 2 constitutes a potential basis for future action. While it goes beyond the FCCC by mentioning PAMs and listing several key areas of action, the language of Article 2 does not legally bind Parties to take any specific action. Despite this limitation, Article 2 may, even in its current form, influence the implementation of the Kyoto Protocol, as the listed key areas for action can provide influential arguments for the related domestic political discussions.

The international process dimension of Article 2 may, in comparison, turn out to be more important. By envisaging a political and cognitive consensus building process on PAMs and their effectiveness (including possible co-ordination) based on Article 2.1(b), as well as by laying the foundation of a potential process of co-ordinating PAMs under Article 2.4, the importance of PAMs under the Kyoto Protocol can grow far beyond the current Protocol text.[29] Much will depend on whether Parties will be able to begin a learning process so as to build up a common understanding of the issues involved.

Emissions from aviation and marine bunker fuels will provide an early test case of the willingness and ability of Parties to meet the demand for common and/or co-ordinated action. The demand for co-ordination is perhaps in no other area as obvious as it is with respect to international transport emissions. However, PAMs co-ordination can also provide considerable mutual benefits and remove barriers in other important areas such as energy taxation, energy efficiency standards of appliances, and removal of subsidies for climate-damaging activities (like burning coal). The co-ordination of PAMs is therefore important for implementing effective climate policies in the next century (see Chapter 25).

---

29 See Weizsäcker/Ott 1998a.

11. Emission Limitation and Reduction Commitments (Article 3)

The negotiations on quantitative targets, the so-called "quantified emission limitation and reduction objectives" (QELROs), were at the core of the pre-Kyoto negotiating process. Lacking legally binding obligations, the Convention had proven inadequate in bringing about a real change of domestic climate policies in almost all industrialised countries (see Chapter 4). Acknowledging the inadequacy of existing commitments, the Berlin Mandate passed at COP 1 set the stage to establish, for the first time in history, specific, legally binding obligations in the Kyoto Protocol to limit or reduce those greenhouse gases primarily responsible for global warming. The Kyoto Protocol to the FCCC contains, for those industrialised countries listed in Annex B, legally binding, differentiated reduction (or limitation) obligations for a basket of four gases (CO_2, CH_4, N_2O, SF_6) and two groups of fluorinated gases that are supposed to amount to an overall reduction in the commitment period from 2008–2012 of at least 5%.

This was the result of complex and intricate negotiations that are described in Section 11.1. Section 11.2 will analyse some of the main features of the agreed targets and timetables, which determine their meaning and effectiveness. One of these features is the treatment of sinks, dealt with separately in 11.3. The assessment, contained in Section 11.4, will shed some light on the likely effectiveness of the targets and timetables included in the Kyoto Protocol.

11.1 Negotiating History

The original protocol proposed by AOSIS in September 1994 and the supplemental "Elements for a Comprehensive Protocol to the UNFCCC", tabled by Germany a few days later, remained, for a considerable time, the only substantive proposals in the AGBM process.[1] While the main element of the German proposal was a list of PAMs that Annex I Parties would have to adopt, the AOSIS proposal incorporated the "Toronto target" of a 20% reduction in the emissions of carbon dioxide by the year 2005. The AOSIS proposal also included a provision concerning the adoption of targets for other major GHGs, including fluorocarbons, but did not indicate a specific target or timetable. The 20% target thus remained the strongest proposal on the negotiating table until the end of the negotiations. Furthermore, it was adopted by almost all environmental NGOs and thus served as a common denominator that kept together the oft fractious environmental groupings.

1 See FCCC/AGBM/1996/MISC.2 for the two proposals.

In October 1995, the EU proposed an "Outline of Possible Protocol Structure", for tabling at AGBM 2, that focused on PAMs (see Chapter 10).[2] The second EU proposal did not go much further than to elaborate on the previously submitted protocol structure.[3] Its main novelty, however, was a new "Annex X" that contained a list of those countries to be bound by "Commitments by developed country Parties and other Parties". This was a first open hint at the possibility to include developing countries' commitments in the protocol – a clear deviation from the Berlin Mandate – and was not received with sympathy by those countries.

In March 1996, Germany attempted to accelerate the negotiating process by proposing a 10% reduction for industrialised countries in the emissions of CO_2 by 2005 and of 15–20% by the year 2010, taking 1990 as a baseline.[4] While the dates 2005 and 2010 had been agreed with the rest of the EU, the level of the proposed targets was solely Germany's initiative. Although no other country officially submitted any commitment proposals, seven other EU Member States orally announced their own preferences during the negotiations.[5] None of the major non-European players, such as Japan or the US, submitted any concrete reduction or limitation proposal until late 1997.

The negotiating process was given another major boost on 3 March 1997, when the Council of Environment Ministers of the European Union adopted, on the first day of AGBM 6, a common target proposal and agreed on an internal burden-sharing arrangement (see Table 6.1). This proposal was for a 15% reduction in 2010, taking 1990 as a base year, for a basket of three gases (CO_2, CH_4, N_2O).[6] Later in June 1997, the EU Environment Ministers were able to agree on a 7.5% reduction target in 2005 for the same basket of gases.[7]

Japan, which was one of the most fervent supporters of a differentiated approach, was next in tabling its proposal in September 1997. Differentiation of commitments had been supported by countries with special circumstances such as Japan, Australia, Norway and Iceland; in the European Union, this position was taken by France because of its low per capita emissions on account of the high percentage of nuclear power in its electricity supply.[8] Whereas Australia and Norway were the first to demand differentiated targets in general,[9] Japan's proposal to differentiate was concrete. Its starting point was a flat-rate target of a 5% reduction of CO_2, CH_4 and N_2O by 2010, that was then adjusted according to three criteria:

- emissions per capita,
- emissions per GDP, and

---

2 FCCC/AGBM/1996/MISC.2, p. 18.
3 FCCC/AGBM/1996/MISC.2/Add.2, p. 19.
4 FCCC/AGBM/1996/MISC.2, p. 30.
5 See the list in the European proposal, FCCC/AGBM/1996/MISC.2/Add.2, p. 20.
6 FCCC/AGBM/1997/MISC.1/Add.2, p. 35.
7 EWWE, 20 June 1997 (special issue), p. 2.
8 FCCC/AGBM/1997/MISC.1, p. 22.
9 FCCC/AGBM/1996/MISC.2/Add.2, pp. 3 and 25.

- population growth.[10]

As a result, instead of the initial 5% reduction nominally suggested, Japan's own commitment would have been a mere 2.5%, Australia's 1.8% and Germany's 3.7%.[11] Further flexibility in other provisions meant that actual commitments would even have been lower, whereas the legally binding part of the Japanese proposal referred only to the stabilisation of emissions on 1990 levels. Nevertheless, with this proposal by Japan, an approach based on differentiated commitments was firmly placed on the table. This ultimately proved decisive for the successful conclusion of the negotiations in Kyoto.

The notion of differentiated targets was not supported by the United States, mainly because under every conceivable rational criteria it would have had to accept higher obligations than the rest of the industrialised world. In December 1996, the US instead opted for a "flexible" approach that included:

- multi-year targets,
- banking (transfer of unused emissions into the next budget period) and borrowing (transfer from the next budget period) over several budget periods,
- a comprehensive approach for six gases,
- emissions trading, and
- Joint Implementation.[12]

The success of the US strategy is evident by the fact that the only element from this list missing from the Kyoto Protocol is the concept of "borrowing" emissions from the next budget period.[13]

President Bill Clinton announced the US's proposed numerical target late in the negotiating process, on 22 October 1997, in an address to the National Geographic Society. The call for "reaching" 1990 emission levels by 2008–2012, and reductions below 1990 levels in the five-year period afterwards, disappointed the EU, most developing countries and environmental groups. It also sounded hollow compared to the eloquent speeches on the threat of climate change that the President had given in the course of the White House Initiative on Global Climate Change. Furthermore, the proposal was linked to developing country participation, contrary to the Berlin Mandate. This condition was a direct response to the Byrd-Hagel "Sense of the Senate" Resolution, unanimously passed, that made US ratification of any legal instrument contingent upon the "meaningful participation" of devel-

---

10 FCCC/AGBM/1997/MISC.1/Add.6.

11 This is in contrast with Germany's own commitment to reduce CO_2 emissions by 25% in 2005. For the calculations see Greenpeace 1997.

12 FCCC/AGBM/1996/MISC.2/Add.4, p. 26.

13 However, "penalty borrowing" might find its way back into the Protocol through a suggested consequence for "non-compliance", see Section 16.3.

oping countries (see Section 6.2).[14] At least, the announcement did not include the idea of a "relief mechanism" or "cost cap" that had been previously considered by the White House. This would have allowed governments to sell "paper credits", not supported by actual emission reductions, under an emissions trading regime if the world price of carbon emission rights exceeded a predetermined level (US $ 25 to US $ 50 per ton of carbon).

President Clinton's announcement had been narrowly preceded by a G-77 and China proposal calling for gas-by-gas reduction targets for CO_2, CH_4 and N_2O for industrialised countries of 7.5% by 2005, 15% by 2010 and 35% by 2020 (Section 4.6). Developing countries thus effectively supported the EU's position on targets. Two other important players in the JUSSCANNZ group had waited for the US and announced their targets only at Kyoto.[15] Canada, whose Prime Minister Chrétien had declared that his country "must do better than the US" in climate policy, proposed a 3% reduction from 1990 levels in "greenhouse gas emissions" by the year 2010, and a further 5% by 2015. New Zealand proposed, one day later, a reduction target of 5% below 1990 levels, to be achieved within a five year budget period beginning no earlier than 2005. The timing of these proposals gave the impression of being part of a carefully orchestrated performance by JUSSCANNZ. Whereas the EU appeared to be isolated with its proposal to reduce by 15% in 2010, the US took the opposite extreme position by merely proposing "stabilisation". Japan, meanwhile, was given the honour of proposing a medium-sized target of nominally 5% (but actually closer to 3%) that could be presented as the much needed compromise. Canada and New Zealand, for their part, submitted targets that occupied the middle ground.

These were the main positions when negotiations began to get serious in Kyoto. It was quite difficult, however, to assess the actual level of allowable emissions under the various proposals, since they differed with respect to the number of gases included, the treatment of sinks and the role of the flexibility mechanisms. In order to improve its negotiating position, the US, at AGBM 8 in October 1997, circulated a calculation which alleged that the EU proposal (minus 15% in 2010) amounted only to a *de facto* reduction of 5.2% if account was taken of the three additional gases in the US proposal and if sinks were included in the calculation.[16] The Japanese proposal, according to this paper, would actually allow for an increase of emissions of between 5.4 and 7.9%. These calculations were later cor-

[14] The term "meaningful" was missing in the President's written proposal, although it did appear in his address to the National Geographic Society.

[15] Switzerland, which was part of this group, had already announced in March 1997 that it would support differentiated commitments leading overall to a reduction of 10% in Annex I emissions in 2010 (FCCC/AGBM/1997/MISC.1/Add.2, p. 67). Australia, in an apparent attempt to ridicule the European burden sharing arrangement, called for differentiated targets in the range of "between a 30% reduction by 2010 and a 40% increase by 2010" (FCCC/AGBM/1997/MISC.1/Add.2, p. 6).

[16] "An Analysis of Effects Of Different Proposals on Annex 1 GHG Emissions", anonymous, undated (on file with the authors).

rected by the US after considerable criticism and the EU target was alleged to be between 8.5 and 11%, if the additional gases were taken into account. This confusion illustrates the rather arcane nature of those targets even to the people most closely involved, although they were the prime focus of public attention during these negotiations.

Most of the basic obligations were still unresolved when the delegations met in Kyoto. During 1996 and 1997, the formal elaboration of the negotiating text had been promoted by the Chairman of the AGBM, the Argentinian Ambassador to China Raúl Estrada. With the help of the UNFCCC Secretariat, he compiled several versions of the negotiating text, which were gradually refined over time (see Chapters 4 and 7).

However, progress was slow and the number of options tended to increase over time rather than diminish. As stated by Ambassador Estrada after AGBM 8, the text contained a "mosaic of different positions".[17] The real negotiations were facilitated by him behind the scenes. Several delegations, among them the EU, the US and AOSIS, expressly urged the Chair at the close of AGBM 8, to continue seeking a solution and "prepare alternative text".[18]

As regards the question of a point target versus a budget approach, agreement was reached at the ministerial meeting of key delegations in Tokyo on 8 November 1997 (see Section 7.1). Contrary to its stated position up until that point, a budget period of five years was agreed to by the EU. The possibility of allowing Joint Implementation between industrialised countries was undisputed as well. No agreement existed, however, with respect to the starting date for the obligations (2003 versus 2008), the level of the target, the question of differentiation, banking, the basket of GHGs (three vs. six), the inclusion of sinks, emissions trading and Joint Implementation with developing countries.

In the early days of COP 3, the US had changed its position on differentiation and was prepared to accept "limited, carefully bounded differentiation".[19] This paved the way for informal talks behind the scenes predominantly between the "big three" – the US, the EU and Japan.

In the evening of 9 December 1997, Chairman Estrada put some numbers on the table that supposedly reflected the state of the negotiations.[20] The numbers for the EU and other European countries (minus 8%) and New Zealand (stabilisation) remained unchanged afterwards. The targets for Russia, the Ukraine (minus 5%), Australia (plus 5%) and a number of Eastern European countries, however, were stricter than the final result, whereas the targets for the United States and Canada (minus 5%), Japan (minus 4.5%) and Norway (plus 5%) still represented lower figures than those accepted in the final deal. These reduction and limitation commitments would relate to the commitment period 2006 to 2010 and would cover a

17 BNA International Environment Reporter, 12 November 1997, p. 1037.
18 Authors' observation.
19 Authors' observation.
20 FCCC/CP/1997/CRP.4, Annex B.

basket of the three main gases (CO_2, CH_4 and N_2O), with a second basket containing the other three gases and groups of gases (HFCs, PFCs and SF_6) to be adopted at COP 4 with a separate timeline.[21] The latter gases were not included at this point because Germany, France and Austria insisted on a three-gases basket. Agreement at this stage had also emerged that "borrowing" emissions allowances from future commitment periods would not be part of the deal.

Eventually, agreement on a six-gases basket was reached during the same night and the trilateral negotiations entered the final stage. The EU was under pressure on account of a Russian proposal to come to an umbrella agreement with JUSSCANNZ. This was a direct response to the EU bubble and the insistence of the European countries on the need for strict rules for trading. This umbrella agreement would be in the context of Article 4 on bubbling (Chapter 12), whereby non-EU industrialised countries would enter into an agreement on the "joint fulfilment" of their obligations, exclusively using the Russian and Ukrainian "hot air" available due to the steep fall in emissions that had taken place there since 1990.[22]

At this stage, negotiations within the triad made up of Japan, the US and the EU had narrowed down to a limited range of options. The main reduction targets discussed were, for the EU, the US and Japan respectively, 8, 7 and 6% (the final outcome), 7, 7, 6% and 6, 6, 5%. Although Russia was not involved, these numbers for the triad relied to a large extent on the availability of Russian "hot air".

However, whereas the targets for the EU, the US, Japan and Canada were the result of negotiations and bargaining, the targets for the remaining countries mainly resulted from "voluntary pledges" based on their "willingness to pay".[23] Although a certain amount of diplomatic pressure was applied by the triad on other industrialised countries (see Chapter 7), their targets do not rely on any particular rationale and reflect basically the degree of intransigence and *chutzpah* employed by these countries. Even months after Kyoto, Chairman Estrada was thus "still looking for the basis of these figures".[24]

The predominance of the trilateral talks over other negotiating groups was one of the main features of the final negotiations in Kyoto. The final deal was a compromise reached within the triad and did not involve other participants. When the Chairman issued the draft for the final session at Kyoto at 1 am in the morning of 11 December, he carefully avoided including any numbers in its Annex B.[25] Russia and Ukraine reacted immediately and declared that they would not accept any numbers going beyond stabilisation. Given the voluntary nature of the process, Russian intransigence prevailed in the end, and, to the disappointment of many involved in the negotiations, the numbers contained in the final document released

---

21 FCCC/CP/1997/CRP.4, Article 3.1/2/3 and Annex A.

22 For a more detailed account of the "hot air"-problem see Section 15.3.3.

23 See Ott 1998c.

24 Raúl Estrada: Keynote Address on the Kyoto Protocol Negotiations; OECD Climate Change Forum, Paris, 12–13 March 1998 (transcript on file with the authors).

25 FCCC/CP/1997/CRP.6.

at 10 am on 11 December required Russia and the Ukraine only to stabilise their emissions at 1990 levels. Even Chairman Estrada confessed his surprise after Kyoto that the EU did not apply more pressure.[26] The Dutch Minister Margaret de Boer showed extreme surprise and annoyance when the numbers came out and assumed that the British Minister John Prescott, who had conducted negotiations on behalf of the European Union, had made a secret deal with the US on this issue.

As a result of this low Russian obligation, the overall target of reducing Annex B emissions by 6% below 1990 levels, contained in the Chairman's last draft,[27] had to be corrected to "at least 5%" (Article 3.1 of the Kyoto Protocol). Since the triad had chosen the highest option of 8, 7, 6%, the final result represents an overall reduction of Annex I emissions of about 5% despite the Russian and Ukrainian insistence on stabilisation. This is almost equal to the fall in emissions between 1990 and 1995 arising from the decline in emissions in the formerly communist countries and thus represents a stabilisation of emissions at 1995 levels.

11.2 The Kyoto Targets: What are the Limits to Emissions?

The "Kyoto Protocol to the United Nations Framework Convention on Climate Change" contains differentiated quantified emission limitation and reduction commitments for a basket of four gases and two groups of gases. These are legally binding for industrialised countries listed in Annex B to the Protocol, and would amount to an overall Annex B reduction in the commitment period from 2008 to 2012 of at least 5% (Article 3.1).[28] There are a number of features of these targets that need to be taken into account when assessing their meaning. Section 11.2.1 will be an overview of the QELRCs. More detail is then provided in subsequent sections. The issue of sinks is dealt with separately due to its special importance and complexity (Section 11.3). The Chapter concludes with an overall assessment of the commitments and an outlook towards the future (Section 11.4).

11.2.1 Overview

The main aspects of the Kyoto targets can be grouped as follows:

- *The legally binding nature of commitments (see Section 11.2.2)*
  Binding commitments have been agreed for those Parties to the Kyoto Protocol that are listed in Annex B (all of whom are industrialised countries) (Article 3.7).

---

26 Raúl Estrada: Keynote Address on the Kyoto Protocol Negotiations; OECD Climate Change Forum, Paris, 12–13 March 1998 (transcript on file with the authors). He thought that this failure was perhaps due to a deal on Emissions Trading.

27 Draft Article 3.1, FCCC/CP/1997/CRP.6.

28 For an overview and assessment of those provisions see Ott 1998c; Yamin 1998a; Smeloff 1998; Coenen/Sardemann 1998; Simonis 1998; Bail 1998; Centre for Science and Environment 1998 and Grubb et al. 1999.

- *The basket approach (see Section 11.2.3)*
  The legally binding targets apply to a basket of four gases and two groups of gases, all of which are listed in Annex A to the Protocol. They thus follow the so-called "six-gases approach", including CO_2, CH_4, N_2O, SF_6, hydrofluorocarbons (HFCs) and perfluorocarbons (PFCs).
- *The commitment period 2008–2012 and applicable baselines (see Section 11.2.4)*
  - The targets relate to a five-year period from 2008 to 2012 (the so-called "commitment period") (Article 3.7).
  - To calculate the limit to emissions applicable to a Party for the first five-year commitment period, the percentage assigned to it in Annex B of its gross emissions of those gases in 1990 is multiplied by five (Article 3.7).
  - For HFCs, PFCs and SF_6, Parties may choose the year 1995 as base year instead of 1990 (Article 3.8). In practice, 1995 is very likely to become the common baseline chosen since emissions of these gases for all countries concerned have risen from 1990 to 1995.
  - Parties that are countries with economies in transition (CEITs) may choose different baselines (Article 3.5).
- *The differentiated commitments (see Section 11.2.5)*
  - Each industrialised country Party has been allocated a specific "assigned amount" of GHGs that is inscribed in Annex B for the first commitment period (Article 3.7). These obligations are differentiated and range from an obligation to reduce emissions by 8% (for the EU and many Eastern European countries) to permission to increase emissions by 10% (Iceland) and 8% (Australia) (see Table 11.1).
  - Taken together, those obligations should amount to a reduction in GHG emissions of all Annex B countries by at least 5% (Article 3.1).
  - If a Party's emissions are below the amount assigned to it in Annex B, the surplus amount may be transferred to a subsequent commitment period (Article 3.13).
  - While CEITs generally enjoy a certain flexibility as regards their obligations, this does not apply to their targets contained in Article 3 (Article 3.6).
- *Sinks (see Section 11.3)*
  - Certain activities that cause net changes in GHG emissions through direct human induced land-use change and forestry, limited to afforestation, reforestation and deforestation, may be used in the calculation of emission targets (Article 3.3).
  - Any Party for whom land-use change and forestry constituted a net source of emissions in 1990, shall include these in their baseline calculations (Article 3.7, second phrase).
  - Should the COP/MOP decide that additional human-induced activities might be added or subtracted from the assigned amount of each Party, this

> decision should, in principle, apply to the second commitment period at the earliest. A Party may decide, however, to apply such a decision already in the first period, provided that these activities have taken place since 1990 (Article 3.4).

The specific obligations of the Kyoto Protocol cannot be fully understood without taking the inclusion of "sinks" (see Section 11.3) and the so-called "flexibility mechanisms" into account. Four such mechanisms have been incorporated in the Protocol: Emissions Trading (Article 17), Joint Implementation between industrialised countries (JI, Article 6) and the Clean Development Mechanism for projects with developing countries (CDM, Article 12); a fourth mechanism allows for the "joint fulfilment" of obligations ("bubbling") by two or more Parties (Article 4). These mechanisms and their impact on Parties' obligations are dealt with in the chapters on the respective articles of the Kyoto Protocol.

11.2.2 The Legally Binding Nature of the Commitments

After COP 2 there was a clear understanding by almost all Parties, industrialised and developing alike, that a voluntary approach as taken by the FCCC had proven to be inadequate and that hard, verifiable and enforceable obligations were required for the next stage.[29] The individual targets contained in Article 3.7 and Annex B are therefore legally binding, in contrast to the soft law approach taken by the FCCC and advocated by some prior to Berlin and Kyoto.[30] There are many examples in the environmental field where "soft", non-binding, instruments and commitments have paved the way for stricter obligations. As such, these "soft" instruments can be important steps in the development of environmental regulation.[31] However, especially in areas like climate change that require changes in economic policy and therefore involve hotly contested domestic decision-making, the existence of legally binding obligations is a necessary component of any internationally co-ordinated strategy.

In addition to legally binding targets, the Kyoto Protocol contains a "soft" provision that requires Parties included in Annex I to make, by the year 2005, "demonstrable progress in achieving its commitments" under the Protocol. Specifying the year 2005 has an important symbolic character, since this was the date which AOSIS countries, the European Union and environmental groups had urged to be the starting time for obligations. No indicator is provided to determine "demonstrable progress", but this provision nevertheless may prove important in the negotiations on subsequent commitments. Article 3.9 states that the COP/MOP

29 On 22 October 1997, President Clinton stated: "The industrialised nations tried to reduce emissions to 1990 levels once before with a voluntary approach, but regrettably most of us – including especially the United States – fell short".

30 See e.g. Victor/Salt1994a.

31 See for example Ott 1998a.

is to initiate consideration of further commitments at least seven years before the end of the first commitment period, i.e. in the year 2005. A failure to show "demonstrable progress" by 2005 could thus influence the negotiations on future commitments. It could, for example, result in stronger obligations for a second commitment period after 2012.

Stronger obligations may also be the result of implementing Article 9.2, which requires the second COP/MOP to review the Kyoto Protocol in the light of new information (see Chapter 19). Depending on when the Protocol enters into force, this review might coincide with the consideration of further commitments and the target date for showing "demonstrable progress". If this review takes place much before the year 2005, it could provide an incentive to step up implementation efforts so as to show demonstrable progress.

The Kyoto Protocol also includes provisions that allow countries to slip under an "umbrella" and thus establish a kind of "collective responsibility" (see Chapter 12). Prior to Kyoto, some Parties (e.g. Russia) advocated a so-called "Annex I bubble". This would have meant setting a quantified target for all industrialised countries, with the final "burden-sharing", i.e. the allocation of the individual commitments, to be agreed upon later amongst themselves.

The Kyoto Protocol does not formally establish such an "Annex I bubble", although it would allow for the same effect. Article 3.1 contains an overall aim to reduce the overall emissions of GHGs in Annex B countries by at least 5%. Furthermore, Article 4 allows Annex I Parties to enter into agreements with each other "to fulfil (their) commitments under Article 3 jointly" (see Chapter 12). Nothing could preclude the Parties listed in Annex B, therefore, to conclude an agreement on the "joint fulfilment" of their quantified targets under Article 3 and to agree on an "internal" burden sharing arrangement. This might allow some Parties to exceed their "assigned amounts" as long as this is compensated by steeper reductions of other Parties to such an agreement. Whereas Article 4 is usually considered only in relation to a limited number of Parties, it could also serve to establish a global industrialised country bubble. However, in addition to establishing certain rules for accountability, Article 4 requires Parties to notify the terms of any such agreement to the Secretariat at the time of ratification of the Protocol. This, in effect, will probably preclude an application of this concept for the first commitment period.

11.2.3 The Basket Approach

Prior to Kyoto, all of the major industrialised players adopted a so-called "basket approach". This approach is not without its difficulties, in particular because of the uncertainties that exist concerning sources and sinks of the different GHGs. Consequently, some FCCC Parties, such as Germany, had urged a so-called "single gas approach" as long as possible.[32] Under the latter approach, reduction targets would have been separately established for each of the three main GHGs. This approach need not preclude comprehensiveness, since the Protocol could have established different schedules for each of the gases (and groups of gases like HFCs and PFCs) in different annexes. For example, the Montreal Protocol on Substances that Deplete the Ozone Layer regulates eight baskets of ozone-depleting substances that contain between one and more than 30 compounds. In principle, climate negotiators could thus have adopted two or more baskets of GHGs with different schedules.

The "one basket approach", by contrast, is "comprehensive" in a second sense: it establishes a single obligation to reduce the overall emissions of those gases, without a preference for any one of them. Each Party is thus in a position to choose how to concentrate its reduction efforts according to its specific emission profile. This provides flexibility, in that the reduction target must only be achieved for the basket as a whole. For example, a Party that in 1990 had very high emissions of CH_4 might choose to achieve a larger part of its obligation through the reduction of this gas if this can be done more cost-effectively than the reduction of CO_2. In any event, since the share of CO_2 in most industrialised countries is more than 80% of their overall GHG-emissions, reducing CO_2 emissions will, in practice, have to be at the centre of all efforts.

The calculation of emissions of the different GHGs in the basket have been made comparable by a weighting system known as the "Global Warming Potential". The basis of this system is the respective greenhouse forcing of other gases relative to CO_2 as determined by the Intergovernmental Panel on Climate Change (IPCC) (see Section 1.2). A country's total GHG emission is thus computed in tonnes of CO_2 equivalent by multiplying the emitted amount of each GHG by its GWP value.

There were two different "baskets" on the negotiating table before Kyoto. Whereas the EU and Japan in 1997 had adopted a "three-gases basket", comprising the main gases CO_2, CH_4 and N_2O, the US and Canada favoured a larger basket of four gases, as well as two groups of gases that included the fluorinated GHGs sulphur hexafluoride (SF_6), hydrofluorocarbons (HFCs) and perfluorocarbons (PFCs).[33] The main reason for advocating this larger basket was to achieve more flexibility in the implementation of the Kyoto agreement. Furthermore, the inclusion of those gases made sense because they are very powerful GHGs and their emissions have been rising rapidly in most countries, notably in Japan and the US.

Consequently, Japan, large parts of the EU and certain components of US industry were opposed to the inclusion of fluorinated gases. While rifts within the

32 This was also the approach recommended by the Wuppertal Institute prior to COP 1 in Berlin, see Jäger/Loske 1994.

33 New Zealand demanded the inclusion of "at least" the three main GHGs CO_2, CH_4 and N_2O.

EU were visible – with the UK and the Netherlands tacitly supporting the "six-gases approach" – it was the chemical industry, the producers of the gases under discussion, that tipped the balance. In a strategic move (and after intense internal discussions), most of this industry decided to lobby consumer industries and governments to support the "one basket" approach. At the same time, they advocated the deletion of any reference to fluorinated gases in Article 2 on policies and measures successfully (PAMs, see Chapter 10). They calculated that including all GHGs in one basket would avoid policy-makers focusing specifically on artificial fluorinated gases and would thus reduce pressure for their total phase-out. Including them in one basket with the main GHGs could then facilitate later arguments in favour of their relative usefulness because of the services they support (refrigeration, foam manufacturing etc.).

As a result, the US and Canada succeeded and the agreed reduction target comprises four gases and two groups of fluorinated gases, as stipulated in Annex A to the Protocol. Whereas 1990 remains the baseline for the first three main gases (CO_2, CH_4, N_2O), a concession to Japan was made that any Party may use a 1995 baseline for calculating the emissions of the other gases (HFCs, PFCs and SF_6). As mentioned above, it is to be expected that all Parties will choose this 1995 baseline because emissions of these latter gases will almost invariably have risen from 1990 to 1995.

11.2.4 The "Commitment Period" 2008–2012

Another central feature of the obligations in the Kyoto Protocol is the "commitment period" of five years. This concept was originally proposed by the US, which used the term "budget period".[34] It is designed, first, to avert the risks that a single-year target may pose arising from fluctuations in economic performance or from extreme weather conditions. Overshooting the target in a single year is thus allowed under a budget-approach, if this is compensated for in other years of the period. Second, the commitment period provides Parties with additional temporal flexibility. Third, the "commitment period" was meant to create a "budget" of emissions large enough to form the basis for a system of emissions trading that had been already envisaged by the US at a very early stage in the negotiations (see Chapter 4).

The US proposed a five-year "budget-period" starting in 2008 and ending in 2012. This proposal prevailed after the European Union accepted the concept in principle. Its attempt to push for an earlier commitment period from 2003–2007 failed. The only concession on the timing that the EU was able to get from JUSSCANNZ was the soft provision in Article 3.2 that "each Party ... shall, by 2005, have made demonstrable progress in achieving its commitments under this Protocol" (see Section 11.2.2 above).

---

34 See Dudek 1996; Environmental Defense Fund 1997 for the origins.

The calculation of each Parties' allowed emissions (the "assigned amount") within a commitment period is not more complicated than for a single-year target (Article 3.7). Baseline for a Parties' assigned amount from 2008–2012 are the GWP-weighted emissions of the "six gases" of the basket in 1990. For any CEITs that had chosen another base year under the FCCC (see Table 3.1), this will also be the baseline of its obligations under the Kyoto Protocol (Article 3.5). As mentioned above, a general exception has been provided for the fluorinated gases. According to Article 3.8, a country may choose 1995 as a baseline for HFCs, PFCs and SF_6.

The allowed emissions for one year of the commitment period are equal to the percentage of the amount of 1990 (1995) emissions assigned to each country in Annex B of the Kyoto Protocol. For example, the reduction obligation for the US is 7% of its 1990 emissions, since the amount inscribed for it in Annex B is "93". To calculate the assigned amount for the five-year commitment period, the amount for a single-year is multiplied by five.

The following example illustrates this. If emissions were 100 Mt CO_2 equivalent in 1990 and the amount assigned to a Party in Annex B equals 92 (i.e. an 8% reduction), the allowed emissions for each year of the first commitment period would equal 92 Mt CO_2-equivalent. Multiplied by five, this would amount to a total of allowed emissions of 460 Mt CO_2 equivalent for the entire commitment period.

Thus, although the budget concept has merit, given the special features of the climate regime, it also has the less desirable effect of considerably delaying any definite assessment of compliance with the obligations in the Kyoto Protocol. This may complicate negotiations on the next step of commitments. For example, if 2010 had been chosen as a single target year, compliance could have been monitored and verified by 2011/2012. The commitment period of 2008–2012, however, will probably not allow an assessment of a Party's compliance until 2014.

In case of the single target year option, any obligations for 2015 fixed prior to 2011/2012 could have been adjusted on the basis of a compliance assessment, and new obligations for, e.g., 2020, could then be negotiated. This would have been similar to the procedure that has been followed in the flexible development of the Montreal Protocol, where Parties frequently adjusted control measures in light of new information.[35] By contrast, under a commitment period of several years, half of the following period will have passed by the time the results of the compliance assessment of any given period become available. Adjustments to the existing obligations for the running period, e.g. in response to an incomplete fulfilment (or an over-fulfilment) of the overall target, will therefore hardly be possible. The compliance assessment can then only support negotiations on the obligations of the third commitment period. Rather than supporting flexible adjustments of the obligations, the Kyoto Protocol is thus likely to be a rather clumsy and slow to develop

---

35 See Benedick 1998a; Ott 1998a and Oberthür 1997.

treaty – not only because of the complicated nature of the issue as such, but also because of its inflexible structural design.

11.2.5 Differentiated Targets

Negotiations regarding a "flat rate" target, i.e. a uniform obligation for all industrialised countries, proved to be very difficult during the AGBM process. Already in July 1997, Chairman Estrada was convinced that "differentiation was the only way out".[36] Consequently, he had started negotiating bilaterally with some major players in the summer of 1997. It was not until the first week in Kyoto, however, that he publicly proposed a list of differentiated targets for certain groups of countries.

When the US accepted the concept at the start of COP 3, the stage for agreeing differentiated targets was set. The targets subsequently agreed for the US, the EU and Japan were the result of intense negotiations. By contrast, the targets for most other Parties are based primarily on pledges or a "willingness to pay" – a "willingness" which the big three tried to support through diplomatic efforts during the last hours at Kyoto.

Eventually, the targets of most of the other industrialised countries reflect more their domestic interests, intransigence and *chutzpah* than any sense of compromise in the international process.[37] While, *prima facie*, special circumstances need to be taken into account, the targets that allow an increase in emissions or require only stabilisation (Australia, Iceland, Norway, Russia, Ukraine, New Zealand, see Table 11.1), have annoyed many environmentalists and are considered particularly good deals for the countries concerned.[38] Notwithstanding these increases, the main industrial polluters in the industrialised West have, however, assumed reduction obligations. The EU and its Member States as well as most Eastern European countries have to reduce by 8%, the US by 7%, Canada, Hungary, Japan and Poland by 6% and Croatia by 5%.

The country with the lowest target, Iceland, was still not satisfied after the conclusion of the Protocol. It claimed that because of the small size of its economy a single industrial project would significantly increase its emissions above any set limit. There would be no option to avoid this emissions increase – unless one wished to forgive the chances of economic development. Iceland succeeded in having COP 3 mandate COP 4 to decide on any special conditions that may be granted to economies for whom single projects would have a significant proportional impact on emissions.[39] COP 4 subsequently requested the SBSTA to report

---

36 Raúl Estrada: Keynote Address on the Kyoto Protocol Negotiations; OECD Climate Change Forum, Paris, 12–13 March 1998 (transcript on file with the authors).

37 See Ott 1998c.

38 The Russian and Ukrainian "hot air" problem is discussed in more detail in Chapter 15.

39 Decision 1/CP.3, para.5(d) (the Iceland provision) in FCCC/CP/1997/17/Add.1.

additional information and decided to "take a conclusive decision on this matter at its fifth session" by the end of 1999.[40]

Table 11.1: The Differentiated Quantitative Obligations of Annex B Parties

Target (percentage reduction from base year or period)	Party
- 8%	Austria, Belgium, Bulgaria[a], Czech Republic[a], Denmark, Estonia[a], European Community, Finland, France, Germany, Ireland, Italy, Latvia[a], Liechtenstein, Lithuania[a], Luxembourg, Monaco, Netherlands, Portugal, Romania[a], Slovakia[a], Slovenia[a], Spain, Sweden, Switzerland, United Kingdom of Great Britain and Northern Ireland
- 7%	United States of America
- 6%	Canada, Hungary[a], Japan, Poland[a]
- 5%	Croatia[a]
stabilisation	New Zealand, Russian Federation[a], Ukraine[a]
+ 1%	Norway
+ 8%	Australia
+ 10%	Iceland

[a] Countries that are undergoing the process of transition to a market economy

Source: Annex of Decision 1/CP.1 in FCCC/CP/1997/7/Add.1.

The history of the special targets for Hungary and Poland illustrates the conditions under which the targets of the Parties outside the triad (US, Japan and EU) emerged and especially the problems the EU faced during the final days in Kyoto in linking up even with its close allies. Both CEIT countries hope to join the EU early in the 21st century and they had generally aligned themselves quite closely with the Union in the negotiations leading to Kyoto. Consequently, Chairman Estrada, in his proposal of 9 December, put in the same figures for Hungary and Poland as for the EU, i.e. minus 8%.[41] However, the EU was fully absorbed by internal quarrels and by organising its relations with the US and Japan. Consequently, communication with its CEIT allies virtually broke down during the last days in Kyoto. When Poland and Hungary realised that Russia and Ukraine would only be committed to stabilisation, they decided to insist on lower figures as well.

40 Decision 16/CP.4 in FCCC/CP/1998/16/Add.1.

41 FCCC/CP/19997/CRP.4.

As one delegate stated, this decision was not made on the basis of facts but politics. There can be little doubt that the Polish and Hungarian commitments would look different if the EU had been more proactive in its diplomatic efforts towards CEIT countries (Russia and Ukraine as well as Poland and Hungary themselves).

11.2.6 Banking of Parts of Assigned Amounts

The Kyoto Protocol allows the "banking" of emissions (Article 3.13). If a Party's GHG emissions stay below its assigned amount, the surplus will, upon request of that Party, be added to its (not yet determined) emission allowance for subsequent commitment periods. This provision has positive and negative implications. On the positive side, transferring those emissions may constitute a further incentive for countries to go beyond their obligations in the first commitment period and may thus encourage early action. Also, it provides additional temporal flexibility and may encourage implementation of longer term strategies to reduce GHGs.

On the negative side, however, banking of emissions may increase the overall amount of GHGs emitted into the atmosphere, since any surplus emission allowances that otherwise would have "become void" and permanently removed can now be kept for future use. This future use is quite probable because the banked emissions may enter the emissions trading market. Consequently, banking of emissions means that even if a quantitative limit is applied to the use of Russian and Ukrainian "hot air", this will not permanently remove these emissions: any unused emissions will accumulate and may be sold onto the market at a later stage, thus allowing other countries to assume lower efforts in later commitment periods. Indeed, the full amount of any banked emission allowances from the first commitment period will only become known once the second commitment period has started and the applicable assigned amount for that period will thus have to be fixed without knowing the amount of banked emissions. Finally, the banking of emissions perpetuates the allocation of emission rights into the future and exacerbates the equity questions already raised by the procedure of "grandfathering".[42]

11.3 The Issue of Sinks (Land-use Change and Forestry)

The issue of "sinks" in the Kyoto Protocol was – and still is – probably the most complicated in a technical sense.[43] In the negotiations leading up to Kyoto this was aggravated by the fact that serious negotiations on the inclusion of sinks only started shortly before COP 3. Contrary to this apparent lack of urgency, the issue was of greatest importance, with serious implications for the credibility,

---

42 See Chapter 15 on Emissions Trading. This issue has been raised for instance by the Centre for Science and Environment 1998.

43 See IPCC 1996b; for a concise overview see Pearce 1998; see also Lashof et al. 1997; WWF 1999.

transparency and verifiability of the emission targets and with great impact on the eventual size of those commitments.

The Parties, in short, settled for a compromise formula for the first commitment period that does not include sinks in the calculation of the baselines, but allows the calculation of commitments to take account of "direct human-induced land use change and forestry activities, limited to afforestation, reforestation, and deforestation since 1990" (Article 3.3). Further activities or categories of direct human-induced land use change may be added to this list for subsequent commitment periods after further consideration (Article 3.4).

This section first provides a short overview of the basic scientific facts relating to carbon storage in ecosystems. It then describes the history of the negotiations and the role of carbon sequestration in the Kyoto Protocol. Finally, some problems connected with the accounting of sinks and sources to meet Parties' commitments and future challenges are explored.

11.3.1 Carbon Storage in Biospheric Systems

Carbon is stored globally in different living biotic and abiotic "systems", namely the atmosphere, the oceans, the terrestrial system and the fossil fuel deposits. Together with deep sea ocean carbonaceous sediments, only the latter keeps the carbon out of the natural terrestrial carbon cycle for geological timescales. All other systems exchange – i.e. store and release – carbon on a more or less continuous level. The transfer from one system to another is called a "carbon flux". A system that stores more carbon than it emits is labelled a "sink". Generally, in nature, there is a equilibrium in the biological carbon cycle between the seasonal changes of CO_2 absorption from and release to the atmosphere, provided all environmental and climate factors are in steady state.

Currently, this stability is disturbed because additional carbon dioxide is released into the atmosphere by the burning of fossil fuels. Terrestrial plants and marine algae undergoing the process of photosynthesis for biomass production/growth take up half of this surplus input. This additional uptake of carbon is alleviated by the high level of anthropogenic nitrogen input into the natural system through fertiliser and nitrous emissions from power plants and cars. A major part of the naturally fixed carbon is released into the atmosphere again through natural processes. A certain part remains within plants for their lifetimes and so contributes to the building of biomass. While much of this is released again to the atmosphere after the plant dies, a portion of the carbon remains fixed. Decomposing plant and tree litter as well as dead roots contribute to soil carbon on land, while dying marine biomass contributes to deep sea sediments in the oceans.

Whether an ecosystem is in fact a sink of atmospheric carbon depends on the kind of ecosystem it is and a number of other factors. Anthropogenic interventions and external natural influences both have to be considered. Apart from active enhancement of sinks, human exploitation can have positive (side) effects on the storage potential of an ecosystem. For example, commercial timber harvests may be accompanied by reforestation, so that in the long run the forest might experience an enhancement of its sink capacity. This will, however, depend on various conditions, most importantly the fate of the harvested timber. As all timber decomposes eventually, releasing the major part of its carbon to the atmosphere again, the potential net contribution of reforestation after timber harvesting to sink enhancement is rather limited.

In general, forests have the highest sink potential, depending, however, on age and condition of the forest. For example, old growth forests have a negligible sink potential and may in some cases even be net sources of CO_2. Events like forest fires or clearing release the main part of the stored carbon quite rapidly into the atmosphere again. Other ecoystems, like wetlands or farmland, store much less carbon but also have the capacity to act as sinks. Finally, external influences must be taken into account. For example, the carbon storage rate of the ecosystems is expected to grow because of the higher CO_2-concentration in the atmosphere. On the other hand, global warming will intensify the respiration and thus diminish the storage potential. Furthermore, altering climatic conditions and changing precipitation patterns pose the danger of turning many carbon sinks into sources of greenhouse gases.[44] Nevertheless, the IPCC concludes that until 2050, by slowing deforestation, promoting natural forest regeneration in the tropics, and implementing a global forestation program, 12–15% of cumulative fossil fuel emissions could be offset.[45]

11.3.2 *The Provisions on Sinks in the Kyoto Protocol*

Although the issue of sinks always loomed in the background of the negotiations prior to Kyoto, serious negotiations on this only began at AGBM 8 in October 1997. At this meeting, many Parties realised that there was almost no factual basis on which to take a decision. The whole issue of LUCF (Land-use Change and Forestry), as it was finally referred to, was surrounded by uncertainties. Many Annex I Parties did not report on sinks in their national communications, and the methodologies used were not in any way comparable. Furthermore, great uncertainty existed with regard to the storage rate of biospheric sinks and Parties lacked a common understanding of how activities affecting these sinks should be taken into account. This situation has persisted since Kyoto.

Following AGBM 8, the Secretariat compiled a document on the issues[46] and Parties submitted their views prior to the Kyoto Conference.[47] There was, generally speaking, a clear dividing line between the majority of Parties, arguing against

[44] See WBGU 1998, p. 19. El Niño may already have turned the rain forests of Amazonia into a net carbon source, see Nature, Vol. 396 (1998), p. 619.

[45] IPCC 1996b.

[46] FCCC/TP/1997/5.

[47] FCCC/AGBM/1997/INF.2 and FCCC/AGBM/1997/MISC.4, FCCC/AGBM/1997/MISC.4/Add.1.

the inclusion of sinks at least for the first commitment period (the EU, Japan and developing countries), and most non-European OECD countries arguing in favour of it (USA, Canada, Australia, New Zealand). Whereas Australia favoured a so-called "net/net"-approach, i.e. the inclusion of sinks for the determination of the baseline as well as for the calculation of the commitments, New Zealand strongly argued in favour of a "gross/net"-approach that would not take sinks into account for determining the baseline but would do so for the commitment period.

After a workshop held immediately prior to COP 3, the chairman reported that four basic positions existed:

- non-inclusion for the first commitment period,
- waiting until further guidance from the IPCC,
- excluding LUCF except for "new activities", and
- adopting commitments on a "net"-basis, thus including sinks.[48]

Calculating the 1990 baseline emissions by including sinks was quickly discarded by most Parties – except by Australia – as inherently uncertain because of the incomplete data available. Developing countries, especially China and Brazil, argued strongly against the inclusion of sinks. In their view, the targets should be clear, transparent and identifiable, whereas the inclusion of LUCF would have the potential to thwart the intention to establish legally binding obligations.

The EU, on the other hand, slowly moved towards accepting certain LUCF-activities, partly for tactical reasons and partly because of internal opposition to the common EU-position from France and Finland. This change in position may have also been affected by the EU's wish to push forward the negotiations on targets – because only once the issue of sinks was resolved, could negotiators be able to calculate the numerical targets for each country.

The Parties, after the first week of intense negotiations, finally settled on a limited gross/net-approach and Article 3.3 provides that not only changes in emissions from sources but also certain removals from sinks shall be used to meet the targets of Annex B countries. LUCF are thus excluded from determining the emissions of the base year 1990, but a limited range of activities is taken into account when calculating emissions in the first commitment period. These activities comprise "direct human-induced land use change and forestry activities, limited to afforestation, reforestation, and deforestation since 1990". Other existing sinks, for example soils or the agricultural sector, are not included.

In view of the rather obscure language of this paragraph, the Parties at COP 4 adopted a clarification, based on the conclusions of the SBSTA at its eighth session, that the meaning of Article 3.3 is as follows:[49]

---

48 Authors' observation.

49 Decision 9/CP.4, para.1 in FCCC/CP/1998/16/Add.1.

"the adjustment to a Party's assigned amount shall be equal to verifiable changes in carbon stocks during the period 2008 to 2012 resulting from direct human-induced activities of afforestation, reforestation and deforestation since 1 January 1990. Where the result of this calculation is a net sink, this value shall be added to the Party's assigned amount. Where the result of this calculation is a net emission, this value shall be subtracted from the Party's assigned amount."

Article 3.4 furthermore establishes a process to add further categories of human induced activities resulting in the absorption of greenhouse gases. The Parties are required to provide further data on their sinks and sources to the SBSTA in order to enhance the knowledge base and diminish existing uncertainties. The COP/MOP is then requested to decide, at its first session, or as soon as practicable thereafter, on

"how, and which, additional human-induced activities related to changes in greenhouse gas emissions and removals in the agricultural soil and the land-use change and forestry categories shall be added to, or subtracted from, the assigned amounts for Parties included in Annex I…." (Article 3.4).

This decision should be based on the methodological work of the IPCC, the advice of SBSTA and decisions by the COP of the Convention.

Two last-minute changes resulted in new uncertainties regarding the determination of the emissions in the base year and the first commitment period. First, whereas any decision by the COP/MOP to include further categories under Article 3.3 would only apply to the second and subsequent commitment periods, a Party may choose to apply such a decision to the first period (Article 3.4). As a result, the extent of such a Party's obligation will not be known exactly for some time to come. Second, upon the insistence of Australia, the last sentence of Article 3.7 allows a Party to add the "net" LUCF emissions to its baseline for 1990, if those emissions were positive. This provision effectively inflates the baseline and thus diminishes the efforts required to reduce GHG emissions to meet the target. At the moment, only Australia and, to a limited extent, the United Kingdom and Estonia, would likely benefit from this clause.[50]

11.3.3 Open Questions and Future Challenges

The language on sinks finally incorporated into Article 3 of the Kyoto Protocol has significantly reduced the danger of a large potential loophole – without eliminating it completely.[51] Negotiations on the open questions surrounding the use of sinks began immediately after COP 3. At the next SBSTA session, in June 1998, the Parties requested the IPCC to prepare a Special Report on LUCF in order to

---

50 See Clive Hamilton: Land-use Change in Australia and the Kyoto Protocol, at <http://www.tai.org.au/> as of 7 December 1998; Ian Noble and Detlev Schulze: The Kyoto Protocol; at <http://www.igbp.kva.se/protocol.html> as of 30 June 1999.

51 See Greenpeace 1998, p. 48.

enable the COP to take decisions.[52] Following the adoption of the Special Report by the IPCC, the Parties decided at COP 4 in Buenos Aires, in November 1998, that the Convention's COP should forward a draft decision to the first session of the COP/MOP on the inclusion of additional categories of LUCF activities according to Article 3.4.[53]

As the IPCC has decided to prepare the Special Report by 2000, further decisions on this have been, as a matter of practice, deferred until COP 6. The Parties will thus be able to base any decisions on the inclusion of further LUCF activities, as well as the specification of the existing ones, on sound science. This may render it possible to remedy, and avoid a repetition of, the uncertainty caused by the political decisions concerning sinks taken in Kyoto. In particular, the COP will be enabled to exclude any additional sinks categories that entail too high levels of uncertainty in their calculation and may thus become loopholes for Parties attempting to evade the implementation of their targets under the Kyoto Protocol.

In 1998, the Parties submitted additional views on the issue of LUCF.[54] Mainly at the insistence of non-European OECD countries, the SBSTA initiated a process parallel to the deliberations of the IPCC. This included a SBSTA workshop in September 1998 on data availability based on definitions used by Parties and international organisations in relation to Article 3.3 of the Kyoto Protocol.[55]

However, these activities have not yet resulted in significant clarification of the intricate questions raised by the inclusion of sinks in the calculation of commitments. Despite the partial interpretation of Article 3.3 by COP 4, referred to above,[56] the question of what constitutes "human-induced" land-use change has remained largely unanswered. Furthermore, the language in Article 3 does not indicate whether carbon stored below the earth's surface is to be considered if changes to this occur as a result of human-induced land use changes. There is no clear definition of what constitutes a "forest" and many divergent definitions exist even of a seemingly simple term like "reforestation".[57]

Another serious problem with the provisions on sinks in the Kyoto Protocol is their disconnection from other regimes pertaining to the protection and sustainable use of forests. These missing "forest linkages" could undermine efforts related to sustainable forest management, since the issue is defined solely in terms of "carbon sequestration" under the Kyoto Protocol (see Chapter 23). As a first step towards establishing the necessary links, COP 4 decided "that issues of mutual concern to the [UNFCCC] and the Convention on Biological Diversity should be

52 FCCC/SBSTA/1998/6, para. 45.

53 Decision 9/CP.4 in FCCC/CP/1998/16/Add.1.

54 These submissions are reproduced in FCCC/CP/1998/MISC.1 and Add.1-2; FCCC/CP/1998/MISC.9 and Add.1-2.

55 The report is contained in FCCC/CP/1998/INF.4.

56 Decision 9/CP.4 in FCCC/CP/1998/16/Add.1.

57 See Schlamadinger/Marland 1998.

taken up by the subsidiary bodies at their tenth sessions" in 1999.[58] Further efforts are needed to avoid a conflict between different environmental priorities relating to forests and other ecosystems that may serve as sinks.

While these problems may be solved in the foreseeable future, the inherent difficulties in calculating the storage of carbon in biospheric sinks will continue to exist for a considerable time to come. For example, in 1998 the terrestrial carbon absorption of the US was reported to be higher than its CO_2-emissions - an enormous difference from previous estimates.[59] At COP 1, in Berlin in 1995, New Zealand was projected to reduce its net emissions by 50% by the year 2005 compared with 1990, mainly due to sink enhancement. However, as a consequence of inventory reviews concerning its sink capacity it was later projected to increase its net emissions by a similar amount.[60]

A wide net approach, encompassing more categories of sectors and activities, would not only be questionable environmentally but could also lead to unexpected increases of emissions for some Parties. Under a wide net approach, unforeseen releases of stored carbon into the atmosphere resulting from shifts in rainfall patterns, droughts and resulting changes in vegetation due to human-induced global warming might be counted towards the GHG emissions of the Parties concerned. The forest fires experienced in 1998 by some countries in Southeast-Asia, in the wake of the El Niño phenomenon may have been just a warning concerning the inherently unstable nature of carbon storage in biospheric systems.[61]

Finally, even without sudden releases of carbon from systems previously acting as sinks ("carbon time bomb effect"), increasing the amount of carbon stored in biospheric systems in order to compensate for carbon emanating from the burning of fossil fuels constitutes a high-risk strategy. Forests cannot be guaranteed to remain intact for geological time-scales, probably not even for centuries. This would be required, however, if carbon storage is supposed to have the same long-lasting effect as the avoidance of emitting fossil fuel emissions into the atmosphere.

11.4 Assessment and Outlook

The assessment of the obligations for industrialised countries in the Kyoto Protocol resembles the question of whether "the glass is half full or half empty".[62] The result, despite its many flaws, is a revolutionary achievement done in the face of huge obstacles, and may provide a relatively sound basis on which to base climate policy in the coming decades. The adoption of a Protocol containing binding targets clearly represents a watershed in international climate and

---

58 FCCC/CP/1998/16/Add.1, p. 71.
59 See Fan et al. 1998, p. 49.
60 Greenpeace 1998, p. 49.
61 On El Niño and climate change see IPCC 1996a, pp. 163 et seq.
62 See Loske/Ott 1998.

environmental policy. A full evaluation of this achievement is provided in Chapter 22. This section is confined to providing a summary appraisal of the agreed targets and the future challenges that follow therefrom.

The numerical targets – and in particular those of the US, the EU and Japan – represent the main achievement of the EU at Kyoto. It is probably fair to say that without the EU's insistence on high numbers, the targets adopted by COP 3 would have been much lower. The downside of this achievement, however, is the fact that the Union had much less impact on the concrete design features of the Kyoto obligations, including the choice of a "budget period" of five years instead of a single-year target, the differentiation and timing of the targets, the six gases contained in the "basket", the inclusion of sinks (land-use change and forestry), and the flexibility granted by the "flexibility mechanisms" (see Chapters 13–15). These design features were largely determined by the alliance of non-European states assembled in the JUSSCANNZ group, and by the US in particular.

Figure 11.1: Projected Impacts of the Kyoto Protocol on Emissions of Greenhouse Gases from Annex I Parties

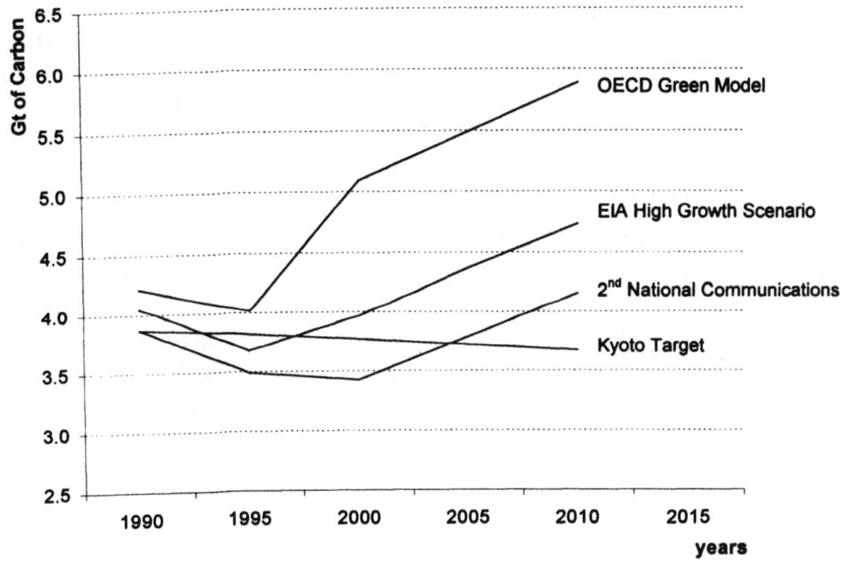

Source: OECD 1993; EIA 1999; 2nd National Communications of Annex I Parties, FCCC/CP/1998/11/Add.2.

The Kyoto targets surely have two main winners: Russia and Australia. Russia's commitment to only stabilise emissions on 1990 levels in the face of its current real reduction of about 30% creates a surplus emission allowance that will have repercussions for many years to come because of its hot air potential. The considerable increase in emissions allowed to Australia – not only in terms of its numerical target but also because of the right to include LUCF emissions in its baseline – induced Prime Minister Howard to declare that he was "pleased" with the results at Kyoto. This has set a bad precedent for future negotiations, especially with regard to developing countries. The targets of Russia and Australia thus clearly illustrate the inefficacy of the pledging procedure applied in Kyoto to arrive at differentiated targets for all countries apart from the US, the EU and Japan.

Overall, the current trend in industrialised countries' emissions nevertheless could be reversed, if these Parties implement the Kyoto targets at least in large part by domestic action. In relation to the 32% increase in CO_2 emissions the International Energy Agency projects for North America until 2010,[63] even the stabilisation of emissions would appear to be a major achievement. Figure 11.1 illustrates that the overall target of reducing GHG emissions by 5.2% until 2008–2012 represents a major deviation and reversal of current trends of emissions of Annex I Parties as a whole. This assessment would not change if, as Parties projected in their national communications, overall emissions would only increase by 8% by 2010 in the second half of the 1990s.[64] Some overly cautious analysts have even warned that the Kyoto Protocol requires reductions that are too steep.[65]

Despite the much steeper reductions that would be required in order to effectively stabilise atmospheric GHG concentrations at safe levels (see Figure 22.1), the Kyoto Protocol may thus be termed a success because it might reverse current emission trends.[66] This cautious positive assessment will, however, only hold to the extent that a number of shortcomings and the environmental necessities are addressed in the future. The following points are of particular relevance:

- The commitment period of 2008–2012 unfortunately lies beyond the usually short-term political horizon. Therefore it might not create enough pressure for political leaders to embark on an effective climate policy, since most governments will no longer be in power by the time the "reality check" comes. Since moving the target years forward does not appear to be politically feasible, there is a concrete danger that the necessary measures will not be taken in time to achieve the Kyoto targets. The focus of most governments on the elaboration of the Kyoto Mechanisms, which are designed to ease the pressure to achieve the targets by climate policy measures domestically, does not bode well in that respect. The business sector, on the other hand, seems to be much more forward

63 IEA 1998.

64 See Decision 11/CP.4 in FCCC/CP/1998/16/Add.1.

65 See for example Richard Benedick: "How Workable is the Kyoto Protocol? How to Salvage the Kyoto Protocol?" on the website of Resources for the Future, <http://www.weathervane.rff.org/pointcpoint/pcp5/index5.html>.

66 See also Coenen/Sardemann 1998.

planning. In particular, some multinational corporations have already started to take the Protocol seriously and adapt their strategy to this new reality (see Sections 2.5, 6.5 and 24.5).

- The inclusion of "sinks" (LUCF) in the Protocol reduces considerably the obligation for a number of Parties to control emissions. The US State Department, for example, announced after Kyoto that "altering the accounting method for carbon absorbing activities ... accounts for about 3% of the 7% reduction" when compared with the original stabilisation proposal of President Clinton.[67] Whether this holds true will depend on the definitions and accounting methods yet to be elaborated. In addition, the inclusion of more carbon absorbing activities under Article 3.4 could further erode the targets. Any such additional sinks categories need to be limited and easily verifiable. The decision of COP 4 to wait for the results of the IPCC Special Report on LUCF before taking further decisions on this issue adopts this approach (see Section 11.3).
- The failure to agree on emission reductions for Russia and the Ukraine (and, to a lesser extent, Poland) has resulted in a huge amount of "hot air". The unlimited use of these cheap reductions through Emissions Trading could considerably reduce the pressure on industrialised countries to act domestically. The CDM for achieving certified emission reductions through JI projects in developing countries will have the same effect. In both cases, future decisions might limit the use of these Kyoto Mechanisms to acceptable levels and provide safeguards against abuse (see Chapters 13–15).
- Some emissions have remained outside of the control regime, most importantly the growing GHG emissions from international aviation and marine bunker fuels. Furthermore, the United States, in a last minute move that has remained largely unnoticed, succeeded in exempting "emissions resulting from multilateral operations pursuant to the UN Charter" from being counted towards national emissions.[68] The need, especially to control emissions from international aviation and marine bunker fuels is likely to be an increasingly important item on the international agenda (see also Section 10.3.2).
- In face of the deep cuts in global GHG emissions eventually required (see Chapter 22), the current targets of industrialised countries need to be strengthened in subsequent commitment periods. In the medium term, commitments by developing countries will also be needed. Industrialised countries' obligations for the second commitment period 2013–2017 will, according to Article 3.9, be on the agenda in 2005 at the latest, the year when the Parties to the Protocol should have made "demonstrable progress" in achieving their commitments (Article 3.2). This might provide the opportunity for interested Parties and *civil*

---

67 See The Kyoto Protocol on Climate Change. State Department Fact Sheet of January 15, 1998, p. 2 (on file with the authors). According to the fact sheet, changing the baseline for the three synthetic gases from 1990 to 1995 accounts for a 1% further reduction.

68 Decision 2/CP.3 in FCCC/CP/1997/7/Add.1.

society[69] to lobby for a strengthening of industrialised countries' targets. Negotiations on commitments for developing countries might then also become possible if industrialised countries succeed in demonstrating leadership by example by that date. Devising commitments of developing countries will without doubt represent a major challenge (see Chapter17).

- The design of the obligations in terms of "commitment periods" instead of a single-year target does not serve a dynamic development of the Kyoto Protocol. By delaying the assessment of compliance with the targets by several years, the structure of the commitments may inhibit their dynamic evolution. Considerable time will pass between the realisation that targets may be insufficient and thus warrant further strengthening and the possibility to react to these needs. Although the budget design does provide some stability to policy-makers and businesses, the downside weighs heavily. It may only be compensated if the precautionary approach is consistently stressed and high targets sought.

69 See von Weizsäcker/Ott 1998b and Talaab 1998.

12. Joint Fulfilment of Commitments (Article 4)

With the inclusion of Article 4 into the Kyoto Protocol, the European Union achieved one of its major goals: the possibility to "bubble". In combination with Article 3.1, Article 4 allows any group of Parties to jointly fulfil their commitments under the Protocol. It specifies that the combined assigned amounts in Annex B may be redistributed through internal agreement and notified to the Secretariat.

A "bubble" is a concept used in national pollution laws, where, for example, emission limits are not set for an individual factory, but for all company sites. The company is then able to comply with the regulations in the most cost-effective manner, because it can choose to reduce emissions at those sites where it is cheapest to do so.[1]

Although the joint fulfilment of obligations was a crucial part of the EU's strategy, the EU's internal struggles prevented it from having a straightforward position on the issue early on in the negotiations to communicate clearly to its negotiating partners. The end result was one not originally intended by the Union, in that this instrument is now open to all Parties, and as such, may ultimately be an alternative to Emissions Trading but without any safeguards.

This chapter will start with the history of negotiations, paying special attention to the particular circumstances of the EU. This will be followed by an analysis of the Protocol's provisions. Finally, the chapter will describe some post-Kyoto developments and provide an outlook to the future.

12.1 Negotiating History

Article 4.2(b) of the Convention allows Parties to act "individually or jointly", but the lack of concretion left wide room for interpretation. To most negotiators and observers this phrase primarily referred to the concept of "joint implementation", a project-based mechanism for the acquisition of emission allowances (see Chapter 13). The Member States of the EU, however, maintained that they would be allowed under this provision to agree on an internal "burden-sharing arrangement", i.e. the internal distribution of emission allowances under a common target. The main purpose of such a bubble was not primarily cost-effectiveness, but to allow the Union to act as a whole. It was only in this way that the so-called "cohesion countries" with relatively low gross domestic products (Greece, Ireland, Portugal and Spain) would be able to participate in the Protocol on an equal footing with the rest of the Union. These countries are trying to catch up economically with the rest of Western Europe and – very much like developing countries – wanted their emission levels to be allowed to grow.

---

1 On these concepts see the contributions in Sorrell/Skea 1999.

However, there are more reasons why the EU insisted on a special provision so as to be counted as one entity under the Kyoto Protocol. For example, EU competence to regulate many areas complements or even replaces the sovereign power of the Member States.[2] International trade, for example, is generally outside the competence of individual Member States. Negotiations with trading partners or disputes in the framework of the GATT/WTO (World Trade Organization) are therefore dealt with primarily by the European Commission. Most environmental treaties, including the FCCC and the Kyoto Protocol, cover areas of "mixed competence", i.e. where both the EU and its Member States possess competence in this field. Consequently, the Union (i.e. the European Community represented by the European Commission)[3] is a member to those treaties as well as the Member States.

With respect to implementation of the Kyoto Protocol, participation of the Union is indispensable since a number of the policies and measures affecting the emission of GHGs in the Member States are decided in Brussels. Under the EU Structural and Cohesion Funds, for example, well-financed programmes influence the economic development or the structural adjustment of less developed regions and provide for some redistribution of resources. Important decisions impacting GHG emissions are furthermore taken in the area of the EU Common Agricultural Policy, as well as in the development of the internal market in the energy sector and many other policy areas.[4]

One important aspect that distinguishes the EU from other international organisations is its "supra-national" character. The Council of the European Union, composed of the respective Ministers of the Member States, has the power, often together with the European Parliament, to adopt legislation that has a direct effect on the citizens and companies in each country. If the Council adopts "Regulations", for example, they create rights and duties effective immediately, and enforceable by any law court in any Member State. "Directives", by contrast, generally have to be implemented by the Member States. Parts of these, however, may also directly create rights for citizens or companies in Europe. Furthermore, the EC possesses a very strong judicial system and the European Court of Justice may impose financial penalties on Member States that do not comply with European legislation.[5]

Despite the importance of the matter, the Union failed to effectively communicate these special features during the pre-Kyoto negotiations. To illustrate this

---

2 On the role and legal competence of the EU in the context of international environmental treaties, see especially Macrory/Hession 1996; Hession/Macrory 1994; Haigh 1991; Heinegg 1998; Oberthür 1999c.

3 On the difference between the European Union and the European Community, which is a Party to the FCCC, see Chapter 2.

4 See, for example, Oberthür/Bär 1998.

5 See generally Craig/de Búrca 1998. On the use of these instruments for environmental purposes see Krämer 1998.

failure, it was only when environmental NGOs at a press conference at Kyoto provided background information on the EU bubble concept, that many media representatives confessed to having understood the issue for the first time. Moreover, for a long time the EU did not submit a textual proposal for an EU bubble. In its proposal on the draft protocol structure submitted by Ireland at the end of 1996, the only hint of its wish to participate under a bubble was the ambiguous phrase that "the (industrialised) Parties shall, individually or jointly, abide by quantified objectives..".[6]

With the informal agreement of the Council of Environment Ministers on burden-sharing in March 1997, just before the start of AGBM 6 (see Section 6.1), the EU opened the debate on the topic. The ambitious Dutch EU presidency had managed to get Member States to agree to a compromise that would result in an overall reduction of EU emissions of 9.2% from 1990 levels by 2010 for a basket of three GHGs. The accepted burden sharing formula was based on a proposal introduced at an informal workshop in January 1997 by a research-team from the University of Utrecht that had been commissioned by the Government of the Netherlands.[7]

This analysis served as a useful starting point for the political bargaining on the share that each Member State was to shoulder.[8] The final agreement in March 1997 contained high reduction targets of 25% by Europe's biggest emitter Germany, accompanied by Denmark and Austria, as well as a ceiling of up to 40% further growth for the economically less-advanced countries (see Table 12.1). The overall reduction added up to little more than 9%, which fell short of achieving the EU's own proposal of a flat-rate target of -15% for all Annex I Parties, which was agreed on at the very same meeting.

This internal agreement was harshly criticised by other Parties. The US, in particular, accused the EU of having double-standards and attempting to establish an unfair advantage. The contradiction in advocating differentiation for itself, while demanding flat-rate targets for the rest of the industrialised countries was perceived as provocative. In addition, the EU did not come up with any proposals on how to ensure transparency and accountability. Neither did the proposal address the future enlargement of the Union by the "accession countries" of Central and Eastern Europe (in particular the Czech Republic, Estonia, Hungary,Poland, and Slovenia). Non-European industrialised countries also charged that the EU's progressive stance on emission reductions was enabled partially by Germany's so called "wall-fall-profits": i.e. decreased emissions due to economic break-down and restructuring in the former German Democratic Republic.[9] Taking the accession candidates into the bubble would result in even more such "hot air".

---

6 Article 2(c) in FCCC/AGBM/1996/Misc.2/Add.2, p. 19, and later in FCCC/AGBM/1997/Misc.1, p. 43.

7 See Blok et al. 1997.

8 See Ott 1997.

9 However, the German emission reductions are not only based on "hot air", but also on a general improvement of energy efficiency and infrastructure, see Ziesing et al. 1998.

As the Union failed to come up with a concrete textual proposal, Australia submitted the first proposal in January 1997 for dealing with the possible enlargement of regional economic integration organisations and a further elaboration on this in March 1997.[10] This was submitted in time to be included in the negotiating text for COP 3. In an apparent attempt to ridicule the Union's agreement on burden sharing, Australia furthermore proposed a range of targets from minus 30 to plus 40% as the basis for all Annex I Party commitments.[11]

The Australian text dealt explicitly with the most critical aspects of the bubble-concept: first, the question of who would be responsible for fulfilling the obligations in cases of non-compliance, and second, the prevention of hot air through accession of more countries to the bubble. As regards the first element, the Australian proposal foresaw a detailed declaration of the division of competence between the organisation and its member states, and made the ratification of the organisation dependent on the approval of this declaration by a meeting of the highest decision-making body. As to enlargement, existing obligations would only apply to the membership of the organisation as from the date of adoption of the protocol, and the implications of any change of the membership would then be subject to review and approval by the meeting of the Parties.

Following the vigorous attacks on its initial proposal, and with the strict Australian proposal on the negotiating table, the EU submitted an elaborate text at AGBM 8 in October 1997 that at last met some of the criticism.[12] Most importantly, it opened the door to joint fulfilment not only for the EU but for all Parties. Furthermore, the member states of a regional economic integration organisation that entered into a bubble would be jointly responsible for fulfilling the obligations under this proposal – a disadvantage compared with other bubbling groups where each party would be responsible only for its own target under the respective bubble agreement.

The new EU proposal again attracted considerable criticism mainly for the reason that the text did not provide for the implications of EU enlargement. In fact, the proposal allowed the burden-sharing agreement to be changed up until "five years before the expiration of the (commitment) period". This would have permitted the Union to change its agreement up until right before the start of a five-year commitment period, which was thought to be somewhere between 2005 and 2008. At that time, negotiations on the accession of Eastern European countries were expected to have finished by then and at least some of them would have already joined. Another criticism referred to the fact that an agreement on internal burden sharing should be required at the time of ratifying the Protocol and not at such a late stage.

10 FCCC/AGBM/1997/2, pp. 40–41 and FCCC/AGBM/1997/MISC.1/Add.2, pp. 9–12.
11 FCCC/AGBM/1997/MISC.1/Add.2, p. 6.
12 FCCC/CP/1997/2, p. 10.

At Kyoto, Norway chaired the informal negotiations among industrialised countries on the issue. The EU was under pressure from JUSSCANNZ, in particular from the US, to give up its position on limiting the use of the flexibility mechanisms, especially on Emissions Trading. The US insisted that there was no difference in principle between bubbling and Emissions Trading and that the same limitations on their use for achieving compliance with the targets should apply. In the end, both provisions were left without strict language on limitation: the EU may make unlimited use of Article 4 and Article 17 contains a weak formulation on "supplementarity" of Emissions Trading (see Chapter 15). Furthermore, the Australian proposals on review and scrutiny of the internal agreement by the COP/MOP were not incorporated into Article 4. The outcome at Kyoto can thus be termed a success for the EU. It came at a price, however – the unrestricted possibility to make use of the provision on joint fulfilment for any two or more countries when becoming Parties to the Kyoto Protocol. Noting this price, Samoa declared upon the adoption of Article 4 in the final negotiations at Kyoto that it was agreeing to this provision on the understanding that only the EU would take advantage of it.

12.2 The Rules Applying to Bubbling

Article 4 of the Kyoto Protocol sets out the rules for entering into the joint fulfilment of obligations in conjunction with Article 3.1, according to which "the Parties included in Annex I shall, individually or jointly" comply with their assigned quantified targets as listed in Annex B. Any two or more industrialised countries that intend to bubble must enter into an agreement and notify the Secretariat of the terms of their agreement on the date of ratification, which will then be forwarded to all other Parties (Article 4.2). The Union was thus unsuccessful in its demand for temporal flexibility as regards the timing and notification of the agreement. According to Article 4.1, the agreement must contain the exact allocation of reduction or limitation obligations for each Party as well as the terms and conditions.

Uncontested in the negotiations was that the agreement must remain in operation during the full commitment period 2008–2012 (Article 4.3). The group's overall achievement is calculated by the sum of their total GHG emissions over the five-year budget period, which must not exceed the sum of their individual assigned amounts inscribed in Annex B over the same period (Article 4.1). Should the total GHG emissions during the commitment period exceed the added amounts assigned to these Parties, each Party will be responsible for its own level of emissions as set out in the agreement (Article 4.5). Thus, Article 4 establishes individual state responsibility as the general rule; i.e. only the Party/Parties exceeding their revised assigned amounts under the bubble agreement will be deemed to be in non-compliance. Responsibility thus lies exclusively with the "seller", the Party that agrees to a higher target than stipulated in Annex B. This "seller liability" might work well in domestic systems, but is problematic at the international level (see Section 15.3.4).

This general rule of individual state responsibility contrasts with the provision in Article 4.6 on regional economic integration organisations, which establishes joint responsibility of member states and the organisation. The EU will be the only such organisation in the foreseeable future. Should the member states fail to achieve the required total emission level assigned to it, the organisation shall be responsible as well as the member states individually for their respective emission levels as notified in the agreement. Although this arrangement appears to go against the idea of joint fulfilment within the framework of a regional economic integration organisation, it was the EU that originally proposed it. This joint responsibility in fact only applies to the EU, even if other countries should decide to bubble within the framework of such an organisation, since the special responsibility rule of Article 4.6 only applies to regional economic integration organisations that are Parties to the Kyoto Protocol.[13]

In contrast, the Union was largely successful in fending off the demand of the JUSSCANNZ group that an enlargement of the EU should not affect the bubble. Article 4.4 provides that "any alteration in the composition of the organization[14] after adoption of the Protocol shall not affect existing commitments under this Protocol". This ensures that the 8% reduction target of the EU according to Annex B will remain after accession of new member states (and also after a hypothetical decrease in membership). At the same time, the commitment of any new EU Member State would, according to that provision, also remain unaffected, which leads to some complication in cases where the commitments of the EU and any accession country differ. For example, if Hungary and/or Poland that have a commitment of -6% joined the EU bubble, the EU would still be committed to -8% (i.e. its common target would not be relaxed).

The second phrase of Article 4.4. further elaborates on this and further strengthens the notion that accession countries might become part of an EU bubble if accession occurs before ratification (by the EU, its current Member States and any new Member States): "Any alteration in the composition of the organization shall only apply for the purposes of those commitments under Article 3 that are adopted subsequent to that alteration." There is certainly some scope for interpretation of this rather cryptic formulation, but the provision only refers to commitments according to Article 3, which may be redistributed according to Article 4.[15] Whereas the second phrase of Article 4.4 ensures that a country's obligation would

---

13 Article 4.6: "If Parties acting jointly do so in the framework of, *and together with*, a regional economic integration organization..." (emphasis by the authors).

14 Article 4.4. contains the same further condition as Article 4.6 that the organisation must also be part of the agreement, thus referring only to the EU.

15 Article 4 grants Parties permission to modify the Protocol "inter se", and accords with Article 41.1(a) of the Vienna Convention on the Law of Treaties (1969); see Brownlie 1990, pp. 625 et seq.

not change just by its accession to the EU, it thus still allows the country to be included in an EU bubble under Article 4. In fact, if negotiators at Kyoto had intended to exclude this possibility, they would have had the opportunity to do so explicitly in Article 4.4, for example along the strict language proposed by Australia. It follows, then, that Central and Eastern European countries acceding to the Union might be part of the bubble agreement as long as accession takes place before ratification of the Protocol and notification of the agreement to the Secretariat.

12.3 Post-Kyoto Developments in the European Union

When the EU signed the Kyoto Protocol on 29 April 1998, it made the following declaration: "The European Community and its Member States will fulfil their respective commitments under Article 3, paragraph 1, of the Protocol jointly and in accordance with the provisions of Article 4." With this statement, the first bubble under the Kyoto Protocol had been announced.

The new UK Labour government under Prime Minister Tony Blair, which took over the EU presidency on 1 January 1998, was eager to demonstrate leadership in the field of climate change. It had already announced an increase in its national voluntary reduction target from 10 to 20% from 1990 levels and proceeded to engage in an effort to devise a new burden-sharing agreement that would take account of the final outcome at Kyoto. For example, a new agreement would have to take into account that the Protocol regulates a basket of six gases, instead of the EU's proposed three gases. The Kyoto Protocol also includes sinks for the calculation of the reduction targets, and refers to a commitment period (instead of the target year 2010). Furthermore, whereas the burden sharing of March 1997 amounted to an overall reduction of 9.2%, the Kyoto Protocol requires the Union to reduce by a mere 8%.[16] And finally, some of the large increases granted in the original agreement to the "cohesion countries" (Greece, Ireland, Portugal and Spain) had annoyed other member states.

While there were thus a number of good reasons to adjust the previous burden-sharing agreement, a number of governments used the new situation to achieve a general relaxation of their targets. Denmark and Germany demanded adjustments and the former "green" countries Austria and the Netherlands admitted that they would not be able to meet their ambitious targets of the first agreement. These announcements triggered distinctive resistance from Greece, Ireland, Portugal and Spain, who were now expected to limit their emission growth to a larger extent than formerly agreed. Of the main emitters of the European Union, only the UK declared that it would take on a stronger commitment than before.[17]

---

16 Calls by environmental NGOs to keep the level previously agreed and thus "overcomply" were unsuccessful.

17 According to the "Second Communication from the EC under the UNFCCC", released on 26 June 1998, CO_2 emissions in the UK had decreased by nearly 7% between 1990 and 1995.

The new agreement reached in June 1998 reflected the altered situation and changed interests. The total reduction was lowered to the exact 8% demanded by the Protocol, although the UK had tried to reach a compromise above this minimum. The figures converge more than they did before, and now range from -21 to +27%, instead of the -30 to +40% before (see Table 12.1).[18]

Table 12.1: The EU Burden-Sharing Agreements of March 1997 and June 1998

Member State	March 1997: Emission Reduction by 2010	June 1998: Emission Reduction by 2008–2012
Austria	-25.0%	-13.0%
Belgium	-10.0%	-7.5%
Denmark	-25.0%	-21.0%
Finland	0.0%	0.0%
France	0.0%	0.0%
Germany	-25.0%	-21.0%
Greece	+30.0%	+25.0%
Ireland	+15.0%	+13.0%
Italy	-7.0%	-6.5%
Luxembourg	-30.0%	-28.0%
Netherlands	-10.0%	-6.0%
Portugal	+40.0%	+27.0%
Spain	+17.0%	+15.0%
Sweden	+5.0%	+4.0%
United Kingdom	-10.0%	-12.5%
EU-Total	-9.2%	- 8.0%

This compromise indicates the sober atmosphere surrounding European and international climate policy since Kyoto. Whereas the 1997 agreement partly reflected the momentum of the pre-Kyoto process, a number of Member States were on the retreat post-Kyoto – as were other countries active in 1998 and 1999 in the international process (see Chapter 24). Against this backdrop, the 1998 burden-sharing agreement still marked a considerable success, as it settled early on one of the most difficult questions of the implementation of the EU target under the Kyoto Protocol. However, this will not suffice to achieve the target itself, which will require implementation of (national as well as European) policies and

[18] Council Conclusions on Climate Change, European Council of Environment Ministers, 2106th Council Meeting, Doc. 9702/98 of 16/17 June 1998.

measures. Without these, total GHG emissions of the EU will, according to projections of the European Commission, increase by 6% by 2010.[19]

12.4 Assessment and Outlook

The provision on joint fulfilment under Article 4 will be one of the most interesting elements of the Kyoto Protocol in the future. Its implications are not yet foreseeable, and it is not clear whether all Parties present in Kyoto actually realised what they had consented to. Members of the JUSSCANNZ group, which certainly grasped the importance of Article 4, have already met with Russia and the Ukraine in Kyoto in order to conduct preliminary discussions on a bubble among themselves. As a result, the JUSSCANNZ grouping has largely been replaced by a so-called "umbrella group" that includes Russia and the Ukraine.

Most important for this group is that Article 4 provides a suitable alternative to a yet undefined and unapproved Emissions Trading system under Article 17 of the Kyoto Protocol.[20] Since there is no provision for approval or scrutiny by the COP/MOP or the COP of agreements to jointly fulfil commitments under Article 4, any group of Parties is allowed to reallocate its assigned amounts independently of whether an Emissions Trading system is in place. For the members of the "umbrella group", a key advantage of bubbling is that the reallocation of obligations under Article 4 is not conditioned by a requirement for supplemental domestic implementation (as exists for Emissions Trading). Under Article 4, Russia could thus transfer/sell all its hot air to one or more other countries without restrictions that might be imposed in relation to the Kyoto Mechanisms. The transparency rules elaborated in the context of Article 17 on Emissions Trading also would not apply. A country with supreme economic, political or military power (notably the US) might thus be in a position to monopolise the excess emission allowances – realising exactly those fears that drove Japan to insist on rules ensuring transparency of Emissions Trading (see Section 15.1).

Eventually, Article 4 would even allow countries to bubble in order to establish their own trading regime, independent of any Emissions Trading under Article 17, as long as they achieve "their total combined level of emission reductions" (Article 4.5). If their combined level of assigned amounts would be exceeded due to the failure of any of the participating Parties, however, each Party would be responsible for its own level of emissions set out in the bubble agreement. Each party to a bubble agreement that provided for internal Emissions Trading would be highly dependent on the other parties to such an agreement. Thus, pursuing Emissions Trading under a bubble agreement might constitute a high risk strategy for the participating countries.[21]

---

19 European Commission 1999.

20 "Trading without rules", see Ott 1998c, p. 41.

21 This danger is less severe for the EU Member States, since they operate in a framework of a long-established economic, political and legal tradition.

As against the opportunities Article 4 opens up, the "umbrella group" does face certain disadvantages and difficulties in using Article 4 as an alternative to Emissions Trading under Article 17. First, there is the general uncertainty of entering into agreements with governments and countries with rather unstable governance structures (e.g. Ukraine, Russia). This weighs particularly heavily if an internal trading arrangement under Article 4 was to be established, in that it would require stable and reliable long-term co-operation. Secondly, if such trading was not pursued, then the flexibility under any bubble agreement provided by Article 4 would be severely constrained, since the emission allowances would be fixed after ratification of the Kyoto Protocol. Third, the initial allocation of emission allowances under any bubble will involve complex negotiations that at the very least will have to deal with the price of any reallocation (whereas this price would be set by the international market under Emissions Trading according to Article 17). And finally, any bubble agreement going beyond the EU and its Member States will meet with fierce opposition from environmental NGOs and possibly the wider public.

Whether Parties other than the EU, such as the "umbrella group", are seriously considering entering into a bubble agreement is unclear, but the possibility puts considerable pressure on the EU with regard to the negotiations on Article 17. Indeed, this is consciously used as a bargaining chip in these negotiations. Eventually, the provision on the joint fulfilment of commitments in Article 4 could well be used as an alternative to Emissions Trading, thereby avoiding any quantitative as well as qualitative (transparency etc.) restrictions that apply to the other Kyoto Mechanisms. It is not clear whether the EU had been in a position to achieve an "EU-only" bubble in Kyoto. In any event, this objective was not furthered by the Union's negotiating strategy on this issue.

13. Joint Implementation (Article 6)

The concept of Joint Implementation (JI) is based on classic economic theory: measures to limit GHG emissions should preferably be taken where they are cheapest or even profitable. Mitigation costs differ among countries, mainly due to differences in the efficiency of energy use. Thus, emission reductions in developing countries and CEIT countries are thought to be cheaper than in OECD countries. To save on resources and maximise emission reductions, Parties from the OECD (or their industries) are striving to receive credit for carrying out climate protection projects abroad. Benefits would accrue to both project partners: the recipient country would receive additional funds, modern technology and know-how, whereas the investing country would acquire CO_2 credits at a lower cost than taking action at home. Although having intellectual appeal, many practical and political difficulties exist that must be solved in order to actually reap the benefits of this instrument.[1]

The term Joint Implementation is not used in the Kyoto Protocol. Nevertheless, the principal concept is underlying two instruments of the Protocol. First, Article 6 regulates the transfer of "emission reduction units resulting from projects" aimed at GHG emission reductions or enhancement of sinks among Annex I Parties. Second, Article 12 on the Clean Development Mechanism (CDM) establishes basic rules for "project activities resulting in certified emission reductions" in developing countries. Project activities under both instruments are supposed to be financed by (private business in) Annex I Parties.

It has become common to refer to Article 6 as "the" JI provision in the Kyoto Protocol, since Article 12 establishes a distinct and innovative multilateral mechanism.[2] This Chapter deals exclusively with Article 6, the evolution and meaning of Article 12 on the CDM are discussed separately in more detail in Chapter 14 below. In this chapter, the analysis of the negotiating history of Joint Implementation is followed by an interpretation of the rules and an analysis of the future challenges in this area.

13.1 Negotiating History

The negotiations on JI have their roots in Articles 3.3, 4.2(a) and (d) of the Convention. The JI concept remained contested throughout the international negotiations since 1992. A number of concerns have raised doubts about whether the assumed benefits of the JI concept will translate into real advantage when implemented. A first concern related to the transaction costs involved in working out the agreement between the partners in a JI project, which might significantly

---

1 For an early discussion of JI in general see Loske/Oberthür 1994.

2 See FCCC/SB/1998/2, p. 8.

reduce its financial benefit. Second, the calculation of the project related emission abatement requires the determination of a so-called "baseline scenario"; i.e. what would have happened in the absence of the JI project and how would emissions have developed without additional measures. And if the project actually resulted in emission reductions, was this result offset by any "leakage effects", i.e. the reallocation of emission-intensive activities to other sites? Clear answers to these question are in many cases difficult, or even impossible, to give. Furthermore, implementation of JI projects is prone to lead to collusion between the project partners so as to deliberately overestimate the emission reductions achieved ("paper emissions"), because of benefits for both partners.[3]

Third, JI might not be cost-efficient in a long-term perspective, because it might ease pressure on industrialised countries to change their development path and introduce innovative technologies. With respect to North-South relations, this also raised an equity concern, since industrialised countries, while escaping own domestic action, would tap the cheap reduction potentials in developing countries, whereas developing countries, later on, would have only the more costly options available for fulfilling their own possible future commitments.

Finally, a highly politically charged concern exists in developing countries relating to the implications of JI on North-South equity. The terms of specific JI projects would have to be negotiated between, and implemented by, one (or more) industrialised and a developing country (or another industrialised country, i.e. a CEIT). Frequently, however, the power differences between these potential partners will favour the industrialised investor country. As a result, the terms of such agreements might more closely reflect the interests of the investor country and could even contradict the national development priorities of the host country. In short, JI was sometimes alleged to lead to a form of "neo-colonialism".

It was this last concern, in particular, that led most large developing countries to object to JI prior to COP 1 in 1995. At that conference, the Parties finally agreed to establish a pilot phase of "activities implemented jointly" (AIJ) covering all GHG sources, sinks and reservoirs.[4] The term "activities implemented jointly" was agreed as a political compromise so as to ensure that this would not represent any predetermination as regards the question of establishing full-fledged JI (including crediting etc.) as proposed by industrialised countries.

While no credits were to be granted during the AIJ pilot phase, it was open to all Parties on a voluntary basis. The financial resources spent on project activities had to be additional to official development assistance (ODA) ("financial additionality"). Three substantive criteria for AIJ projects were agreed, i.e. such projects should:

- be compatible with national environment and development priorities,
- have prior approval by the participating governments, and

---

3 See e.g. Ott 1998b; Roland/Haugland 1995.

4 Decision 5/CP.1 in FCCC/1995/7/Add.1; Oberthür/Ott 1995, pp. 146-147.

- result in real, measurable and long-term benefits that would not have occurred otherwise ("environmental additionality").

Since the AIJ pilot phase was to be reviewed before the end of the decade,[5] JI appeared to be off the agenda of the negotiations on the Kyoto Protocol, which, according to the Berlin Mandate, were supposed to conclude before the end of 1997 (see Chapter 4). However, given the high degree of political interest of industrialised countries, the concept found its way on the agenda of the protocol negotiations.

Negotiations on the issue did not begin in earnest before 1997, when all the major players had put their positions on the negotiating table.[6] Two questions in particular caused some disagreement among industrialised countries: the timing for including developing countries in JI activities (before or after evaluation of the AIJ pilot phase) and the principles that should govern Joint Implementation under the Protocol. The US (and JUSSCANNZ in general) favoured a *laissez-fare*-approach without restrictions on participation. In contrast, the EU favoured more stringent monitoring and verification requirements and was opposed to the inclusion of sink-related projects. The Union furthermore attempted to accommodate the concerns of many developing countries and proposed to wait for the evaluation of the AIJ pilot phase before taking a decision on their participation.[7]

This was particularly important for some larger developing countries like China and India, which wished to preserve the unity of the G-77 on this issue. Most developing countries thus vividly opposed participation of developing countries in JI under the Kyoto Protocol – at least until the AIJ pilot phase had been evaluated. Concerns about "neo-colonialism" merged with environmental concerns about shifting action away from the industrialised countries where it was most needed. Developing countries were able to point to the slow start of the pilot phase and the incomplete picture arising thus far from the experience in AIJ projects.[8]

This common position could, however, hardly conceal large internal differences. On the one hand, OPEC countries opposed JI in principle, because depriving OECD countries of this flexibility mechanism would further undermine the prospects for ratification of the treaty. On the other hand, Costa Rica most outspokenly aired positive views about JI and developing country participation.[9] As had already been the case at COP 1 in 1995, Latin American countries tended to have a more favourable stance towards JI. A large number of AIJ projects were already implemented in this region and especially in Costa Rica, which was actively push-

---

5 Ibid.

6 See the compilation in FCCC/AGBM/1997/2, paras. 108-117 and FCCC/AGBM/1997/MISC.1/Add.4 (US); FCCC/AGBM/1997/MISC.2/Add.1 (EU); also FCCC/AGBM/1997/MISC.1/Add.2 (Costa Rica).

7 See FCCC/AGBM/1997/2.

8 A synthesis report on the status of the AIJ pilot phase was presented by the Secretariat in document FCCC/SBSTA/1997/12 and Add.1.

9 See FCCC/AGBM/1997/MISC.1/Add.2.

ing for the right to sell emission credits (see Table 13.1). In addition, some of the bigger and economically more advanced Asian developing countries were not generally opposed to the idea of JI.

Table 13.1: AIJ Projects by October 1997 and June 1999

Region	1997				1999			
	Total		Sinks projects		Total		Sinks projects	
	number	impact[a]	number	impact (% of total)	number	impact[a]	number	impact (% of total)
CEITs	45	47,854	2	21%	76	52,806	3	21%
LAC	15	71,050	8	79%	25	97,431	11	80%
Asia	1	134	1	100%	7	7,327	1	2%
Africa	1	1,450	-	0%	1	1,450	-	0%
Total	62	120,488	11	55%	109	159,015	15	56%

[a] Impact in thousand metric tons of CO_2 equivalent as reported/projected in 1999.

Source: FCCC/SBSTA/1997/12/Add.1 and UNFCCC – CC:INFO/AIJ – List of Projects (5 June 1999) at <http://www.unfccc.de/program/aij/aijproj.html>.

Nevertheless, the G-77 and China managed to preserve its unity and its principal opposition in the negotiating process up to Kyoto, although the consolidated negotiating text by the Chairman of October 1997 already foresaw JI with developing countries after a COP decision following the evaluation of the AIJ pilot phase.[10] Disagreement between industrialised countries remained on whether sink projects should be eligible and the degree to which guidelines for reporting, calculation of credits and verification should be specified.[11]

Progress towards an agreement was driven by the negotiations on the CDM. First, the CDM negotiations dealt with developing country participation in a JI-type mechanism, effectively removing one of the major points of disagreement stalling the discussions on Article 6 of the emerging Protocol. Second, after accepting the basic concept of a project-based mechanism in the context of the CDM, it would have been inconsistent for developing countries to further oppose JI among industrialised countries. Agreement among industrialised countries was finally secured by deleting most of the controversial language and following the

10 FCCC/AGBM/1997/7, Article 6.

11 This was indicated by the brackets included in the Non-Paper of the Chairman of the Committee of the Whole of 7 December 1997; see FCCC/CP/1997/CRP.2, Article 7.

precedent set by the AIJ pilot phase in allowing sink projects. As a consequence, Article 6 was agreed in principle after the first week of negotiations in Kyoto[12] and was adopted without discussion in the final plenary of the Committee of the Whole in Kyoto (see Chapter 7).

13.2 The Rules Applying to Joint Implementation

According to Article 6.1 of the Kyoto Protocol, any Annex I Party "may transfer to, or acquire from, any other such Party emission reduction units" resulting from projects for the purpose of meeting its quantified targets. Projects may be carried out in any sector of the economy, including the enhancement of sinks. By limiting the use of JI to industrialised countries that are subject to emission ceilings under the Protocol, Article 6 to a large degree ensures that "paper emissions" created by JI projects will not have negative effects on the climate system. Any emission reduction units (ERUs) transferred under Article 6 will, according to Article 3.10 and 3.11, be subtracted from the transferring Party's allowed emissions (and added to the acquiring Party's assigned amount). At the same time, Article 3.10 and 3.11 provides for the full integration of ERUs in the Emissions Trading market under Article 17 (see Chapter 15).

As a result of Article 3.10 and 3.11, JI under Article 6 leaves the total emission allowance of industrialised countries unchanged. This alleviates the problem of cheating by inflating the emission reductions from JI projects. If such crediting of "paper emissions" was to occur, it would reduce the assigned amounts of the host country (possibly reducing its potential to sell emission allowances on the Emissions Trading market) (see Figure 13.1).

The reference to Annex I Parties, however, opens up JI for all industrialised country Parties to the Convention, instead of limiting participation to Annex B Parties of the Protocol. The implications are twofold. First, there are two Annex I Parties to the Convention that are not included in Annex B of the Protocol, Belarus and Turkey, since they were not Parties to the Convention at the time of COP 3 (and there might be more such countries in the future). These might participate in JI under the Protocol once they ratify the Convention. Second, even those Annex I Parties to the FCCC that do not ratify the Kyoto Protocol will nevertheless be allowed to participate in JI.

These Parties to the Convention would thus not be restricted by Article 3 of the Protocol and could in principle transfer ERUs accruing from JI project activities without being limited by a quantified amount of emissions assigned to them. However, most of these countries have a large incentive to sell "part of their assigned amount" under Article 17 on Emissions Trading, since most CEITs have excess emission reductions to sell ("hot air"; see Chapter 15). As this requires to be Party

---

12 Thus, the Chairman's paper of 9 December 1997 already contained, more or less, the final language on JI; see FCCC/CP/1997/CRP.4, Article 7; see also Chapter 7.

to the Kyoto Protocol, the problem described with respect to JI should be of limited relevance in reality.

Figure 13.1: The Kyoto Mechanisms

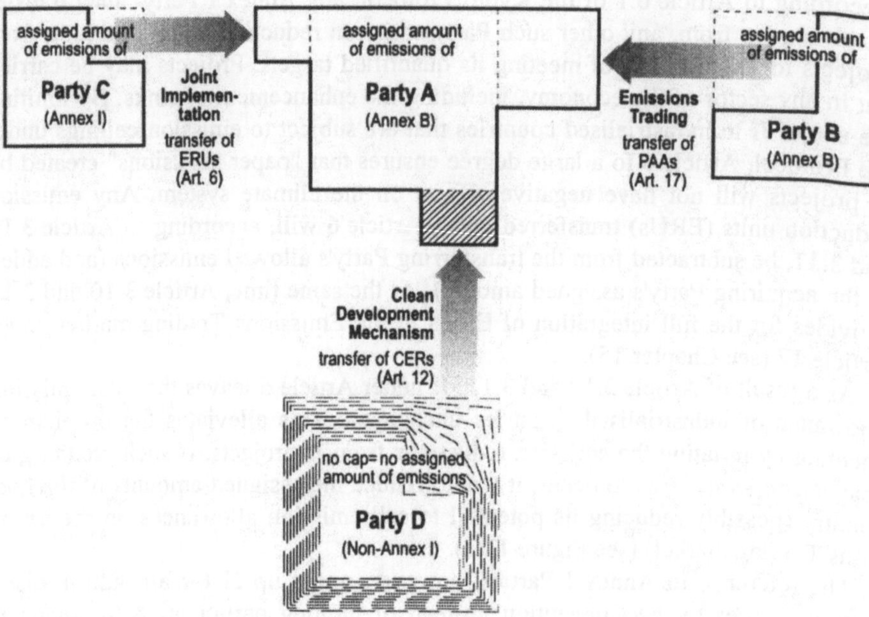

Going beyond the criteria for projects under the AIJ pilot phase, Article 6.1 places four conditions on Parties engaged in JI activities:

- JI projects need the approval by the Parties involved (Article 6.1(a)),
- JI projects must lead to additional emission reductions or enhancement of removals by sinks ("environmental additionality") (Article 6.1(b)),
- acquiring Parties have to be in compliance with their obligations to establish national systems for the estimation of net GHG emissions in accordance with Article 5 and to report relevant information in accordance with Article 7 (Article 6.1(c)), and
- acquisition of ERUs shall be supplemental to domestic actions for the purpose of meeting the quantified targets (Article 6.1(d)).

The first two substantive criteria are clearly inspired by those applying to the AIJ pilot phase (see Chapter 13.1 above). Notably missing from the list in Article 6 is the requirement that projects must be compatible with national environment and development priorities. This criterion was included under the AIJ pilot phase (as well as the CDM, see Chapter 14) mainly in response to developing country concerns. As Article 6 of the Protocol does not relate to developing countries, the compatibility criterion was left out (as was the requirement of "financial additionality").

The third criterion of Article 6.1 reflects concerns that inadequate reporting by Parties involved in JI projects might endanger the credibility and reliability of the system. It provides a further incentive for proper reporting by Parties, a core precondition for the functioning of the treaty (see Chapter 16). The fourth condition, regarding "supplementarity" to domestic actions, is ambiguous, and subject to varying interpretations. It reflects the request by the EU to limit the use of JI, Emissions Trading and the CDM for achieving the quantified targets (the so-called "cap" on the use of the flexibility mechanisms), but deliberately does not contain any specific amount due to US and JUSSCANNZ opposition (see Chapter 15).

Further guidelines for the implementation of JI, in particular for verification and reporting, could not be agreed.[13] This was partly due to a disagreement over whether such guidelines would need to be elaborated at all. Thus, Article 6.2 merely contains an enabling clause: the COP/MOP "may, at its first session or as soon as practicable thereafter, further elaborate guidelines (...), including for verification and reporting". The "Buenos Aires Plan of Action", adopted at COP 4 in November 1998, provides for the elaboration of such guidelines.[14] A decision on this is expected at COP 6.

JI projects are intended to be conducted not by governments but by private actors, most notably businesses looking for cheap opportunities for emission reductions. Industrialised country governments have therefore sought to engage private enterprises directly. The underlying assumption of that approach is that the free market will be more efficient than governments in identifying and implementing least-cost reduction potentials. It also reflects the general trend towards deregulation in industrialised countries. As a consequence, an Annex I Party "may authorise legal entities to participate" in JI projects (Article 6.3).

Certain problems may arise from the fact that private entities are given a prominent role in the framework of an intergovernmental treaty. This raises the issue of responsibility and liability in case of project failure, since at present only Parties to the Protocol are legally responsible, represented by their governments. Article 6.3 thus determines that private entities authorised by a Party will still operate "under its responsibility". Further liability rules may have to be elaborated, however. For example, there are liability issues related to the relationship between the private entities involved in a JI project and the governments concerned.[15]

---

13 These elements were contained in the Non-Paper by the Chairman of the Committee of the Whole of 7 December 1997; see FCCC/CP/1997/CRP.2, Article 7.

14 See Annex of Decision 7/CP.4 in FCCC/CP/1998/16/Add.1.

15 See Ott 1998b, pp. 26 et seq. The required agreements might possibly build on existing foreign investment treaties (see ibid.).

Article 6.4, finally, is a result of the remaining concerns that JI projects might create "paper emissions". Although this danger is to some extent alleviated (see above), Parties agreed that ERUs may not be used for meeting the targets, where the implementation of the criteria mentioned above is questionable. If such a problem is identified by the in-depth review process in accordance with Article 8 of the Protocol (see Chapter 16.1), transferred ERUs shall not be used for meeting a Party's quantified target "until any issue of compliance is resolved". This restriction refers to the transferring as well as acquiring Party. Since any transfer of ERUs without proper accounting and reporting should in principle not be possible, the provision may help avoid the transfer of emission allowances over and above those that have been achieved in the JI project ("paper emissions"). The acquiring (most probably OECD) Party may thus not offset any ERUs acquired, if there is an unresolved question regarding implementation of the JI criteria related (1) to itself or (2) to the transferring (most probably CEIT) Party.

Article 6 is silent about the starting date for JI projects under the Protocol and crediting. This, in combination with Article 3.10 and 3.11, however, suggests that JI projects may only be used to meet the quantified obligations once the commitment period has begun, from 2008 onwards. This does not preclude an earlier start for projects intended to be recognised as JI activities.If Parties had reached agreement to allow banking of emission reductions from JI projects achieved before 2008, they would have expressed it explicitly, as in the case of the CDM (see Chapter 14). However, the issue was reopened after Kyoto[16] and the "starting date for Article 6 projects" is to be considered as part of the Buenos Aires Plan of Action adopted at COP 4.

13.3 Assessment and Outlook

In the aftermath of Kyoto, Joint Implementation has received the least attention of the three Kyoto Mechanisms. This is due in part to the fact that discussions on JI have a long history. They started in 1993 on the basis of the Convention, whereas the CDM and Emissions Trading were only introduced into international climate policy by the Kyoto Protocol (and may have a greater potential). Extensive work has already been devoted to the elaboration of JI and some experience has been gained in the context of the AIJ pilot phase.[17] Nevertheless, a number of issues are still contentious or need clarification.

One of those issues concerns the elaboration of more specific rules for Joint Implementation. Article 6 enables the COP to elaborate guidelines for its implementation, including verification and reporting. COP 3 provided for a prompt start

16 Indeed, the Secretariat suggested after Kyoto that Parties might wish to express their views on the issue; see FCCC/SB/1998/2.

17 For some of the relevant work, see FCCC/SBSTA/1997/12/Add.1; OECD 1999; Ott 1998b; Hamway 1998; Luhmann et al. 1997; Parson/Fisher-Vanden 1997; Nordic Council 1997.

of the discussions,[18] and negotiations are underway as part of the Buenos Aires Plan of Action adopted by COP 4 (see above). Work on the elaboration of rules for JI could in principle be closely linked to the related discussions on the CDM and could draw on the AIJ experience referred to above.[19]

There are two principal models of JI, which diverge in particular on the degree of oversight and control of projects in the framework of the climate regime's institutions. One concept is largely bilateral and the other multilateral and more institutionalised.[20] The first model might also be called a free-market model that leaves most of the decisions to the Parties involved in JI projects. The bilateral model of JI has been favoured by the US and the so-called "umbrella group" (Australia, Canada, Iceland, Japan, New Zealand, Norway, Russian Federation, Ukraine, United States) in order to minimise costs and provide for maximum flexibility in the use of the mechanism.[21] According to this model, participating Parties (that can number more than two[22]) would largely decide on how to verify emission reductions and share the benefits of JI projects. They would also be primarily responsible for the role of any private "legal entities" involved in JI projects.

Other Parties have been supporting a multilateral model of JI activities with a much bigger role for institutions operating under the international climate regime.[23] One of the main reasons for this position in the general debate was to balance power differentials between investor and host country participants in projects. Under this approach, the COP/MOP could create a subsidiary body to take on tasks such as approving projects, deciding on crediting and resolving conflicts. Furthermore, the COP/MOP could elaborate provisions for the verification and possible certification of credits achieved by project activities. A clearing-house function could be built into the system, possibly operated by the UNFCCC Secretariat.[24] This is the concept that, to a large degree, underlies Article 12 on the CDM (see Chapter 14).

While the multilateral model of JI has been supported by developing countries in the general conceptual debate about project-based joint implementation, they should still have a profound interest in a more institutionalised design of Joint

18 Decision 1/CP.3 in FCCC/CP/1997/7/Add.1.

19 The Annex to Decision 7/CP.4 mentions "Implications of the AIJ pilot phase" as one of the issues to be considered in elaborating JI under Article 6; see FCCC/CP/1998/16/Add.1, p. 28.

20 See Ott 1998b; Yamin 1998b on the CDM; see also Chapter 14.

21 For this approach see Responses to G-77/China Questions on Flexibility Mechanisms (Australia, Canada, Iceland, Japan, New Zealand, Norway, Russian Federation, Ukraine, United States) in FCCC/CP/1998/MISC.7 (Paper No. 6).

22 In particular, more than one investor country might be involved. Despite this fact, the model is referred to as "bilateral", since two sides generally face each other (the investor(s) and the host) with no/little oversight by third Parties/multilateral institutions.

23 See the contributions in FCCC/SB/1998/MISC.1 and addenda.

24 See for an elaboration of the institutionalised model Ott 1998b.

Implementation under Article 6. Although they are not involved directly, very low transaction costs would make this mechanism more attractive than the CDM, which imposes a fee to cover administrative expenses and finance adaptation costs in developing countries (see Section 14.2).

Nonetheless, the formulation of Article 6 – which, for example, does not mention certification – appears to lean more towards the bilateral option. Proponents of a strict bilateral approach claim that separate arrangements for certification and verification would not be needed. Since any ERUs transferred by a Party are subtracted from the Party's assigned amount in accordance with Article 3.11, this Party should have a strong incentive to verify the reported emission reductions and transfer only the ERUs actually achieved by a project. For example, an overestimation of ERUs of JI projects might diminish the assigned amounts available for Emissions Trading under Article 17.

However, as in the case of Emissions Trading, reality is more complicated. Because of the large amount of hot air available in some CEITs (see Section 15.3.3), these countries may not have such a strong incentive to prevent the transfer of paper emissions, even though such transfers would diminish the parts of assigned amounts available for sale. If the size of the market for emissions trading does not match expectations of today or a JI investment offers special incentives, Russia or the Ukraine may still be prepared to transfer some of their excess emission allowances as "paper emissions" via JI. This might also hinder the realisation of additional emission reductions in JI projects (because those engaging in the extra-effort needed would be at a competitive disadvantage as compared to those getting credits for "paper emissions"). Furthermore, a common procedure for project execution and evaluation might provide a level-playing field for investing and host countries, whose negotiating power might otherwise diverge widely and lead to inequitable agreements. Such considerations speak in favour of an enhanced oversight of JI projects.

Independent verification and certification therefore serve an important function under Article 6 of the Protocol. Overall, accountability and transparency of project activities are fundamental for the credibility and viability of Joint Implementation. Climate protection demands that JI activities lead to additional emission reductions. To attract serious investments and ensure public acceptance of the mechanisms, certainty and stability will be required. Furthermore, certain standards might be required because ERUs created in JI projects under Article 6 will be allowed to enter the Emissions Trading market (see above). Therefore, reporting on acceptance of JI projects by Parties, baseline emissions, projected emissions of the project (and the assumptions from which these were derived) and actual emissions could be subject to the in-depth review process under Article 8 of the Protocol. This would provide the basis for identifying questions of implementation in accordance with Article 6.4. The reporting procedures and requirements could draw on the (limited) experience gained in the AIJ pilot phase.[25] In addition, considerable experience exists with regard to the monitoring, reporting, verification and certification of certain types of projects.[26]

As a second option, verification and certification functions could be assigned to private actors, as has been considered in the case of the CDM (see Chapter 14). Because of possible collusion of interests, these obviously would have to be different from the private businesses which are expected to be involved in the execution of JI projects as investors. The COP/MOP might adopt minimum requirements regarding such private actors.

One of the main issues for any project-based mechanism is the determination of baselines. ERUs will have to be calculated on the basis of assumptions about what would have happened in the absence of the JI project in question (and, in the project planning phase, projections of what will happen given the implementation of this project). Emission reductions are determined by deducting actual emissions from these baseline emissions. One way of achieving accountability and reliability of JI might be to develop guidelines and "best practices" for calculating baselines. This might include defining benchmarks or reference cases for certain types of projects, regions and sectors which might avoid difficult case-by-case consideration.[27] Such benchmarks might then be used in reporting and possibly verifying and certifying ERUs, which would involve checking whether the guidelines have been followed properly.[28]

Furthermore, the COP/MOP might wish to establish procedures for determining compliance with the requirements of Article 6, including the provisions of national GHG inventory systems (Article 5) and reporting obligations (Article 7). As these tasks are rather specialised, it might be useful to delegate the compliance check to some subsidiary body, be it one of the existing Subsidiary Bodies or any other body that may be established in the future under the multilateral consultative process or the non-compliance procedure (see Section 16.3). At the national level, incentives will be required for business to engage in JI projects while ensuring that failure of such projects will not lead to non-compliance by states (e.g. by obliging investors to buy equivalent emissions in an emissions trading system in the case of project failure, or insure projects against failure).

As in the case of the other flexibility mechanisms, a clearer definition of the supplementarity provision of Article 6.1(d) will be on the agenda of future nego-

25 See the uniform reporting format for AIJ (FCCC/SBSTA/1997/4) as adopted by Decision 10/CP.3 in FCCC/CP/1997/7/Add.1; see also Ott 1998b for reporting formats of private entities vis-a-vis national authorities.

26 See e.g. Vine/Sathaye 1999; Jackson/Begg 1999. See also the International Performance Measurement and Verification Protocol of the US Department of Energy, December 1997, available at <http:// www.ipmpv.org> as of June 1999.

27 E.g. FCCC/CP/1998/MISC.7 (Paper No. 6); Luhmann et al. 1997.

28 See the report on the Technical Workshop on Mechanisms under Articles 6, 12 and 17 of the Kyoto Protocol organised by the FCCC Secretariat in Earth Negotiations Bulletin, Vol. 12, No. 98, 19 April 1999.

tiations. In addition, a number of rather technical and procedural questions will need to be answered for JI under the Kyoto Protocol to function properly. For example, the Parties have to specify the information any ERU certificate should carry so as to minimise the risk of forged ERUs being introduced into the system. Further, some criteria on the authorisation of "legal entities" by Parties to take part in JI projects might be considered necessary to prevent precarious businesses from getting involved.[29]

The relevance of Article 6 on Joint Implementation in the Kyoto Protocol remains uncertain, considering the range of other flexibility mechanisms that may be used to close any compliance gap. Neither the exact design of the Kyoto Mechanisms nor their relative cost efficiency is known and will not likely be known until some time into the next decade. From an environmental point of view, however, JI project activities have the advantage of leading to real emission reductions in CEITs.

Evidence for the relevance of JI under the Kyoto Protocol was provided by the Japanese and Russian governments already shortly after the adoption of the Kyoto Protocol. In April 1998, Russia's President Yeltsin and Japanese Prime Minister Hashimoto agreed that Japanese firms would conduct feasibility studies for JI projects on the modernisation of 20 Russian power plants and factories. Indicating the importance of JI to Japan, Japan's trade ministry allocated US $ 20 million for the fiscal year 1998/99 to conduct such studies. JI projects could contribute to a much needed modernisation of the Russian energy sector, which is 3–7 times less energy efficient than that of OECD countries, while at the same time providing an option for Japan to fulfil its ambitious emission reduction target of 6% under the Kyoto Protocol.[30]

Two more aspects point to the continuing relevance of JI under the Kyoto Protocol. The first is that a large number of projects were already implemented in CEITs within the framework of the AIJ pilot phase. 76 of the 109 AIJ projects reported by June 1999 were implemented in CEITs (Table 13.1). Investors thus have an incentive to make use of the experience gained in these projects and get credit for emission reductions achieved in AIJ projects (after the AIJ pilot phase) under the Kyoto Protocol. The emission reductions projected are as yet rather limited, but significant. In June 1999, the total (as opposed to annual) reductions were projected at about 160 Mt CO_2 equivalents (roughly equalling 1% of annual emissions of industrialised countries) (see Table 13.1).

The second aspect pointing to the continuing relevance of Article 6 is the uncertainty over whether the other Kyoto Mechanisms will be able to deliver the required emission allowances and at what price. The extent of participation of

29 See Annex of Decision 7/CP.4 in FCCC/CP/1998/16/Add.1.

30 See "Japan and Russia Conclude Landmark Greenhouse Gas Swap", Reuters News Service, 19 April 1998. According to one undisclosed source, this deal was already envisaged during the final days of COP 3 in Kyoto so as to re-ensure Japan that it would be able to draw on the Kyoto Mechanisms for meeting part of its quantified target.

large developing countries in the CDM remains uncertain. With regard to Emissions Trading, the future price of parts of assigned amounts on volatile markets will always entail some risk. As negotiated agreements, JI projects have the advantage to provide relative certainty as regards the price of future emission credits. Furthermore, whereas both the CDM and Emissions Trading will require further elaboration before they can operate, JI project activities under Article 6 could in principle be conducted and emission reductions credited without further preparatory work. According to Article 6.2, the COP/MOP "may" – but not "shall" – "further elaborate guidelines for the implementation of this Article".[31] On the other hand, credits acquired through JI might turn out to be excessively expensive as compared to traded "parts of assigned amounts". A risk aversion strategy (probably best suited for big industry) might therefore involve using all three Kyoto Mechanisms.

31 See also FCCC/CP/1998/MISC.7 (Paper No. 6).

14. The Clean Development Mechanism (Article 12)

The Clean Development Mechanism (CDM) was the "Kyoto surprise",[1] elaborated with little public debate in the final days of COP 3. Taking many developing countries' concerns into account, the CDM establishes a multilateral framework for project-based joint implementation between industrialised and developing countries. This was only possible because some key negotiators built a winning coalition between AOSIS, several major developing countries and industrialised countries. AOSIS and other developing countries particularly vulnerable to climate change will benefit from a fee levied on CDM activities, developing countries in general will benefit from additional financial and technological resources transferred through the CDM, and industrialised countries were eager to achieve substantial participation of developing countries in the Protocol and to be able to earn emission credits from project-based activities in developing countries.

Only a loose framework for the CDM is defined in the Kyoto Protocol. As with the other Kyoto Mechanisms, most of the details need to be fleshed out in future negotiations. In any case, the CDM represents an important mechanism for involving developing countries in the Kyoto Protocol. To what extent the CDM will succeed in fostering climate protection will, however, depend on its eventual design to be negotiated under the Buenos Aires Plan of Action which is to be concluded by COP 6 in 2000 (or early 2001). Sticking to this deadline is particularly important with respect to the CDM, since the Protocol allows for crediting of emission allowances generated in relevant projects as from 2000.

This chapter deals with the negotiating history of Article 12 of the Kyoto Protocol on the CDM, analyses the agreed rules, and sets out the major unresolved design issues of the mechanism. The negotiating history of Article 12 was closely linked to the negotiation of Article 6 on JI, since the two mechanisms were only separated very late in the process. The innovative CDM proved to be the way out of the deadlock on the JI negotiations as it resolved the issue of developing country participation in joint implementation projects. This chapter only deals with the CDM according to Article 12. The reader interested in the broader context of Joint Implementation under the Convention (and JI under Article 6 of the Protocol) is referred to Chapter 13.

14.1 Negotiating History

The CDM has its roots in the proposal for a Clean Development Fund by Brazil at a late stage of the Kyoto process in May 1997.[2] The Brazilian proposal originally contained two elements. First, it suggested a rather complicated methodology for

---

1 Werksman 1998a; Grubb et al. 1999, p. 226.

2 FCCC/AGBM/1997/MISC.1/Add.3, pp. 3-57.

assigning emission ceilings to individual industrialised countries. Second, Brazil proposed the establishment of a Clean Development Fund financed by contributions from non-compliant industrialised countries at a fixed rate per ton of carbon equivalent above the level of assigned amounts.[3]

Resources generated by this fund were to be used for climate change mitigation and adaptation projects in developing countries. Any developing country would be able to apply for funding of specific projects, but would only be eligible for a limited share of the overall amounts available. These limits were to be differentiated according to the level of emissions. The biggest developing country emitters would thus have been eligible for the largest shares of the funding. Up to 10% of the total resources available were to be used for the adaptation component.[4]

The core element of the Brazilian proposal – the establishment of a fund that would provide additional financial resources to the South – was attractive to other developing countries as well. The criteria for distributing the contributions to recipient countries, however, were subject to criticism, since they favoured the biggest developing country GHG emitters, leaving little resources for small and least developed countries.[5] At AGBM 8 in October 1997, the G-77 and China therefore merely endorsed the core concept. This was the proposal calling for the establishment of a fund to be fed by Annex I Parties found to be in non-compliance with their commitments "to assist the developing country Parties to achieve sustainable development and contribute to the ultimate objective of the Convention".[6]

Although generally not very supportive of the idea of additional North-South resource transfers, the US grasped the chance which the Brazilian/G-77 proposal represented for industrialised countries interested in "geographical flexibility" for the implementation of their commitments. Instead of a penalty for non-compliant industrialised countries, the Brazilian proposal could also be read as a means of avoiding non-compliance and thus fulfilling the commitments under the Protocol by investing in climate change related projects in developing countries. For this, the punitive element of the Brazilian proposal would need to be dropped. Instead, the US idea was that industrialised countries would provide resources to acquire emission credits in advance of any compliance assessment. Thus, the idea of the CDM was born, taking into account many of those concerns that drove the majority of developing countries into opposing Joint Implementation (see Section 13.1).

The US made a first quasi-official attempt to give the Brazilian proposal its own spin at the meeting of key governments in Tokyo held at the beginning of November 1997 (see Chapter 7). Subsequently, intensified bilateral discussions between Brazil and the US prepared the ground for elaborating the CDM at Kyoto. By the

---

3 See FCCC/AGBM/1997/MISC.1/Add.3, p. 8.

4 Ibid.

5 See Werksman 1998a, p. 151.

6 FCCC/AGBM/1997/MISC.1/Add.6, p. 16.

time they reached Kyoto, both countries had come close to a common understanding that leaned towards the US interpretation of the proposed mechanism. On this basis, they were able to steer the discussion in Kyoto towards compromise.

Gylvan Meira Filho, the Brazilian architect of this proposal on the fund, chaired the contact group in Kyoto with the mandate to elaborate the CDM. Only late – at the beginning of the second week of the Kyoto conference – did an Article on the CDM appear in the negotiating text.[7] At that point, the G-77 and China had signed up to the draft text, but a number of elements remained controversial. These included the provision for a fee on resource flows under the mechanism to cover administrative costs and finance adaptation measures in particularly vulnerable developing countries. Disagreement further prevailed on institutional issues, such as whether the mechanism would build on existing financial institutions, and the GEF in particular, or whether a new institution should be established. While the dividing line on these issues was mainly between Annex I Parties and developing countries, this was not necessarily the case on other issues, such as pre-commitment period banking of emission allowances generated by CDM projects and the eligibility of sink projects. Both these options were opposed by the EU but supported by the US and a number of developing countries.

EU negotiators were caught somewhat unprepared by the US's willingness to compromise on a number of G-77 demands in order so save the CDM. Furthermore, some Parties complained that there was insufficient time to work on Article 12 and assess its implications. However, this lack of time might have contributed considerably to the adoption of Article 12, since this "veil of uncertainty" also covered negative implications of the mechanism for some Parties.[8]

During the final plenary, discussions arose on the issue of pre-commitment period banking of emission reductions. Several developing countries as well as the EU pressed for the deletion of the provision that would allow for banking of emission credits generated by CDM projects as from 2000. However, the EU did so only when the decision had already been taken, because EU co-ordination took place concurrently with the plenary and did not produce a common position in time.[9] Chairman Estrada followed his rigid procedure (see Section 7.6). Since the negotiating text contained the provision on banking of emission reductions from 2000 and since developing countries and JUSSCANNZ spoke in its favour, he decided to stick to the draft text. Some discussion also arose as a result of a request by Saudi Arabia to channel some funds intended for adaptation into mechanisms for developing countries dependant on production and exports of fossil fuels, i.e. OPEC members. This request, however, did not find much support.

---

7 Compare FCCC/CP/1997/CRP.2 of 7 December 1997 with FCCC/CP/1997/CRP.4 of 9 December 1997 (Article 14).

8 On the significance of the veil of uncertainty in international environmental negotiations see Young 1989.

9 As France noted, the delay was due to "EU organisational issues" (authors' notes).

To conclude, the agreement on Article 12 was one of the most notable developments at Kyoto. It was the result of an unlikely combination of motivations and interests, which produced a dynamism that even the strongest opponents to the participation of developing countries in joint implementation ultimately could not resist. The US and other industrialised countries thereby succeeded in including a key flexibility mechanism and possibly an element of "meaningful" developing country participation into the Protocol (as demanded by the US, see Chapter 17). The EU had no particular difficulties to accept JI with developing countries, as its former opposition had partly been a concession to developing countries. Developing countries succeeded in creating a new source of funding for sustainable development, although the establishment of a new, independent fund for this purpose was not agreed. However, the agreement of developing countries on the CDM and Emissions Trading enabled them to achieve the deletion of the proposed article on voluntary commitments of developing countries.[10] AOSIS finally dropped its former opposition to developing country participation in project-based JI in exchange for the provision of additional money for adaptation projects through the CDM.

14.2 The Rules Applying to the CDM

Article 12.1 "defines" a "clean development mechanism". This compromise language conceals the different approaches of developing and industrialised countries to the CDM.[11] Industrialised countries were determined to avoid the establishment of a new financial institution like the Multilateral Fund for the Implementation of the Montreal Protocol on Substances that Deplete the Ozone Layer, as they are generally striving to concentrate environmental North-South financial transfers in one institution, the GEF. Developing countries opposed the GEF because they had been dissatisfied with it since its inception and hoped to gain more influence in a new institution.

The language in Article 12.2 and 12.3 reveals divergent purposes and objectives of the CDM with respect to industrialised and developing countries. The CDM is to assist developing countries "in achieving sustainable development and in contributing to the ultimate objective of the Convention" (Article 12.2). Even more unspecific is the provision in Article 12.3(a) that developing countries "will benefit from project activities" under the CDM. As regards the benefits for industrialised countries, the language is more precise: the CDM shall assist Annex I Parties "in achieving compliance with" their quantified targets under Article 3 (Article 12.2), and certified emission reductions (CERs) accruing from CDM projects can be set off against "part of" their obligations, as determined by the COP/MOP (Arti-

---

10 Chapter 7; on the proposed article on voluntary commitments by developing countries see Chapter 17.

11 See also Article 11 FCCC, which "defines" the Financial Mechanism of the Convention.

cle 12.3). Article 12 does not refer to the enhancement of anthropogenic removals by sinks as Article 6 on Joint Implementation does.

As a further contrast to JI under Article 6 and Emissions Trading under Article 17, CERs will be added to the assigned amount of the acquiring Party (Article 3.12*) but not subtracted from the assigned amount of another Party* (see Figure 13.1). This is due to the fact that developing countries are not subject to quantified targets under the Kyoto Protocol. CERs acquired by an industrialised country thus increase its assigned amount, i.e. its emission allowance. As this is not balanced by a corresponding reduction of the assigned amount of another Party, the overall amount of GHG emission allowances of Annex B industrialised countries increases. This is one of the most important features of the CDM and implies that particular diligence is required in elaborating its modalities (see below).

The formulation that industrialised countries may use the CDM to fulfil "part of" their obligations has its roots in the attempt of the EU to limit the use of the Kyoto Mechanisms (see Chapter 15). As with JI and Emissions Trading, the Union unsuccessfully pushed for a quantification of the "part" of the target which could be met by acquiring CERs under the CDM. In the context of the CDM, the EU did not only meet with opposition from JUSSCANNZ countries, but also from developing countries because a quantitative cap would have limited the potential resource flows under the CDM.

Article 12.4 recognises the COP/MOP as the highest CDM authority and a still-undefined "executive board" is to supervise the mechanism. The vague language is again due to disagreement over whether to rely on existing institutions for the executive board or whether to set up a new one. Industrialised countries were successful to avoid any indication that the executive board might be composed of Parties, which would also have been implied if the paragraph had referred to an executive *committee* (as in the case of the Montreal Protocol's Multilateral Fund; see below).

Similarly unspecified "operational entities" to be designated by the COP/MOP shall certify CERs on the basis of three criteria (Article 12.5):

- voluntary participation approved by each Party involved,
- real, measurable, and long-term benefits related to the mitigation of climate change, and
- "environmental additionality", i.e. CDM projects must lead to emission reductions that would not have occurred otherwise.

As in the case of the JI criteria under Article 6, the CDM criteria largely reflect those agreed in 1995 for the AIJ pilot phase. The requirement that any projects be compatible with national environment and development priorities was dropped. This should be guaranteed by the need to get approval by all Parties involved. Notably missing from the list is the requirement of financial additionality, i.e. that resources invested under the CDM should be additional to official development assistance.

According to Article 12.6, the CDM "shall assist in arranging funding of certified project activities as necessary". One possibility for implementing this is to establish a clearing house. This would support the matching of potential investors and hosts of possible CDM projects. Whether or not the provision also means that the CDM will actually channel funds from investors to projects was left open, with most industrialised countries opposing such an interpretation (see Section 14.3).

Further modalities and procedures to ensure "transparency, efficiency and accountability" are to be elaborated at the first COP/MOP (Article 12.7). However, the Buenos Aires Plan of Action adopted by COP 4 envisages agreement on these modalities by the COP at its 6th session.[12] This deviation from the text of Article 12 is due to the fact that, first, all of the Kyoto Mechanisms are to be elaborated as a package. Second, agreement on those rules is indispensable if CDM projects are to operate from the year 2000, as envisioned in Article 12.10. However, any decisions by the COP will have to be confirmed formally by the first session of the COP/MOP of the Protocol. An important part of the rules to be elaborated are those concerning the "independent auditing and verification of project activities" (Article 12.7). This provision neither predetermines who should perform those functions, nor does it specify the process for conducting auditing and verification.[13]

Article 12.8 introduces one of the most noteworthy and innovative elements into the CDM. It determines that "a share of the proceeds from certified project activities" is to be used to cover the administrative expenses of the mechanism as well as "to assist developing country Parties that are particularly vulnerable to the adverse effects of climate change to meet the costs of adaptation". Clearly, this provision lead to AOSIS accepting and even supporting the CDM proposal. AOSIS had been unsatisfied with the funding for adaptation through the GEF. Article 12.8 thus was seen as opening up a source of funding not limited by budgetary constraints of industrialised country governments, as the CDM will primarily deal with private sector investment.[14]

This Paragraph remained controversial up to the last moment in Kyoto, because of industrialised countries' concerns. There was, first, some general opposition, especially by the finance ministries, to the introduction of international taxation, which would set an important precedent and might thus open up related requests in other areas of international co-operation. Second, such taxation of CDM projects would represent an additional burden and would thus reduce the amounts of CERs that could be generated. Even more importantly, several industrialised countries identified constitutional problems with such a provision. As taxation in many of these countries belongs to the exclusive competence of parliaments and is to be used for specified purposes, introduction of some kind of international tax by way

---

12 Decision 7/CP.4 in FCCC/CP/1998/16/Add.1.

13 See also Werksman 1998a, p. 155.

14 See Cameron/Buck 1998a.

of an intergovernmental agreement was considered problematic. As a result, the tax was finally called "a share of the proceeds" instead of a "user fee".[15]

Article 12.9 mirrors Article 6.3 on JI among Annex I Parties. It allows for participation by "private and/or public entities" under the CDM in industrialised investor countries and developing host countries. Reflecting the multilateral character of the CDM, participation is "subject to whatever guidance may be provided by the executive board". This contrasts the bilateral approach to JI among Annex I Parties under Article 6, which provides the Parties to JI projects with a mandate for authorising "legal entities" to participate in the projects (see Section 13.2).

Article 12.10, finally, allows for pre-commitment period banking of CERs from the year 2000. CERs resulting from project activities during the years 2000–2007 may thus be acquired by industrialised country Parties and be used to increase the "assigned amount" of their own emissions of the first commitment period. The Protocol, however, does not specify any conditions for doing so. In particular, Article 12.10 is silent on whether the entry into force of the Protocol, the designation of operational entities for certification or the adoption of appropriate modalities and procedures (as called for by other paragraphs of the Article) is required prior to such banking of CERs. When Chairman Estrada gavelled down opposition to banking in the final night of Kyoto, the EU tried to provide COP 4 with a mandate for revisiting the issue. This was rejected by the US and a number of developing countries. As a compromise, the Parties adopted a decision that scheduled further discussions in the Convention bodies to analyse "the implications" of Article 12.10.[16] Post-Kyoto, however, the EU has not followed up on its request to prohibit pre-commitment period banking, thus indicating its acceptance.[17]

Overall, Article 12 "defines" an innovative instrument of international climate policy. Precedents for the CDM do not exist. Since the Kyoto Protocol merely establishes a loose framework for its development, most of the design features remain to be determined. The CDM therefore is not only the "Kyoto surprise", but also one of the still-undetermined elements of the Kyoto Protocol and part of its "unfinished business".

14.3 Assessment and Outlook

Post-Kyoto, the CDM has drawn much attention from the climate policy community, researchers as well as politicians.[18] The design and development of the CDM was given priority in the Buenos Aires Action Plan adopted by COP 4 in

15 Compare FCCC/CP/1997/CRP.4 of 9 December 1997 with FCCC/CP/1997/CRP.6 of 10 December 1997.

16 Decision 1/CP.3, para. 5(e), in FCCC/CP/1997/7/Add.1.

17 See Non-Paper on the Clean Development Mechanism by the EU and others in FCCC/CP/1998/MISC.7.

18 See, for example, contributions in Goldemberg/Reid 1999 and Goldemberg 1998; Grubb et al. 1999; Werksman 1998a; Cameron/Buck 1998a and b; Michaelowa/Dutschke 1998.

November 1998.[19] The following sections will first analyse the options available for setting-up the general approach and governance system of the CDM (Section 14.3.1), and will then investigate the options for determining the operational criteria and policies of the mechanism (Section 14.3.2). This is followed by analyses of the relationship of the CDM to the other Kyoto Mechanisms (Section 14.3.3) and the special issue of the "share of proceeds" to be used for administrative expenses and adaptation (Section 14.3.4). The section closes by exploring the likely future relevance of the CDM (Section 14.3.5).

14.3.1 General Approach, Governance, Functions and Actors

As a result of the political compromises in negotiating the text, Article 12 of the Protocol provides scarce guidance on the governance system of the CDM. The roles of, and the relationship between, the COP/MOP as the supreme body, the executive board, the mentioned private and/or public entities, the not yet defined "operational entities" for certification and some equally unspecified auditors and verifiers are in need of clarification. Furthermore, the appropriate arrangements to collect, administer and disburse the "share of the proceeds" used for administrative expenses and adaptation activities remain to be determined.

In taking the needed decisions, Parties could follow *two principal models* that are discussed in the literature: (1) a multilateral and institutionalised model and (2) a bilateral or clearing-house model.[20] The multilateral model is inspired by the example of the Multilateral Fund for the Implementation of the Montreal Protocol and has generally been favoured by developing countries. As with that Fund, the executive board would closely oversee project selection and implementation. Within such a multilateral framework, host countries could, in principle, have a greater say in the selection and design of projects than if negotiating with donor countries bilaterally.[21] The Executive Committee of the Multilateral Fund under the Montreal Protocol plays a similar role in co-operating with designated "implementing agencies" responsible for project implementation. Under the Multilateral Fund, these are UNDP, UNEP, UNIDO and the World Bank.[22] The executive board of the CDM might in particular also oversee auditing and verification of project activities and handle relations with (and supervise) any "operational entities" responsible for the certification of CERs.

Project activities under the multilateral approach might be conducted on a bilateral basis and/or within the framework of one or several trust fund(s), in which

---

19 Decision 7/CP.4, para. 1 in FCCC/CP/1998/16/Add.1.

20 Others have compared a bilateral with a portfolio approach (Yamin 1998b) and a marked-based, laissez-faire with an interventionist approach (Werksman 1998a) towards the CDM in general. See also Michaelowa/Dutschke 1998; Ott 1998b.

21 See Yamin 1998b; Michaelowa/Dutschke 1998.

22 On the Multilateral Fund for the Implementation of the Montreal Protocol see DeSombre/Kauffman 1996; Biermann 1997.

donor resources intended for CDM projects could be pooled. Possibilities to combine both modes of implementation are again exemplified by the Multilateral Fund of the Montreal Protocol, which has a component that allows for bilateral implementation of projects (i.e. without going through the implementing agencies mentioned above). In contrast to the Multilateral Fund of the Montreal Protocol, any portfolio funds under the CDM would for the largest part not rely on government money but on private-sector resources, since the concept of joint implementation underlying the CDM aims at mobilising private investment for climate protection (see Chapter 13).

The executive board might administer a trust fund itself and create a portfolio of CDM projects, for which host developing countries and interested investors could submit applications. It might also oversee the implementation of a number of sub-funds operated by various organisations. For example, the World Bank is planning to establish a global Carbon Investment Fund for the development of a project portfolio.[23] Although this does not currently aim at creating a fund under the CDM (but rather at catalysing the build-up of relevant experience), it illustrates some interesting possibilities. Other public and private actors could become involved in establishing such funds as well.

The bilateral model[24] has been favoured by most industrialised countries. According to this approach, the CDM primarily aims to employ private investment and market forces for reducing GHG emissions (and generating CERs), thus minimising transaction and administrative costs. The CDM would thus focus on the clearing-house function mentioned above (see Section 14.2). It would facilitate matching of potential investors with partners in host countries and in this way "assist in arranging funding of certified project activities as necessary" (Article 12.6). Project selection and implementation would largely be left to the participants (subject to whatever operational criteria adopted by the COP/MOP: see Section 14.3.2). The executive board would then simply supervise resource flows under the CDM, and would handle the certification, verification and auditing (thus exhibiting a more limited control function than under the multilateral approach).[25]

Although frequently contrasted, these two models are not necessarily mutually exclusive, as suggested by the experience under the Montreal Protocol. Rather, they describe two ends of a spectrum ranging from low to high multilateral supervision, control, and involvement in project activities. Thus, a clearing-house

---

23 World Bank, The Global Carbon Initiative of the World Bank, (Handout, on file with authors).

24 Even under the bilateral model, more than one investor country might be involved in a project (thus the project participants might number more than two). Nevertheless, especially from the perspective of the host country the relationship within such a model would be a bilateral one, i.e. one or more investors would face one host country, with no/little oversight by third Parties/multilateral institutions. Thus, the model may be coined "bilateral".

25 For an illustration of the bilateral model see, e.g., FCCC/CP/1998/MISC.7 (Paper No. 6); see also the "bilateral approach" described by Yamin 1998b.

function could and would certainly be part of a "multilateral" CDM, but it need not be confined to this function. The two models mirror the different approaches to JI in general and, more specifically, under Article 6 of the Kyoto Protocol (see Chapter 13). While the latter appears to lean towards the bilateral approach, much of what is contained in Article 12, most importantly the supervision by an executive board, points towards a more multilateral variant for the CDM.

There are a range of further options for the governance system of the CDM, which depend partly on the choice of the basic approach outlined above. These options concern the *composition of the executive board, its powers and functions and its procedures*. Should a new institution be established, Parties (i.e. the COP and later the COP/MOP at its first session) will have to elaborate its decision-making procedures and its composition. Experience with the Executive Committee of the Multilateral Fund of the Montreal Protocol suggests that such an independent executive board might consist of a representative sample of Parties with equal representation of industrialised and developing countries, possibly using some kind of "double-weighted" majority voting.[26] Under the Montreal Protocol, for example, decisions require a two-thirds majority representing a majority of industrialised countries and a majority of developing countries. In contrast, making use of an existing institution like the GEF might basically imply accepting the pre-existing composition and voting rules. However, a separate body with special composition and decision-making procedures could, in principle, also be established under the GEF.

It would be reasonable to transfer extensive powers to oversee the day-to-day operation, determine the policies, procedures, guidelines and criteria applicable to the CDM and possibly even take decisions on single projects, to a new institution created specifically for this purpose (and probably composed of Parties). Such an executive board would provide a relatively continuous and detailed control by Parties over project activities under the CDM, thereby taking account of concerns by developing countries about potential loss of sovereignty and government control. As such, it corresponds with the multilateral model of the CDM (see above). By contrast, assigning the tasks of the executive board to an existing institution might imply granting more limited powers to the executive board (e.g. little more than operating a clearing house) and thus rather corresponds with the bilateral model preferred by industrialised countries.

The institutional options also have some implications for the degree of oversight and control by the COP/MOP. Article 12.4 incorporates the division between industrialised and developing countries on this issue in its formulation that the CDM "shall be subject to the *authority and guidance*" of the COP/MOP (emphasis added). The first term, "authority", corresponds with the establishment of something new that would be subject to the authority of the COP/MOP. In contrast, the term "guidance" corresponds with a model where the respective tasks are assigned

---

26 See Multilateral Fund 1998; Ozone Secretariat 1998, pp. 199 et seq.

to an existing institution like the GEF. This would imply a contractual relationship through the elaboration of a Memorandum of Understanding, and Parties would thus retain less direct control over the operation of the CDM.

In both cases, *secretariat* assistance will be needed, which might be more extensive in the case of an independent executive board with far-reaching powers (preparation of meetings, documents, evaluation of project/fund proposals, administration of a clearing-house etc.). The UNFCCC Secretariat could build upon relevant expertise that it has acquired in providing support under the AIJ pilot phase. However, Parties might also establish separate secretariat capacities serving that institution and financed by the share of the proceeds generated by the CDM (which could nevertheless be related to the FCCC Secretariat).

The *certification, auditing and verification* functions do not suggest any particular form of governance, and could in principle be carried out by a variety of actors and entities. These include international and national, governmental and private actors. Designating a number of actors for these tasks would enhance competition and could thus reduce costs. As certification will build the basis for granting CERs, intergovernmental organisations with relevant expertise could be designated as "operational entities". Options include the implementing agencies under the Montreal Protocol's Multilateral Fund (UNEP, UNDP, UNIDO, the World Bank), all of which except UNIDO are GEF parent organisations. However, private firms might also be considered as such firms fulfil similar functions already in various areas of international trade.[27] Care has to be taken, however, that certifiers are not involved in project implementation so as to avoid conflicts of interest and build up trust in the reliability of CERs.

Auditing and verification (Article 12.7) are of utmost importance for the credibility of the CDM.[28] These functions will involve checking that project implementation meets the established operational criteria (see below).[29] Thus, auditing and verification would need to precede any final certification of emission reductions. As auditing and verification will, in large part, involve rather specialised technical activities (possibly including on-site inspections), it has been suggested that this function may be performed by private actors such as "internationally recognised accounting or consulting firms".[30]

---

27 SGS, a company with international operations, has already been involved in the certification of project activities in Costa Rica, see Goldberg 1998; the International Standards Organisation (ISO) might be involved in elaborating relevant standards.

28 See the discussions at the Technical Workshop on Mechanisms under Articles 6, 12 and 17 of the Kyoto Protocol organised by the FCCC Secretariat, in: Earth Negotiations Bulletin, Vol. 12, No. 98, 19 April 1999.

29 See also the International Performance Measurement and Verification Protocol of the US Department of Energy, December 1997, available at <http://www.ipmpv.org> as of June 1999.

30 E.g. Werksman 1998a, p. 155.

Article 12.9 allows for participation of *"private and/or public entities"* in the CDM.[31] However, their role, functions and responsibilities have to be clarified. It follows from the concept of joint implementation that project activities are expected to be mainly carried out by private companies. Most importantly, the Parties will have to establish the obligations and rights of such private entities involved in project implementation and determine how they will relate to the governmental actors and institutions active in the CDM.[32] Options include allowing direct relations between private actors and the CDM governing bodies (e.g. for project approval, reporting, dispute settlement) or mediated relationships through governments of Parties. The same choice exists with regard to the ability to hold and sell CERs. Furthermore, liability for project failure may be conferred directly upon these private entities or mediated through their respective governments. The potential role of private and/or public actors, however, goes far beyond project implementation, auditing or verification. Private actors might furthermore become involved in the acquisition and selling of CERs (brokering), if these are made "fungible" (see below). These are only some of the complex issues that need to be resolved in the future negotiations leading up to COP 6 and beyond.

Developing suitable rules for liability and dispute settlement becomes even more necessary, since a broad range of both private and public actors are going to be involved in various functions regarding CDM project activities. For example, entities such as insurance companies, regional development banks or the International Finance Corporation (IFC) of the World Bank might become involved in insuring investors against project failure and covering the risks of investments.[33] Although the CDM primarily aims at mobilising private sector resources, "public entities" are likely to get involved in its operation and implementation as well. Some Annex B countries might wish to use their experience in development assistance to implement CDM projects that are financed by public funds. Japan, for example, has already shown a clear interest in this.[34] Using public funds, however, raises the issue of financial additionality, i.e. to what extent such funds may come from existing budgets of official development assistance (ODA) (see below on operational criteria).

As regards developing countries, public entities might get involved in project implementation, especially in countries where the private sector is underdeveloped. It has been suggested that developing countries might self-finance CDM project activities and make the generated CERs available to investors (while following the rules applicable under the CDM). Developing countries could thus, by

---

31 On the role of these actors see Campbell 1998.

32 See generally Ott 1998b.

33 See Cameron/Buck 1998b.

34 See Werksman 1998a, p. 154.

investing extra-money in CDM projects, earn CERs and sell them, for example, through an auctioning.[35]

The fundamental design of the CDM will be subject to political decision-making by the Parties. Halfway through 1999 it appears rather unlikely that the Parties will be able to agree on all relevant issues by COP 6. Since the executive board will not become operational before the entry into force of the Protocol, there will be some extra time to find a political compromise. However, such a delay would create uncertainty for potential investors and thus diminish the potential of the CDM. Other more technical issues are more pressing, especially those relating to auditing, verification and certification (including guidelines for conducting these tasks). In this respect, it will also be necessary to develop guidelines for reporting so as to enable checking whether the rules have been followed properly. Elaboration of such guidelines could build on work undertaken on the general subject of joint implementation and under the AIJ pilot phase.[36]

Without agreement on these basic rules and those operational criteria discussed below, the CDM will not function properly as from 2000, as envisaged in Article 12. Afterwards, these rules will most likely be subject to further refinement by the executive board and the COP/MOP. The COP/MOP will also be required to confirm any decisions on the CDM taken by the COP of the Convention prior to the entry into force of the Kyoto Protocol.

14.3.2 Operational Criteria and Policies

In addition to those functions of the CDM discussed above, the Parties need to develop criteria and policies for the day-to-day operation of the CDM – as other environmental financial institutions like the GEF and the Multilateral Fund of the Montreal Protocol have done.[37] Article 12.5 of the Kyoto Protocol merely establishes three criteria that are to guide implementation of CDM projects, namely voluntary participation, environmental additionality and the achievement of real, measurable, and long-term climate benefits (see above Section 14.2).[38] While voluntary participation merely requires express approval by the Parties involved in project activities, the other two criteria require the establishment of credible baselines against which to measure any emission reductions and/or the establishment of benchmarks/reference cases which could be used to determine "environmental additionality" (see also Chapter 13). These baselines are inherently difficult to establish and the topic was intensely debated prior and post-Kyoto. A procedure that is workable, cost-effective and environmentally sound may involve

---

35 Indeed Costa Rica has already engaged in JI activities of a unilateral kind; see Goldberg 1998; Dutschke 1998.

36 See the uniform reporting format for AIJ (FCCC/SBSTA/1997/4) as adopted by Decision 10/CP.3 in FCCC/CP/1997/7/Add.1.

37 See Multilateral Fund 1998; GEF 1996.

38 See Goldemberg/Reid 1999.

the determination of project types and the establishment of generalised baselines that are continuously updated.[39]

Furthermore, the Parties could elaborate "positive" lists of eligible projects (e.g. renewable energy projects or improvement of energy efficiency) for which environmental additionality can be presumed.[40] However, this does not dispense from the need to set concrete quantitative baselines in order to determine the actual emission reductions achieved and the corresponding amounts of CERs.[41] The Parties might also establish a negative list, which would exclude certain projects from being eligible. Nuclear energy projects, for example, might be excluded from the CDM due to the inherent dangers concerning operation and disposal of nuclear wastes (especially in developing countries).[42]

A further issue in need of clarification is the possible role of official development assistance (ODA) in the CDM. The CDM criteria do not include the requirement of "*financial additionality*". Thus, official development assistance projects would prima facie appear to be eligible under the CDM, although the concept of joint implementation underlying the CDM primarily aims at private rather than public investment. At least Japan is known to be interested in employing official development assistance for this purpose. However, demands from developing countries to exclude such projects have continued post-Kyoto.[43]

Another hotly disputed policy issue revolves around the question whether sink projects should be eligible under the CDM. Article 6 on Joint Implementation *expressis verbis* allows the generation of emission reduction units resulting from the enhancement of anthropogenic sinks. Article 12, in contrast, only refers to "emission reductions" resulting from project activities. Although it may be argued that activities resulting in reduced deforestation might qualify as leading to "emission reductions", this difference in wording suggests that sink projects are not eligible. This is also supported by the negotiating history: throughout the negotiations, the draft texts on the CDM contained a note to the effect that sinks might be included in the scope of the CDM depending on the resolution of the issue in relation to quantified targets. The final text, however, does not contain this footnote anymore, nor does it refer to sinks at all.

Post-Kyoto, opinions differ as to whether Article 12 means to exclude *sink projects* under the CDM.[44] A number of industrialised and developing countries –

39 See OECD 1999; Michaelowa/Dutschke 1999; Luhmann et al. 1997 and the discussions at the Technical Workshop on Mechanisms under Articles 6, 12 and 17 of the Kyoto Protocol organised by the FCCC Secretariat, in: Earth Negotiations Bulletin, Vol. 12, No. 98, 19 April 1999.

40 See Cameron/Buck 1998b; Panayotou 1998; Hassing/Mendis 1998.

41 See Luhmann et al. 1997, pp. 91 et seq. and pp. 167 et seq.

42 Approximately 150 representatives of the nuclear industry participated in COP 4. Demands for nuclear energy projects under the CDM and JI have been aired, but were not officially discussed.

43 See, for example, FCCC/CP/1998/MISC.7/Add.2 (African Group).

44 Yamin 1998b.

most importantly those involved in sink projects under the AIJ pilot phase – have kept this debate alive since Kyoto and have demanded that sinks projects be eligible under the CDM, possibly even going beyond the limited sink categories specified for industrialised countries' GHG inventories under Article 3.[45] As with the general issue of sinks, a definite decision is likely to be taken only after the completion of the IPCC special report on sinks is completed, after the year 2000 (see Section 11.3).

There are good reasons to exclude sink projects from the CDM (as argued by several EU Member States), most importantly the large scientific uncertainties surrounding sinks, the special difficulties to establish quantifiable, reliable baselines and to ensure "environmental additionality" especially in developing countries (see Section 11.3). Furthermore, sink projects lack the important element of transfer of technology, which is meant to be a central component of the CDM (although they may contribute to capacity building). However, considerable political pressure will continue to be exerted to allow sink projects under the CDM. This is partly because investments have already been made in projects under the AIJ pilot phase, which the participating countries would also like to transfer to the CDM as from 2000. Should this line of reasoning prevail and sink projects be included, particularly stringent criteria as to the types of eligible projects, the calculation of CERs and sustainability indicators (environmental, social and cultural) would have to be established for these projects, to deal effectively with the inherent ambiguities. For example, particularly high insurance and liability might be required for forestry projects, since project failure through destruction (e.g. by forest fires) would nullify the CO_2 sequestration achieved in the past and thus invalidate *all* CERs accrued from such projects (as opposed to future CERs should non-sink projects fail). By the same token, the CDM needs to reconcile the potentially conflicting objectives of maximising carbon absorption and achieving sustainable forest management (see also Chapter 23).

A special challenge arises from Article 12.10, which allows for the banking of CERs generated from 2000 to 2007 and their transfer into the first commitment period. COP 4 settled to resolve the outstanding issues at COP 6, i.e. in late 2000 (or early 2001).[46] There is little doubt that, even if agreement is reached, any arrangements will need some time to become fully operational. It is also likely that Parties will have to refine the CDM further after COP 6. The meaning of Article 12.10 therefore requires specification. This concerns the question of whether retroactive crediting will be allowed, i.e. whether emission reductions generated before the elaboration of the relevant rules and guidelines (on auditing and verification, baselines, etc.) may be certified later (thus becoming CERs). Such *ex-post* certification would be difficult and may well prove to be impossible. Under the

45 See papers in FCCC/CP/1998/MISC.7 and addenda.

46 Decision 7/CP.4 in FCCC/CP/1998/16/Add.1.

AIJ pilot phase, participants applied a wide range of rules and coherent data on the calculation of emission reductions achieved are still lacking.[47]

Furthermore, the Parties will have to decide whether emission reductions achieved through CDM projects will be certified at the beginning of a project, at its end or in regular intervals (e.g. annually) over its duration. Certification before actual emission reductions have occurred would *de facto* result in the "borrowing" of emissions, a concept that was rejected early in the Kyoto process (see Section 11.1). The choice of terms (certification) as well as reasons of prudence and transparency thus speak in favour of regular certifications of actually achieved reductions. In addition, however, investors might demand a reliable projection of the amount of CERs that are expected to be generated in a project. This in turn might require some type of auditing and preliminary estimation prior to actual achievement of the emission reduction.

Equity issues are involved with respect to the distribution of CDM projects among developing countries. If left to the free market, CDM investments might well concentrate on those developing countries that are already the focus of foreign direct investment: in 1997, 10 developing countries attracted nearly three-fourths of the total foreign direct investment in developing countries, while many others received only negligible amounts.[48] For example, the share of Sub-Saharan Africa in the total was only around 1%. Thus, African countries, least developed countries and countries with small markets are not likely to benefit much from the CDM.[49]

COP 4 acknowledged this problem and requested the Secretariat to prepare "a plan for facilitating capacity building" for CDM activities in developing countries with particular emphasis on small island states and least developed countries.[50] In addition, it has been proposed that CDM projects should be undertaken equally in all developing country regions or that quotas be defined for certain groups of countries.[51] Precedents for such equity-based approaches – which have to be balanced against efficiency considerations – can be found in the Montreal Protocol's Multilateral Fund. The Parties to this agreement agreed to allocate "sufficient funds" to developing countries with low levels of consumption of ozone depleting substances, because, on the sole basis of established criteria, these countries would not have received the necessary assistance.[52] Such equity-based considerations would be easier to implement under a multilateral approach.

---

47 See synthesis report on the status of the AIJ pilot phase by the Secretariat in FCCC/SBSTA/1997/12 and Add.1.

48 World Bank 1999, pp. 21-30; see also Yamin 1998b.

49 See, e.g., Dessus 1998.

50 Decision 7/CP.4, para. 4 in FCCC/CP/1998/16/Add.1.

51 For example, Yamin 1998b; see also African Common Position on the Clean Development Mechanism in FCCC/CP/1998/MISC.7/Add.2.

52 Multilateral Fund 1998, pp. 4-5.

Additional policies and guidelines will need to be developed, depending on the fundamental approach the CDM takes. For example, conditions for auctioning of CERs and unilateral implementation of CDM projects by developing countries might be elaborated if and when these design features are agreed upon.

14.3.3 Relationship of the CDM to the Other Kyoto Mechanisms

Another far-reaching policy decision to be taken by the Parties concerns the "fungibility" of CERs, i.e. whether CERs that have been generated by a CDM project can only by acquired by those industrialised countries involved in a CDM project or whether they might be transferred to other Parties. The rules for the transfer of emission allowances are not dealt with in the respective Articles on the Kyoto Mechanisms, but in Article 3 of the Protocol: paragraphs 10 and 11 of Article 3 regulate the acquisition and transfer of parts of an assigned amount for Emissions Trading and of ERUs under Joint Implementation. Article 3.12, on the other hand, concerns the acquisition of CERs under the CDM. As such, CERs differ from the other two allowances, since they do not constitute part of an assigned amount of a Party that is subtracted after the transfer (see Figure 13.1). As Article 3.12 only regulates the acquisition but not the transfer of CERs, one plausible interpretation might be that only the Party investing in a developing country might "acquire" such CERs and add them as an emission allowance to its assigned amount.

Prohibiting the free transfer of CERs would, however, preclude the acquisition of CERs and the transfer to another Party by the host developing country. The transferability of CERs thus entails important consequences for the role developing countries might play in the CDM. Full fungibility could provide an incentive for developing countries to acquire (part of the) CERs generated in CDM projects and to participate in the trading of such allowances. Without transferability, there would be no incentive for unilateral implementation of CDM projects by developing countries (see above).

Transferability of CERs would therefore lock the CDM onto the not yet operational Emissions Trading under the Kyoto Protocol and would provide limited access to this mechanism for developing countries. Such participation of developing countries might well enhance their interest in both Emissions Trading and the CDM. However, it would also enhance the incentive to certify as many emission reductions from CDM projects as possible. Transferability of CERs and consequential participation of developing countries in Emissions Trading would thus require particularly stringent guidelines for determining baselines and monitoring actual emissions as well as auditing, verifying and certifying emission reductions.

These considerations and the fact that CERs actually increase the total emission allowances of industrialised countries under the Kyoto Protocol, justifies the more robust "safety belts" for the CDM than for JI, including particularly stringent rules on project eligibility and verification. In addition, establishing a level playing field for the different Kyoto Mechanisms is required if the CDM is to play a major role under the Protocol (see below). Since both JI and the CDM are rooted in the same project-based concept and similar problems will be faced by Parties in their implementation, some degree of co-ordination in the elaboration of these mechanisms would be beneficial.

14.3.4 The Share of Proceeds Used for Administrative Expenses and Adaptation

Many questions are open with regard to Article 12.8 – the provision that a "share of the proceeds from certified project activities" shall be used to cover administrative expenses and adaptation measures for particularly vulnerable countries. The size of this "share of the proceeds" to be used for adaptation to the adverse impacts of climate change (e.g. sea-level rise) and administration has not been determined by the Kyoto Protocol. Further, the basis for calculating this share has yet to be defined: whether this is to be based on the CERs generated or rather on the rents generated by CDM projects.

Another important decision concerns the distribution of those funds among the administration and adaptation functions. The part of the funds used for administration expenses might be determined by reference to comparable institutional costs. UN agencies generally deduct 13% of total project costs for administration ("programme supply and support costs"), although they have increasingly come under pressure to decrease and differentiate this rate (according to the size of the project).[53] Furthermore, the amounts raised within the UN framework usually include actual project implementation, whereas this task would be expected to be taken on by private actors under the CDM.

If several actors were to compete for the administration of CDM projects, the share of the proceeds to be used for this task could be diminished. In fixing the part of the "share of the proceeds" to be used for adaptation, there is a clear trade-off between maximising this "share" and preserving the attractiveness of CDM projects. The higher the "adaptation duty" the greater the incentive for investors to look for other options, be it domestic action in industrialised countries or the other Kyoto Mechanisms.

A separate function of the CDM will be the collection, administration and disbursement of funds for adaptation. As elaborated above, the Parties could establish new institutional structures (within an independent CDM) or use established structures such as the GEF, which is already funding adaptation measures under the Convention. The whole array of governance issues elaborated above thus becomes relevant for the aspect of funding for adaptation as well.

---

53 For example, the Executive Committee of the Multilateral Fund of the Montreal Protocol decided in 1998 to differentiate the percentage of the agency fee according to the financial volume of a project with a range of 11-13% and a provision of case-by-case decisions in the case of projects with a value exceeding US $ 5 million; see Decision 26/41 of the Executive Committee in UNEP/OzL.Pro/ExCom/26/70.

Furthermore, the kind of adaptation projects eligible for funding need to be determined. Under the FCCC, adaptation projects have been pursued in three different "stages". Stage I consists of identifying particularly vulnerable areas and appropriate policy options. Stage II comprises measures to prepare for adaptation, while measures to actually facilitate adaptation, including insurance, are to be carried out in Stage III. At COP 1, financing through the GEF was confined to Stage I activities.[54] Thus far, funding for adaptation was limited largely to preparing national communications of countries particularly vulnerable to the impacts of climate change, including planning for adaptation.[55] COP 4 in Buenos Aires agreed in 1998 that Stage II adaptation activities should be eligible for funding in particularly vulnerable countries and regions identified in Stage I activities.[56]

The extent to which funding for adaptation generated under the CDM will be locked onto this existing system under the Financial Mechanism of the Convention is an open question. Funding adaptation with proceeds generated under the CDM might complement the obligation of Annex II Parties to the Convention to assist particularly vulnerable countries in adaptation (Article 4.4 FCCC). The resources raised under the CDM for adaptation could provide further resources for Stage II and III activities. Because of the obvious substantive links between funding for adaptation under the FCCC and the CDM, a close political link between the institutional arrangements for administering and channelling these funds has its obvious merits.

Finally, the politically most sensitive issues will be of a distributional character: the part of the "share of the proceeds" available for adaptation projects and the eligible countries. A concrete definition of eligible countries has not yet been put forward in the framework of the FCCC. As substantive amounts of resources might be involved in adaptation, this task can be expected to result in heavy political bargaining.

14.3.5 Conclusions

The future role of the CDM in the framework of the Kyoto Protocol and in relation to the other mechanisms is difficult to ascertain. Quite a number of factors are unknown or yet to be decided: the design of the CDM itself and that of Emissions Trading and Joint Implementation, whether Parties other than the European Union will enter into a bubble agreement under Article 4 (see Chapter 12) and, most importantly, whether and to what extent the obligations contained in the Kyoto Protocol will be implemented domestically by industrialised countries. Much will furthermore depend on the interest and willingness of developing countries to use the CDM (perhaps also to demonstrate "meaningful participation" in the Kyoto Protocol).

---

54 Decision 11/CP.1 in FCCC/CP/1995/7/Add.1; Oberthür/Ott 1995, p. 147.

55 See GEF 1996, pp. 31-32; Werksman 1998a, p. 150.

56 Decision 2/CP.4, para. 1(a) in FCCC/CP/1998/16/Add.1.

The total amount of CERs that may be created through CDM project activities will depend on a number of the factors mentioned above, including the determination of various policies and eligibility criteria. The decisions on whether and to what extent to include sinks within the scope of the CDM, introducing a cap on the use of CERs by industrialised countries and the fungibility and tradability of CERs will be particularly relevant in this respect. Fixing regional shares for CDM investments would reduce the overall attractiveness, increase the average cost of CDM projects and thus limit the amount of CERs generated.

In a totally unregulated and uncontrolled situation, virtually any private foreign direct investment and a large part of the official development assistance activities in developing countries might qualify as a CDM project. If unilateral implementation by developing countries themselves were allowed under the CDM and if developing countries would be able to transfer these CERs to third countries, hardly any investment/activity in the South would be excluded *a priori*.

Because of the undetermined character of the CDM, no exact quantitative assessment of its eventual impact can be presented. Available estimates span a wide range, of between US $ 1 and US $ 26 billion for emission credits of between 27 Mt and more than 700 Mt of carbon equivalent that might be generated.[57] This range is equivalent to between less than 1% and more than 10% of the overall GHG emissions of industrialised countries in 1990. The CDM could thus lead either to an "inflation" of emission allowances or to being actually rather negligible. If the higher projections were to become reality, this would definitively "blow wide open" the emission ceiling of Annex B Parties under Article 3.1 of the Kyoto Protocol. Environmental NGOs have thus called for a separate cap on the use of the CDM for the fulfilment of industrialised countries' obligations.

However, there are a number of safeguards and provisions in Article 12 of the Protocol that will limit the amount of CERs that may be available to Annex B Parties, including auditing, verification, certification and the (environmental) additionality that would need to be guaranteed. These safeguards will also require a certain amount of resources, thereby excluding a number of potential projects operating at the cost margin. The share of proceeds to be used for administration and adaptation will furthermore raise transaction costs. The extent to which these factors will succeed in avoiding an inflation of emission allowances depends on their exact design.

With respect to the relative attractiveness of the CDM, the cost differential between reducing emissions in developing countries and in CEITs of Eastern Europe and Russia can be decisive.[58] Competition is already biased in favour of Emissions Trading because of the large amount of "hot air" that exists in the system (see Section 15.3.3). The prospects for CDM projects will thus very much depend on whether any additional restrictions will be applied to Joint Implementa-

---

[57] See Grubb et al. 1999, p. 190.

[58] See on the relationship of the Kyoto Mechanisms Missfeldt 1998.

tion and Emissions Trading. These restrictions might take the form of a quantitative cap on these mechanisms, as is being sought by parts of the EU. The imposition of a transaction fee on all mechanisms constitutes another way to establish a more level playing field[59] and could furthermore generate additional resources for developing countries.

Furthermore, expectations will influence the CDM as much as the real situation. Since CDM projects still need some time to materialise, expectations about the future are the major driving force for attracting investments. Potential investors will also take into account the danger of project failure and available insurance schemes against such failure, which will impact on the attractiveness of CDM projects. Non-economic factors, such as political stability in the host country, will further affect the cost-benefit assessment of investors.

The existing experience with project planning and implementation in developing countries, for example in the framework of GEF and the Multilateral Fund of the Montreal Protocol, but also with the AIJ pilot phase, suggests that several years will pass before considerable emission reductions might be achieved (if environmental additionality is ascertained). Only the inclusion of the rather problematic sink projects would lead to sizeable amounts of CERs soon. This suggests the need for a phased approach in which criteria and policies are refined and fine-tuned as experience with the CDM is gained. For example, certain project types where little question about environmental additionality arises (e.g. renewable energy projects) and where baselines can be established comparatively easily, could be started with, while more types could be added with experience. Such an approach might also provide a suitable tool to avoid an inflation of CERs.

The challenge of the CDM lies in the need to balance the chance of greening and de-carbonising private investment in developing countries, while avoiding the danger of rendering the system unreliable or unsustainable. A balance also has to be found between the demands for meaningful participation of developing countries and the danger that industrialised countries might be relieved of the pressure to take the urgent domestic action required to induce structural technological and deeper lifestyle changes. Because of the likely spill-over effects of such changes in the industrialised world on developing countries, they might be more decisive than the CDM in inducing meaningful change on a global scale.

---

59 See Tietenberg et al. 1998, p. 71 and the proposal by the African Group in FCCC/CP/1998/MISC.7/Add.2.

15. Emissions Trading (Article 17)

Along with "meaningful participation of developing countries", the inclusion of so-called "flexibility instruments" into the "protocol or another legal instrument" to be adopted at COP 3 constituted top priority for the US and was of vital importance for other industrialised countries outside of the European Union and for Russia. These instruments were Emissions Trading[1], joint implementation between industrialised countries and the "Clean Development Mechanism" (CDM), a project-based instrument for offsetting emissions with developing countries (see Chapters 13 and 14 and Figure 13.1). First among those flexibility instruments, however, was the possibility to trade surplus emission reductions with other Parties.

Emissions Trading involves the trading of "unused" GHG emissions, from those countries that stay below their allowed emissions, with those countries that do not meet their obligations and thus exceed their allowed emission levels in a particular commitment period. Such a mechanism requires the setting of limits or "caps" and thus the creation of a commodity. In the case of the Kyoto Protocol, such caps are provided for by the binding limitation and reduction obligations of Article 3. Any country that stays below the limit (or is planning to stay below) is in a position to offer the difference for sale. The "assigned amount" that is transferred is then subtracted from the allowed emissions of the seller-country and added to that of the buyer-country.

Emissions trading therefore creates a market of emission rights, where the costs of different countries to comply with the quantified targets are levelled. As a result, overall abatement costs of achieving a given emission reduction target are minimised or, put another way, emission reductions that can be realised from limited resources are maximised. In general, Emissions Trading is thus an instrument for reaching maximum efficiency of abatement efforts.

The concept of trading of pollution rights has long been explored in the environmental economics literature and has, in some cases, been successfully implemented nationally with regard to certain pollutants.[2] However, while experience at the domestic level is limited, its international application is rarer. Some examples may be found in the Montreal Protocol on Substances that Deplete the Ozone Layer (1987), which allows for the transfer of production quotas for purposes of

1 The term "emissions trading" is the one used in Article 17 of the Kyoto Protocol and thus shall be used here (note that the concept is referred to in the Annex I Expert Group as "emission trading").

2 See Burtraw 1999; Ellerman 1999; Schärer 1999; Solomon 1995; Raufer 1996; Hansjürgens 1998.

"industrial rationalisation".[3] The trading of emission allowances as a part of international climate policy had also been the subject of research by institutions like the OECD[4] and UNCTAD[5], which informed the international negotiations on the way to Kyoto.

15.1 Negotiating History

Negotiations on Emissions Trading were opened at COP 2 in Geneva (July 1996), when the United States announced a change in its position on the nature of targets to be adopted in Kyoto. After the adoption of the Second Assessment Report of the IPCC, the Clinton Administration called for a "binding emission target" (Chapter 4). This move, however, was made contingent upon three conditions, one of them being achievement of those targets "through flexible and cost-effective market-based solutions". The US did not at this time expressly ask for Emissions Trading to be part of an agreement. However, since the use of taxes was not an option seriously considered by the US-administration, the move towards a trading regime was clearly being contemplated.

At AGBM 5 in December 1996, the US circulated a "non-paper" that explored some preliminary ideas and, in January 1997, a formal proposal was submitted to the Secretariat as part of the US protocol proposal.[6] It contained the basic principles of the trading regime, namely that:

- "tonnes of carbon equivalent emissions" could be transferred,
- domestic entities might be authorised by any Party to participate, and
- further guidelines to facilitate the reporting of Emissions Trading information should be elaborated on later by the Parties to the Protocol.

AGBM 6 in March 1997 saw the presentation of the proposal and the first discussions were characterised by the rather slow reaction of the EU and other negotiating Parties. Some research and discussion of trading concepts did, however, take place within the "Annex I Expert Group on the UNFCCC", which was supported by the OECD and the IEA. This group quickly became the main forum for elaborating a system of Emissions Trading.[7]

The move towards Emissions Trading by the US was met with caution from most of the European countries, as well as from many environmental groups, and with some opposition by developing countries. The objection of developing countries, like that of many environmentalists in industrialised countries, was to a great extent based on ethical grounds, since the concept of Emissions Trading carries

3 Some guidance may also be found in the regime of tradable fishing quotas; see Mullins 1997.

4 See, e.g., OECD 1992.

5 See, e.g. UNCTAD 1994; Steward et al. 1996.

6 FCCC/AGBM/1997/Misc.1, p. 78.

7 See Mullins/Baron 1997; see also Section 5.1.

with it the notion of "pollution rights".[8] Second and most importantly, developing countries, and India in particular, raised fundamental concerns with regard to equity, since emission rights were to be allocated based on past emissions – the so-called "grandfathering approach". Thus, industrialised countries would be entitled to comparatively high per capita emission levels, while, by implication, developing countries' per capita emission rights (although not fixed quantitatively in the Kyoto Protocol) would be much lower. Although mainly based on principle, the opposition from developing countries was also used partly as a negotiating ploy to counter other US demands, in particular the inclusion of developing country commitments in a protocol.

Furthermore, European and developing countries feared alike that trading might provide a cheap way for the US, Canada, Australia and New Zealand to "buy" themselves out of their obligations. This concern was raised in particular because of the large emission reductions that had taken place in Russia and other countries with economies in transition (CEITs) due to their economic decline. Given the refusal of Russia and Ukraine, in particular, to accept a meaningful reduction target, a large amount of tradable emissions would be available in the system from these countries. This so-called "hot air" of the CEITs constitutes a serious problem on the practical level, in addition to difficulties inherent in any trading mechanism, that proves to be a major challenge towards a credible trading system in the Kyoto Protocol (see Section15.3.3).

Language on Emissions Trading found its way into the first compilation of proposals as well as all negotiating texts subsequently produced on the way to Kyoto, including the revised text under negotiation of November 1997.[9] The G-77 and China nevertheless objected continuously to this language and requested its deletion, as reflected in a footnote to the negotiating text. This draft Article 6 on Emissions Trading, as contained in the negotiating text, was still rather rudimentary and mandated the first meeting of the Parties to the Protocol to decide upon "modalities, rules and guidelines for emissions trading ... including methodologies for verification and reporting". The text did, however, contain specific rules regarding the potential actors (Annex I Parties and intermediaries), gases (all GHGs covered) and compliance.

Contentious issues that were left in square brackets, apart from the square brackets around the whole text due to developing countries' objections, were whether developing countries with voluntary emission limitations would be able to participate and whether emission reductions achieved before the start of the commitment period might be traded. This would have added enormous amounts of emission rights to the system as a tradable commodity and was unofficially referred to in the negotiations as "superheated air". The first issue constituted an attempt by the US to link developing country commitments – as required by the

8 For some ethical objections see The Corner House 1997; Belliveau 1998.

9 FCCC/CP/1997/2, Article 6.

US Senate (see Section 6.2) – with its second major goal of achieving "flexibility" in the implementation of emission reduction obligations through Emissions Trading. Developing countries would thus be given a special incentive to take on voluntary quantitative commitments in order to be able to sell surplus emission rights on the Emissions Trading market.

One of the main weaknesses of this approach was the fact that developing countries were not involved in the elaboration of the trading provisions, neither in the Annex I Expert Group of the OECD/IEA nor in the context of the AGBM negotiations. This did not change in Kyoto itself. Like the targets, the interconnected issue of the trading provisions was negotiated by the triad formed by the US, the EU and Japan. Japan's main concern was to ensure transparency of the trading regime, as it feared that the US and other countries might buy up the Russian "hot air" for themselves. The EU relied on its "bubble concept" as the main instrument of flexibility and was thus not in the same way dependent on Emissions Trading.[10] However, it did not object to this instrument in principle. Stressing the interrelationship between trading and the size of the target, the Union attempted to raise the US position on targets in exchange for increased flexibility due to the trading regime. The EU also tried to introduce a "cap" on Emissions Trading, which meant a certain restriction on the percentage of the obligation that could be fulfilled through trading, thereby determining what would be achieved domestically. And, finally, the Union subjected any trading regime to strict rules, as opposed to the "leave it to the market" approach of the US and other JUSSCANNZ countries.

Certain "cultural" differences also became visible between a predominantly Anglo-American approach (JUSSCANNZ and the UK) that was favourable towards trading and a certain mistrust of such concepts by continental European countries. On this basis, experts mainly from the EU and the US invested a major effort during the final days at Kyoto to hammer out a compromise. When they finally presented the results to Chairman Estrada and a broader circle, they, as one participant put it, "hit a brick wall".[11] Neither were China and India prepared to even discuss the text nor was Chairman Estrada, on that basis, going to present it to the plenary as part of the negotiating text. Industrialised countries thus had to pay dearly for neglecting the concerns of developing countries and not involving them in the negotiations. Upon learning about the defeat, the US chief negotiator reportedly asked environmental groups for advice and telephoned Washington for further instructions. As all involved knew, Emissions Trading was an essential part of the deal for the US.

Accordingly, when Chairman Estrada submitted the final draft, he deleted the former Article 6 on Emissions Trading but inserted a new paragraph 10 into Arti-

---

10 On Emissions Trading in the EU see Klaassen 1999.

11 Grubb et al. 1999, p. 95.

cle 3 that was stripped down to the bare necessities of an Emissions Trading regime. It merely contained three sentences setting out:

- the permission that Annex I Parties could participate in Emissions Trading,
- that the relevant rules and guidelines for verification, reporting and accountability were to be defined by the Conference of the Parties, and
- that trading should be supplemental to domestic actions.[12]

Even this reduced version of Emissions Trading met with fierce opposition from many developing countries, and China and India in particular. When the issue was taken up during the last night of Kyoto, they again requested deletion of these provisions as they saw their equity concerns disregarded. Instead, they demanded the development of equitable rights and entitlements to trading, preferably based on per capita assignments. As was well known from previous discussions, this was not acceptable to the US and most other industrialised countries. The UK, on behalf of the EU, attempted to find some middle ground. It proposed that the COP should define *all* relevant rules (i.e. not only those referring to verification, reporting and accountability) and that trading should only start "once the COP has defined" such rules. While the US was prepared to accept these changes, China and India still insisted on deletion. The result was a breakdown of negotiations that had to be interrupted for last minute consultations (see Chapter 7).

When the Chairman resumed the session, he removed the text from Article 3 and changed it into a new Article that was eventually numbered 17. In response to the proposal of the UK, however, he had rearranged the order of the first two sentences. Also, the UK request to mandate the COP to define the relevant rules beyond those required for verification, reporting and accountability was reflected. With these changes, and after a reminder by the Chairman that this instrument was a decisive part of the agreement, the Article was finally adopted along with a decision that requested the subsidiary bodies to allocate work for those matters that would be decided upon by subsequent meetings of the COP.[13] During the break, the developing countries who had previously flexed their muscles, reportedly did not discuss trading among themselves, but other agenda items – when negotiations resumed they silently accepted the Chairman's proposal. As a tacit trade off, they succeeded, later on, to have the article on "voluntary commitments" of developing countries deleted from the text of the Kyoto Protocol (see Chapter 17).

15.2 The Rules Applying to Emissions Trading

The compromise that was finally reached in Kyoto stems from the many different positions outlined above. Only several basic rules on Emissions Trading were incorporated into Article 17 of the Kyoto Protocol, as well as paragraphs 10 and

12 FCCC/CP/1997/CRP.6.

13 Decision 1/CP.3 in FCCC/CP/1997/7/Add.1.

11 of Article 3 (see Box 15.1). The latter two paragraphs define the parameters for Emissions Trading. Any emission reduction units, or any part of an assigned amount, which a Party acquires from another Party is to be added to the assigned amount of the acquiring Party and shall be subtracted from the assigned amount of the transferring Party.

Box 15.1: Kyoto Protocol Provisions Related to Emissions Trading

Article 3.10

Any emission reduction units, or any part of an assigned amount, which a Party acquires from another Party in accordance with the provisions of Article 6 or of Article 17 shall be added to the assigned amount for the acquiring Party.

Article 3.11

Any emission reduction units, or any part of an assigned amount, which a Party transfers to another Party in accordance with the provisions of Article 6 or of Article 17 shall be subtracted from the assigned amount for the transferring Party.

Article 17

The Conference of the Parties (to the Convention) shall define the relevant principles, modalities, rules and guidelines, in particular for verification, reporting and accountability for emissions trading. The Parties included in Annex B may participate in emissions trading for the purposes of fulfilling the commitments under Article 3. Any such trading shall be supplemental to domestic actions for the purpose of meeting quantified emission limitation and reduction commitments under that article.

According to Article 17, Parties to the Kyoto Protocol listed in Annex B are to be the participants in a trading regime. These are the Northern industrialised countries of the OECD as well as Central and Eastern European countries and some successor states of the former Soviet Union. Parties without legally binding emission reduction and limitation objectives under the Protocol are thus excluded from participating in trading. These are first and foremost developing countries, but also those industrialised countries that do not ratify the Kyoto Protocol or are not included in Annex B (e.g. Belarus). Article 17 therefore provides an incentive to take on legally binding obligations in order to be able to participate in trading. This is primarily an incentive for those countries that would have surplus emissions to sell, especially CEITs.

Restricting participation to those Parties with legally binding targets will also ensure that the overall amount of units circulating in the system is stable. Indeed, it is difficult to imagine how a credible trading system could be developed without binding obligations on Parties. This "cap" on the overall amount of units circulating in the system is, therefore, not only required for ecological reasons in order to definitely restrict emissions, but also to provide stability and allow a functioning market to develop.

The commodity that may be traded is called "any part of an assigned amount". This term refers to the emission "budget" (assigned amount) of the six-gas basket of GHGs (CO_2, CH_4, N_2O, HFCs, PFCs, SF_6) allocated to the Parties by Article 3.7 and Annex B of the Kyoto Protocol for the commitment period 2008–2012 (see Chapter 11).[14] The assigned amount is defined in "tonnes of CO_2 equivalent" and is calculated by weighing the non-CO_2 GHGs according to their respective "Global Warming Potential", as determined by the Intergovernmental Panel on Climate Change (IPCC). It is thus "tonnes of CO_2 equivalent" that will be the common denominator for Emissions Trading.

A second commodity that is tradable, according to Article 3 paragraphs 10 and 11, is an "emission reduction unit". This term refers to Article 6, where "emission reduction units" may result from Joint Implementation projects, and thus clarifies that the "credits" generated may not only be used by the acquiring Party to meet its own obligations, but that these credits are transferable to other Parties. According to Article 3.10 of the Protocol, the acquired emission reduction units "shall be added to the assigned amount for the acquiring Party" – and can thus be traded as "parts of an assigned amount" under Article 17. It is not yet clear whether a third category of emission rights – the "certified emission reductions" generated by the CDM in accordance with Article 12 of the Kyoto Protocol – will be tradable as well. The Parties will have to decide at future conferences on the "fungibility" of those credits (see Chapter 14).

To what extent a Party can sell parts of its assigned amount of emissions or increase it by acquiring "emission rights" is not fixed in the Kyoto Protocol. The sole remnant of the EU request for a quantitative restriction is the provision that Emissions Trading "shall be supplemental to domestic actions" for the purpose of complying with the targets set.

Post-Kyoto, the Parties had to address "the relevant principles, modalities, rules and guidelines, in particular for verification, reporting and accountability". According to Article 17, this may be decided upon by the COP of the Convention – not the COP/MOP of the Kyoto Protocol. This is the only reference in the Kyoto Protocol to the COP and indicates the extremely important role that the Parties attached to the (early) elaboration of Emissions Trading. Thus, agreement on the relevant rules for Emissions Trading can be established prior to entry into force of

---

14 The term "budget" was specifically objected to by developing country delegates and is used here only in its technical sense, without any inference to property rights etc.

the Protocol. Bearing in mind that many issues were left unresolved, COP 3 adopted a decision that requested the subsidiary bodies to allocate work in view of those matters that would have to be decided by subsequent COPs.[15] The COP did not set a deadline to decide on this matter. At COP 4 in Buenos Aires in November 1998, no agreement on questions of substance was achieved. However, the Parties agreed on a comprehensive work programme for all "Kyoto Mechanisms" (Emissions Trading, JI, CDM) "with a view" to taking decisions on all of them at COP 6 by the end of the year 2000.[16]

15.3 Assessment and Outlook

Emissions Trading was, and still is, maybe the most important issue on the agenda of the post-Kyoto negotiations. Since it has the highest potential of all Kyoto Mechanisms, the establishment of a trading system is a *de facto* condition for most members of the umbrella group to ratify the Kyoto Protocol. However, notwithstanding the basic principles of a trading regime established by the Kyoto Protocol, most issues remain unresolved. These issues are, *inter alia*, the time when trading might start, the definition of participants, which part of the Kyoto obligations may be achieved through trading (hot air and "supplementarity") and the regulations regarding monitoring, verification and, ultimately, enforcement of the rules.[17] These questions shall be explored in the next sections.

15.3.1 When Should Trading Start?

An initial disagreement among the Parties concerns the starting date for Emissions Trading. Whereas the US, Canada and other non-European industrialised countries believe that trading could (and should) start immediately, the EU and others argue that the COP first has to establish the relevant rules of the mechanism. The inherent logic of a trading system and a careful analysis of the content and history of Article 17 tend to support the EU's view.

An effective trading system inherently requires certainty and credibility. As in any stock market, actors involved in Emissions Trading will only be prepared to invest in emission rights if they can expect such rights to have real value. The certainty and credibility simply cannot be achieved without basic rules being agreed upon beforehand, especially as regards the issues of transparency, verifiability, reporting and accountability. A system without adequate rules does not make sense from the economic standpoint. Neither would it be acceptable politically for developing countries.

Furthermore, the negotiating history indicates that Parties agreed to start Emissions Trading only after agreement was reached on its rules. The UK, speaking on

---

15 Decision1/CP.3 in FCCC/CP/1997/7/Add.1.

16 Decision 7/CP.4 and Annex in FCCC/CP/1998/16/Add.1.

17 There are other open questions, however, for example the institutional set-up.

behalf of the EU, introduced new language to make trading dependent on a decision by the COP. In an apparent move away from the common JUSSCANNZ position, the US representative, in a carefully worded intervention made prior to the adoption of the article, assured that "any country has the opportunity to be involved in shaping the rules before the first trading rights can be applied".[18]

The Chairman finally settled the argument using a semantic trick – by changing the order of sentences. The order was simply reversed so that the first sentence now requires the Conference of the Parties to "define the relevant principles, rules and guidelines, in particular for verification, reporting and accountability for emissions trading". The possibility for Annex B Parties to "participate in emissions trading for the purposes of fulfilling their commitments under Article 3" follows. This editorial change expresses a compromise between the positions of the major opponents: first, the COP defines the rules, then Parties are allowed to trade.

The second last-minute amendment to the Article on Emissions Trading at the request of the UK further supports this conclusion. The original language requested the COP to define only the "relevant rules and guidelines for verification, reporting and accountability".[19] The final text enlarges the scope of the COP's mandate by inserting "in particular" after "guidelines" and "principles, modalities" before "rules and guidelines". Article 17 thus requires that the COP establish *all* relevant principles, modalities, rules and guidelines for the trading system, not just those for verification, reporting and accountability. Starting to trade emissions before the COP has defined the relevant rules would mean prejudging the COP's decision.

Consequently, Chairman Estrada, when finally gavelling down any further opposition and declaring the provision for adopted, expressly referred to Article 17 as having "the characteristics of an interim arrangement". He thus expressed the common understanding that the basic rules for Emission Trading were still to be agreed. Eventually, COP 4 in Buenos Aires in 1998 confirmed this understanding by deciding to establish a work programme on mechanisms "with a view" to taking decisions at COP 6. However, as explained in detail in Chapter 12, the "joint fulfilment" of obligations under Article 4 allows any two or more countries to establish its own trading system with limited participation if agreement on Article 17 cannot be reached.

15.3.2 Who Can Trade?

Parties to the Kyoto Protocol are the primary participants in a trading regime and it is not yet clear whether and how private entities may also participate in trading. Article 6.3 on Joint Implementation ("legal entities") and Article 12.9 on the CDM ("private and/or public entities") both provide for the involvement of participants

18 Authors' observation.
19 FCCC/CP/1997/CRP.6, Article 3.10.

other than Parties. Contrary to the final language of Article 17, the draft text contained, until the last day, a provision which gave Parties the possibility to authorise "legal entities" to participate.[20] This question therefore has to be addressed in the process of determining the terms of Emissions Trading in the context of the Buenos Aires Plan of Action adopted by COP 4 in 1998.

Participation of non-governmental actors is conceivable in a variety of forms: first, national governments could allocate parts of their assigned amounts to individual firms which could then be enabled to either sell their unused parts if they overcomply with their obligations, or to buy the required amounts on the emission market if they fail to reach their target. Second, governments could allow certain companies to act as brokers, to buy and sell emission allowances like any other commodity. This would considerably increase the number of actors in the market and might lead to a "real" market where prices truly reflect the costs of abatement (since it would be more difficult to control transactions and thus prices). Furthermore, private companies – and/or non-governmental organisations – might become involved in monitoring and supervising trading.

From an economic point of view, participation by the private sector could enhance the potential for cost-effectiveness and efficiency under a trading regime.[21] It does, however, complicate the system and would thus require much more thoroughly elaborated rules than simply for governmental Emissions Trading.[22] First, the number of participants in the trading system might become too large and pose problems for the monitoring and enforcement of the rules. Second, by involving private entities the Parties might have to create new "certificates". As the "parts of an assigned amount" are the commodity which is traded between states, these cannot be held by privates, since they represent the sovereign obligation of each Party to other Parties. Should a Party wish to involve private entities in Emissions Trading, it will thus have to transform its assigned amount of emissions and create another commodity.

Third, the permission for private actors to participate freely in Emissions Trading has the potential to undermine the targets that governments have assumed for themselves. This could be remedied, however, by the introduction of a prior approval procedure for international transactions by legal entities. This would have the additional advantage of providing greater security for private participants. "Green" governments would be in the position to refuse approval of trades with seller countries that have a high "hot air" potential. Fourth, problems might also arise in relation to competition law in a number of countries, if governments decide to allocate emission rights only to a selection of economic actors (e.g. those that can be particularly well monitored). In this case, others, including competitors,

---

20 FCCC/CP/1997/CRP.4, Article 6.

21 See Dutschke/Michaelowa 1998a, pp. 23 et seq. and Dutschke/Michaelowa 1998b.

22 See also European Commission 1998b, p. 24.

might complain about unjustified and unequal opportunities being given to such businesses, since Emissions Trading may well prove profitable.

Whatever rules on participation in Emissions Trading, the national governments that are Parties to the Kyoto Protocol remain ultimately responsible under international law not only for achieving their targets, but also for the correct behaviour of those private entities under their jurisdiction that they permit to engage in Emissions Trading.

15.3.3 How Much Can Be Traded?

The Kyoto Protocol stipulates that any trading shall be "supplemental" to domestic actions for the purpose of meeting the obligations. "Domestic actions" in this context means measures taken domestically to reduce emissions. This clause is the sole remnant of the attempt by the European Union to limit the possibility for countries to "buy themselves out" of their obligations. Putting limits on trade, although adding inefficiency to the market, is a legitimate objective for those Parties that place an overriding political emphasis on domestic action. The main reason to limit the use of trading in fulfilment of the Kyoto obligations in the present situation is, however, posed by the significant emission reductions in Russia and the Ukraine resulting from their economic decline caused by their transformation to market economies, the so-called "hot air".

Russia's emissions of carbon dioxide in 1997 were some 30% below its 1990 level. Those of the Ukraine were even lower, and in both cases emissions are not expected to rise back to the 1990 level until the end of the commitment period (i.e. 2012).[23] These countries, however, are only obliged to stabilise emissions at the 1990 level in the first commitment period. The difference between the 1990 emission levels of Russia and the Ukraine and their actual emissions during this first commitment period would thus constitute a "reservoir" of hot air from which Western industrialised countries could buy emission credits to prevent any non-compliance.

Moreover, since they have been achieved already for reasons other than climate policy, these "surplus emission reductions" do not require additional effort or involve additional cost to the transferring countries. They could for this reason considerably distort the market, lead to declining prices and make mitigation efforts by Western industrialised countries unprofitable compared to the possibility to buy the required emission permits. This is a serious economic concern, which has implications for the functioning of an emissions market and for the environmental effectiveness of the regime.

From an environmental point of view, the mere prospect of the possibility to buy cheap emission reduction units might already prevent some countries from undertaking serious domestic action – despite the third sentence of Article 17,

---

23 FCCC/IDR.1/RUS of 21 February 1997.

which provides that Emissions Trading "shall be supplemental to domestic action" but remains unspecified. Emissions Trading could thereby take away an incentive for early action and lead to a continuation of business-as-usual policies in these countries. And, finally, using this hot air would increase the overall cumulative emissions of greenhouse gases, because emission rights that remain unused may be "banked" and thereby transferred into the next commitment periods. Limiting the amount of "hot air", on the other hand, would increase the overall reduction target of 5.2% (Article 3.1), since that is calculated by taking the large reductions in Eastern Europe into account.

The extent of the problem is dependent on the future level of Eastern European emissions. By the mid-1990s, the emission reductions in the CEITs resulted in overall emission reductions of Annex I Parties in the range of about 4.5%, which roughly equals the overall commitment of Annex I Parties under the Kyoto Protocol (-5.2%). Under the assumption of stable emissions in these countries, Annex I Parties could thus achieve their commitments by engaging in Emissions Trading and stabilising overall GHG emissions at the level of the mid-1990s. With expected economic recovery, emissions in Eastern Europe may increase. According to the 1997 in-depth review of the Russian national communication projections, its CO_2 emissions, however, are still likely to be some 15% below 1990 levels in 2010.[24] Other projections indicate that the biggest suppliers of hot air, Russia, the Ukraine and Poland, together may be able to offer the equivalent of about 2% of global CO_2 emissions in 1990 for sale.[25] This does not even take fully into account the considerable cost-effective reduction potential that still exists in these countries.

Detailed scenario calculations indicate that without restrictions on Emissions Trading, GHG emissions of the US and Japan might be some 10% above their 1990 levels around 2010, as compared to their reduction commitment of 7% and 6% respectively.[26] Taking into account that, in a "business-as-usual scenario", GHG emissions of the US alone are projected to increase by some 25%, domestic action might still be required. However, taking the above calculations into account, about half of the gap might be closed by buying "hot air", thereby begging the question whether such Emissions Trading would still be "supplemental". In addition, actual domestic action taken might be even further reduced depending on the eventual treatment of sinks by the international regime (see Section 11.3) and the availability of "certified emission reductions" from the CDM (see Chapter 14). The emission reduction path of industrialised countries agreed to in Kyoto would thus be endangered.

These scenarios are of course riddled with uncertainties, since they are based on various assumptions that may or may not turn out to correspond to reality. The

24 FCCC/IDR.1/RUS of 21 February 1997.

25 See Missfeldt 1998, p. 131.

26 See Grubb et al. 1999, pp. 155 et seq.

economic recovery of CEIT countries might be slower or more dynamic than expected, efficient energy technology may become available to varying extents, the development of energy prices is difficult to predict etc. Nevertheless the amount of hot air available will be considerable in virtually all projected scenarios – and it is likely to be even higher with every year that the start of economic recovery in these countries, especially in Russia, does not occur.

The easiest way to solve the "hot air" problem would be to raise the obligations for Russia and the Ukraine. This, however, does not appear to be a politically feasible option. Another method of dealing with the problem would be to limit the seller's capacity. This could include prohibiting the sale of those amounts of emission reductions achieved prior to the start of trading. In this case, either a partial trading with the assigned amounts could be allowed or trading could be limited to the percentage of reductions achieved as a result of domestic policy measures. The latter option might, however, negate the incentive to take early action on the reduction of GHGs. Apart from that, methodologies for determining those reductions which have resulted from climate policy measures would be difficult to elaborate and agree upon internationally.

In May 1999, the EU proposed a quantitative limitation upon the right to sell parts of assigned amounts ("seller cap"), which was primarily aimed at the CEIT countries. According to the formula, the total amount of hot air available for sale in these countries would be limited to about one-third compared to projections of the International Energy Agency (IEA).[27] As to the implications, the IEA study concludes that this would limit the supply of hot air to about 20% of what countries would be allowed to purchase – and would thus impose an effective restriction on the use of Emissions Trading for the purpose of complying with the Kyoto targets. However, since this cap would only apply to Emissions Trading, higher demand could be met by JI and CDM project activities. As for the political prospects of this proposal, apart from the certain rejection of this proposal by members of JUSSCANNZ, it is difficult to imagine Russia and the Ukraine agreeing to such a cap.

In addition to addressing the hot air, the EU raised the issue of "supplementarity" with the political objective of preventing Annex B Parties from buying their way out of the Kyoto obligations. There are several ways to deal with this, but since Kyoto, the negotiations on "principles, rules and guidelines" have not seen much progress with respect to this issue. One possibility is the introduction of a cap on that part of the commitment that may be met through trading. In this case, the question of how to define the "part" – as a percentage of the obligation incorporated in Annex B of the Protocol or as a percentage of the obligation compared

27 "A Preliminary Analysis of the EU Proposals on the Kyoto Mechanisms", by Richard Baron, Martina Bosi, Alessandro Lanza & Jonathan Pershing, <http://www.iea.org/new/releases/1999/eurpro/eurlong.htm> as of 28 May 1999.

to the business-as-usual scenario – remains. Another alternative is to define the cap or "ceiling" as a part of a Party's GHG emissions in 1990.

For a long time, the EU was unable to agree internally on a formula for a so-called "concrete ceiling", partly because of the complexity of the issue, but also because of the fact that only two Member States seriously fought for it. In May 1999, the Council finally approved a complicated formula.[28] In addition to the seller cap introduced above, it contained a "buyer cap", i.e. a limitation on the part of a Party's obligation that could be achieved through all three Kyoto Mechanisms. According to the calculations by the International Energy Agency, the EU under this formula would have been allowed to meet 42% of its target via the flexibility mechanisms.[29] It thus favoured the Union over the US, who would only be allowed to meet 33% through trading and other means. Individual Member States of the EU would be allowed an even higher percentage, for example 56% in the case of Germany.

The US and other non-EU industrialised countries rejected this proposal.[30] It is not known, of course, whether agreement on the term "supplementarity" in Article 17 of the Protocol was a tacit agreement not to introduce quantitative limits or whether this was a compromise formula that everyone accepted in order to save face. It is a different question, however, whether such a cap would achieve reasonable environmental and economical results.[31] In any event, the political prospects of a quantitative ceiling on the Kyoto Mechanisms are not very promising (see Chapters 24 and 25).

A better solution might be to impose a fee on all transactions under the trading regime, possibly with different levels for intra-OECD trading as compared to other trading activities.[32] This would not preclude or limit the use of hot air, but instead would raise the transaction costs of trading and thus improve the comparative advantage of taking domestic action to reduce GHG emissions. A fee of US $ 5 per tonne of carbon equivalent, levied on transfers of 300 Mt of carbon per year, might raise approximately US $ 1.5 billion annually during the first commitment period.[33] These resources could be used to assist developing countries in adapting to climate change (see Chapter 25). Such a fee might also serve to balance the relative disadvantage of the CDM in terms of costs, because a "share of the proceeds" under the CDM will be used to cover administrative expenses and certain costs of adaptation to climate change impacts in developing countries (see Chap-

---

28 Conclusions of the Council on a Community Strategy on Climate Change, Doc. 8346/99, 18 May 1999; see also Global Environmental Change Report of 14 May 1999, p. 3.

29 "A Preliminary Analysis of the EU Proposals on the Kyoto Mechanisms", by Richard Baron et al., see above; see also Global Environmental Change Report of 11 June 1999.

30 "Clinton Accuses EU of Rewriting Global Warming Pact", Wall Street Journal, 18 May 1999.

31 The drawbacks are discussed in detail in Grubb et al. 1999, pp. 213 et seq.

32 See Tietenberg et al. 1998, p. 71.

33 Grubb et al. 1999, p. 223.

ter 14). This option would thus have the advantage of providing a more level playing field as exists between the Kyoto Mechanisms. Given the long-lasting resistance by many industrialised countries to accept international "taxation", framing such a solution in a politically acceptable form will require some creativity.

15.3.4 How are the Rules Going to be Enforced?

The enforcement of its rules is one of the most crucial elements of any trading system. This is evident for example from the first analyses of the system established in the US for the trading of sulphur dioxide emission rights.[34] Under the supervision of the Environment Protection Agency, a very complex system for verification, tracking of allowances and enforcement was put in place. This system has proved to be instrumental in ensuring the credibility and proper functioning of the system. Unfortunately, under the conditions of today's international politics and law, such an elaborate and strict regime is not a realistic option. This does not, however, dispense with the necessity to establish a rigorous regime for the international trading of GHGs.

Procedures and institutions to monitor, verify, assess compliance and enforce the rules in cases of non-compliance must have top priority in the establishment of the trading regime for two reasons.[35] First, non-compliance with the commitments of the Kyoto Protocol resulting from a flawed trading system would undermine the climate regime, lead to a slower reduction in the emissions of greenhouse gases and cast serious doubt on the willingness of industrialised countries to tackle the threat of climate change. This in turn would not bode well for the much needed involvement of developing countries at a later stage. Second, the credibility of the trading regime itself is at stake. Since trust in the market and the correct behaviour of participants is one of the main pillars of any trading regime, the failure to detect and sanction instances of non-compliance would prevent the establishment of a stable and thriving market.

An effective compliance system requires several components. First, monitoring and reporting procedures must be in place. The Kyoto Protocol, in Articles 5 and 7, contains general monitoring and reporting procedures that could be adapted to the needs of a trading regime. Information that would need to be provided for the close tracking of trading could, for example, include the exact amounts of emissions traded, the Parties and/or legal entities involved and the dates when such trading occurred. Compliance with these requirements is essential for the creation of a credible market and would thus have to be a precondition for participation in the Emissions Trading regime.

---

34 See for example Solomon 1995; Burtraw 1999; Ellerman 1999; Hansjürgens 1998.

35 For an overview see Werksman 1998b.

Second, the correctness of the data submitted would need to be assessed. One of the cornerstones of such a procedure, the verification function, has already been incorporated in Article 8 of the Protocol (see Section 16.1). The expert review teams established under this article might also be enabled to check the data submitted on trades, nationally as well as internationally. Additionally, certified auditors linked to the Secretariat could perform some of these functions.

Third, the appropriate reaction to a presumed infraction must be determined. Following the general practice in international environmental agreements, this function will ultimately lie with the Conference of the Parties serving as the meeting of the Parties (COP/MOP) to the Kyoto Protocol. The supreme body of the Protocol could base its decisions on recommendations of a smaller body that could make recommendations in a focused manner. The Montreal Protocol on Substances that Deplete the Ozone Layer, where an Implementation Committee discusses and prepares decisions on possible cases of non-compliance, provides a pertinent example of such an approach.[36] Article 8 of the Protocol already provides a basis for establishing such a procedure by empowering the COP/MOP to consider information submitted by the Parties and to "take decisions on any matter required for the implementation" of the Kyoto Protocol.

The more specific Article 18 of the Kyoto Protocol on non-compliance does not lend itself well to the establishment of a sanctions regime for Emissions Trading. Although it requires the Parties to "approve procedures and mechanisms to determine and address cases of non-compliance", it contains a clause that makes "binding consequences" dependent on a formal amendment of the Protocol (see Section 16.3). Since an amendment to the Protocol requires ratification by at least three-fourths of the Parties, its entry into force would be very time-consuming and might not even occur at all.

Since appropriate sanctions under an Emissions Trading regime would certainly have to include those with "binding consequences", they could instead be incorporated directly into the trading regime (i.e. bypassing Article 18). This is possible because Article 18 only refers to binding consequences "under this article" – strict enforcement rules might thus well be adopted under other provision of the Protocol. Including sanctions into the Emissions Trading regime might thus have the beneficial side effect of providing the Parties to the Kyoto Protocol with a much needed partial enforcement regime, should the Parties fail to adopt (or ratify) a strict non-compliance procedure.

Non-compliance with the rules of the trading system is conceivable in a variety of ways. Participants may fail to fulfil the reporting or monitoring obligations, they may violate the trading rules inadvertently or with intent, and, ultimately, they may fall into non-compliance because at the end of the commitment period they have exceeded their allowable level of emissions.

---

36 See Ott 1998a, pp. 225 et seq.; Gehring 1994; Werksman 1996.

A number of possible punishments for infractions of the rules of trading – e.g. personal civil and criminal liability – fall within domestic jurisdiction. Should a Party decide to involve private actors, the provision of adequate sanctions at the national level should be compulsory. This is in the interest of any Party that ultimately will be held responsible internationally for the activities of such private actors. At the international level, however, financial sanctions are extremely difficult to impose. Nonetheless, in order to facilitate the extraction of such fines from non-compliant States, a certain financial deposit could be required prior to receiving permission to trade. In case of non-compliance, such deposit might be (partly) lost and go into a fund to finance mitigation and adaptation measures in developing countries. In case the fines exceed the deposit, future participation in Emissions Trading might be made dependant on the additional payment.

As part of such a sanctions regime, Emissions Trading itself can also be the best means for imposing penalties, in that parts of assigned amounts can be devalued, either partly or completely. Such inherent sanctions could be determined and imposed without any interference with Article 18 on non-compliance. There are basically two approaches, depending on whether liability lies with the seller or with both buyer and seller (shared liability).[37] Under the traditional rules of international law, only the seller would be responsible for the violation of rules and regulations of the trading regime (seller liability). Sanctions would thus be directed towards the country that has sold part of its assigned amount in breach of the rules (e.g. because its remaining emission allowance does not cover its actual emissions). In the framework of domestic legal systems, this approach is usually sufficient, since penalties for non-compliance can be severe and enforced by the judicial and executive branches of government.

Such an approach was introduced by the US during the negotiations on "borrowing",[38] but was not accepted by other Parties because it could have provided a disincentive to comply. According to this concept, the amount of emissions exceeded by a country in a given commitment period would have been subtracted from the assigned amount of the next period. An additional 20% would have been added to the amount to be subtracted as a "penalty". Whereas this might certainly have induced some motivation not to exceed the "budget", it would have entailed some risks as well. First, the mere possibility of such borrowing could lead a government to the conclusion that a "wait and see"-strategy might be worth following, maybe waiting for some grand technological breakthrough that could abate emissions. Second, this kind of sanction does not help in cases of successive non-compliance. After the second instance of non-compliance, some more traditional sanctions would be needed.

37 Other mechanisms like joint and several liability may also have a potential that should be explored.

38 As part of the US proposal, FCCC/AGBM/1997/Misc.1, p. 78 (see Section 11.1).

Under international law, therefore, this principle of "seller beware" might not work well due to the relative absence of enforcement powers as against delinquent states. In contrast to other trade transactions, buyers may be indifferent to whether the seller has obeyed the rules, as long as any emission rights acquired remain valid. In this case, a "buyer beware" principle creates an environment where the buyer has an interest in observance of the rules by the seller. For example, if a selling country is in danger of exceeding its assigned amount at the end of a commitment period and its parts of an assigned amount may be devalued afterwards, buying them will entail a great risk. The buyer would thus abstain from acquiring emission rights from this country or would do so at a much lower price. Observing the rules and staying within the limits of assigned amounts allocated to it thus becomes of prime interest to any country engaging in Emissions Trading.[39]

A "buyer beware" approach demands transparency of trade conditions and transactions. Such transparency and assessment of a potential seller's capacity to provide emission rights of value could be ensured by an early warning system, where a supervisory institution raises "yellow" or "red" flags once it becomes apparent that a country might exceed its assigned amounts. Based on annual reports, this institution (this might be the Climate Secretariat) would be in a position to assess the performance of all participants and issue a "rating". Any country that stays within the "safe zone" might freely sell parts of its assigned amount. Should a country enter the "danger zone", the supervisory institutions will issue a warning and raise the "yellow flag". This will indicate to potential buyers that its emission rights are in danger of being devalued after the commitment period. Any country that enters the "forbidden zone" might furthermore be barred from trading.

15.3.5 Conclusions

Not all questions on the establishment of a system of Emissions Trading could be dealt with in this chapter. The elaboration of the institutional set-up of such a mechanism, for example, will be a tremendous task once the more basic structural issues have been settled. More so than in the case of the other two Kyoto Mechanisms, the agreement upon a global system of Emissions Trading is a step into the unknown. It touches upon complex issues like global governance in a rapidly changing world and the relationship between private actors vis-á-vis sovereign governments. In fact, the introduction of this instrument designed to alleviate fears of economic decline appears to be as bold a step as the enterprise of phasing out fossil fuels itself.

At the same time, the fate of the Kyoto Protocol very much hinges upon the establishment of a trading system. Neither the US nor the other members of the JUSSCANNZ group will ratify the Protocol without the possibility of meeting a substantial part of their quantified targets by trading. Russia's ratification, on the

[39] See also Grubb 1998, p. 141.

other hand, is dependent on the possibility to sell its hot air and thereby generate substantial foreign earnings. In the end, most Member States of the European Union may also have to rely on Emissions Trading and the other mechanisms because of the slow implementation of their obligations (see Chapter 24). For this reason, the Parties should attempt to make use of the potential inherent in the trading regime, without, however, compromising the fundamental base of the system. This requires transparency and verifiability of transactions as well as the establishment of a credible and effective enforcement mechanism.

16. Implementation Review and Compliance (Articles 5, 7, 8, 16, 18, 19)

The Kyoto Protocol contains several means for addressing non-compliance and enforcement of its obligations. None of them is particularly innovative and much will depend on subsequent elaboration by the COP/MOP. This is in line with the process-oriented character of the climate regime, i.e. its potential to develop and evolve over time.[1] The basis for determining non-compliance is the system of national communications and the review procedure for those reports contained in Articles 7 and 8 of the Protocol, which will be based upon methodologies for the emission inventories to be elaborated under Article 5 of the Protocol.

To deal with possible difficulties encountered by Annex I Parties in implementing the Protocol's obligations, three basic mechanisms can be distinguished. First, the Protocol requests the COP/MOP to consider, and modify as appropriate, the multilateral consultative process elaborated under the FCCC (Article 16). Second, Article 18 calls for the approval of "procedures and mechanisms to determine and to address cases of non-compliance with the provisions of this Protocol". To further this aim, the fourth Conference of the Parties, November 1998 in Buenos Aires, established a "joint working group on compliance" to consider issues relating to non-compliance and to report to COP 5.[2] And third, Article 19 incorporates the dispute settlement provision of Article 14 FCCC into the Protocol. In addition, the Emissions Trading regime under Article 17 may in the future contain elements of compliance control and an enforcement mechanism of its own (see Section 15.3.4).

This chapter will provide an overview of how the review of national communications acts as the basis for any compliance mechanism. Then, the various provisions of the Kyoto Protocol dealing with compliance will be presented and discussed: the multilateral consultative process, the procedure on non-compliance and the dispute settlement mechanism.

16.1 The Review of National Communications (Articles 5, 7, 8)

Obligations for reporting and reviewing national communications are important elements of both the FCCC and the Kyoto Protocol. Precise information on sources and sinks of greenhouse gases, policies and measures being taken, and their effect on emissions provides the basis for assessing the individual performance of each Party. As such it may serve as the basis for a procedure that is designed to deal with cases of individual non-compliance. Eventually, this

1 For a general treatment of the "process-oriented character" of international environmental regimes see Ott 1998a, pp. 268 et seq.

2 Decision 8/CP.4 in FCCC/CP/1998/16/Add.1.

information helps assess the overall effectiveness and the adequacy of the Protocol's obligations.

The Protocol builds upon the Convention, and contains more elaborate requirements for national reporting (Article 7) and a process for the review of this information by an "expert review procedure" (Article 8). National communications will only provide a realistic basis for the envisaged expert review, if the Parties apply common methodologies. To this end, Article 5 of the Protocol stipulates rules for developing such common methodologies.

16.1.1 Negotiating History

The Parties to the FCCC developed a system of national communications and a review procedure for the Convention that has been widely considered useful and effective.[3] The FCCC requires certain information to be included in regular national communications from industrialised countries.[4] Guidelines adopted by COP 1 specify that Annex I Parties have to report on sources and sinks of GHGs, provide a detailed description of the policies and measures taken to implement the commitment contained in Article 4.2 FCCC (i.e. the "aim" to return GHG emissions to 1990 levels) and provide a specific estimate of the effects that these steps will have on anthropogenic GHG emissions.[5]

The Conference of the Parties is to review the information received and assess the overall effectiveness of the Convention (Articles 4.2(b) and 7.2(c) FCCC). The Convention, however, does not specify the process for reviewing these communications. In order to elaborate this process, COP 1 adopted an "in-depth review procedure"[6] that provides the basis for evaluating the information submitted by the Parties. For this purpose, the Secretariat may, *inter alia*, compare the data reported with data gathered from other organisations. In-depth review teams may – with the approval of the Party concerned – conduct country visits in the course of their assessments. The Parties at COP 3 further decided that "in-depth reviews (...) should, as a general rule, include visits of review teams co-ordinated by the secretariat".[7] COP 4 decided that a third national communication will become due on 30 November 2001 and thereafter on a regular basis to be decided at a future session of the COP.[8]

Encouraged by the rather successful implementation of this system in the FCCC, the Secretariat suggested to AGBM 3 in March 1996 that it might be

---

3 See Corfee Morlot 1998.

4 Articles 12.1, 12.2, 12.3 and 4.2(b) FCCC; see Kinley 1994.

5 Decision 3/CP.1 in FCCC/CP/1995/7/Add.1; see the report of INC 9, UN Doc. A/AC.237/81, Decision 9/2 and report of INC 11, UN Doc. A/AC.237/71/Add.1 for the development of the procedure.

6 Decision 2/CP.1 in FCCC/CP/1995/7/Add.1; see Ott 1994; Oberthür/Ott 1995, p. 149.

7 Decision 6/CP.3 in FCCC/CP/1997/7/Add.1.

8 Decision 11/CP.4 in FCCC/CP/1998/16/Add.1.

adapted to fit the needs of a protocol or, at least, that there be co-ordination between the two systems.[9] This was met with approval by most Parties, with some of them suggesting a stronger review procedure than that contained in the Convention.[10] Others suggested to incorporate the required information under the Protocol in the communications to be submitted under Article 12 FCCC. The US submitted in its proposal to AGBM 6 a detailed "review and compliance process" that already included a review of communications by "expert review teams", as they were finally referred to in the Kyoto Protocol.[11]

As the envisaged system of national communications and their review largely built on the practice developed under the FCCC no major disputes arose during the negotiating process. The same is true for the rules concerning the methodologies to be applied in calculating GHG emissions, which constitute the basis of the review system. As a result, the three articles on methodologies, national communications and their review contained in the revised text under negotiation submitted by the Chairman in November 1997 remained largely unchanged during the final stages of the Kyoto conference.[12]

16.1.2 Reporting and Review under the Kyoto Protocol

The reporting system under the Protocol will be closely linked with the one existing under the Convention. According to Article 7 of the Kyoto Protocol, industrialised country Parties (Annex I) will have to fulfil their reporting obligations under the Protocol by submitting additional information in their communications required under Article 12 of the Convention. Thus, Article 7.1 of the Kyoto Protocol requires Annex I Parties to include in their annual inventory of GHG emissions and sinks "the necessary supplementary information for the purposes of ensuring compliance with Article 3" of the Protocol. This is not only information on GHGs other than CO_2, which was not required by the reporting requirements of the FCCC, but furthermore data on "land-use change and forestry" and background information in order to assess its validity.

Furthermore, "each Party included in Annex I shall incorporate in its national communication submitted under Article 12 FCCC the supplementary information necessary to demonstrate compliance with its commitments under this Protocol" (Article 7.2). This requirement refers to information on the following:

- activities relating to provisions of the Protocol other than the targets and timetables agreed under Article 3, namely on policies and measures (Article 2),

9 FCCC/AGBM/1996/4, para. 18.

10 FCCC/AGBM/1996/10; FCCC/AGBM/1997/2 and Add.1.

11 FCCC/AGBM/1997/MISC.1, p. 78.

12 See FCCC/CP/1997/2, Articles 5, 8 and 9; only the references to "Parties acting under Article 10" were deleted after this article that had allowed developing countries to take on voluntary commitments failed to get adopted, see Chapter 17.

- the joint fulfilment of obligations according to Article 4 ("bubble"),
- Joint Implementation under Article 6,
- the general commitments contained in Article 10,
- activities under the Clean Development Mechanism (Article 12), and
- activities under the system of Emissions Trading elaborated according to Article 17 (see the respective Chapters).

The information on the inventories shall be submitted annually, beginning with the first inventory due under the Convention for the first year of the commitment period after the Protocol has entered into force for that Party (Article 7.3).

Furthermore, the COP/MOP will have to adopt guidelines for the preparation of information at its first session and review them periodically thereafter, taking into account the guidelines adopted by the COP to the Convention (Article 7.4). This provision has to be seen in the context of Article 5, which stipulates the methodological foundations of the emission inventories that need to be prepared by industrialised countries and will represent the backbone of the system. Accordingly, Annex I Parties have to establish national systems for estimating GHG emissions and removals by sinks ("inventories") by the end of 2006 at the latest. The first COP/MOP will have to decide on related guidelines (Article 5.1).

These guidelines will include the applicable methodologies for estimating GHG emissions by sources and removals by sinks, which shall generally be those elaborated by the IPCC and adopted by COP 3 (Article 5.2).[13] The second sentence of Article 5.2 stipulates that, where such methodologies are not used, appropriate adjustments shall be applied according to methodologies agreed upon by the first COP/MOP. The COP/MOP will also have to review the applied methodologies and adjustments, and may revise them. Any such revision will then, however, only take effect for any commitment period adopted subsequent to such a revision (Article 5.2). The same applies to any later changes to the Global Warming Potentials (GWPs) of the various GHGs, that serve to weigh gases other than CO_2 in the GHG basket controlled by the Kyoto Protocol. For the first commitment period, and until any future revisions, the GWPs contained in the Second Assessment Report of the IPCC and agreed upon by COP 3 are applicable (Article 5.3).[14] As GWPs were not fixed for all HFCs at that point, some uncertainty exists as to how emissions of these gases will be treated in the first commitment period. Another difficulty may arise if the Parties to the Convention agree on revised methodologies prior to the first commitment period. Although the reporting system has been carefully crafted so as to harmonise reporting under the Convention and the Kyoto Protocol, this could lead to a situation in which Parties to the Protocol have to use

13 Decision 2/CP.3 in FCCC/CP/1997/7/Add.1. On uncertainties of methodologies and inventories see IPCC 1998; Lanchbery 1998.

14 Decision 2/CP.3 in FCCC/CP/1997/7/Add.1.

different methodologies for reporting under the Convention (revised methodologies) than under the Protocol (methodologies agreed at COP 3).

The review procedure of Article 8 of the Protocol follows in large part the "in-depth review procedure" developed under the Convention (see above). The information on inventories is to be reviewed by "expert review teams" pursuant to the relevant decisions of the COP and in accordance with guidelines to be adopted by the COP/MOP. The "supplementary information" submitted according to Article 7.2 shall be reviewed as well as part of the procedure established under the Convention (Article 8.1).

Motivated by the desire of many developing countries to keep as much power in the hands of the Convention bodies as possible, the expert review teams, which will be co-ordinated by the Secretariat, shall be composed of experts nominated by the Parties to the Convention or intergovernmental organisations in accordance with guidance provided by the COP (Article 8.2). These expert review teams shall "provide a thorough and comprehensive technical assessment of all aspects of the implementation by a Party of this Protocol" (Article 8.3).

They shall furthermore prepare a report to the COP/MOP, assessing the implementation of the commitments of a Party and identifying any potential problems in, and factors influencing, the fulfilment of commitments (ibid.). As is common practice under the Convention, these reports shall be circulated to all Parties to the Convention.[15] An expanded role has been foreseen for the Secretariat by Article 8.3. It is requested to "list those questions of implementation indicated in such reports for further consideration" by the COP/MOP. This provision will considerably enhance the status of the Secretariat in the implementation of the Protocol and may be used in the context of the procedure on non-compliance to be elaborated under Article 18.

Paragraphs 5 and 6 of Article 8 contain rudimentary provisions for a non-compliance procedure. The COP/MOP is requested to consider, with the assistance of the subsidiary bodies, the information submitted by the Parties, the reports of the expert review teams, the questions of implementation listed by the Secretariat as well as any question raised by Parties (Article 8.5). In light of these considerations, the COP/MOP is furthermore requested to "take decisions on any matter required for the implementation" of the Protocol (Article 8.6).

Overall, the system of reporting and reviewing information under the Kyoto Protocol remains closely related to the system operating under the Convention, so as to avoid overlap and double-work. On this basis, Articles 7 and 8 provide, without explicitly referring to non-compliance, for the first and the last step of a multilateral procedure for the implementation and treatment of non-compliance under the Protocol. The COP/MOP will receive any information required for the assess-

15 The Secretariat went one step further, however, and published the in-depth review reports on its website, <http://www.unfccc.de>. According to the FCCC, "publication" only refers to the communications themselves, not the in-depth reports, see Article 12.10 FCCC.

ment of individual compliance by the Parties themselves, by other Parties, by the expert review teams and by the Secretariat. As the supreme body of the Protocol, the COP/MOP is, on this basis, in the position to decide on any step that it deems necessary for the functioning of the Protocol.

16.2 The Multilateral Consultative Process (Article 16)

Article 16 provides a mechanism for finding co-operative solutions for difficulties Parties may have in implementing their obligations and for detecting early on possible conflicts. Without establishing a binding mandate and without setting a concrete timetable, the Protocol requests the COP/MOP to consider, as soon as practicable, the application to the Kyoto Protocol of the multilateral consultative process elaborated in the framework of the FCCC.

16.2.1 Negotiating History

Suggestions for a multilateral compliance procedure of a future climate regime had already been considered in the late 1980s by the first political gatherings on this issue.[16] During the negotiation of the Convention, elements of compliance control were discussed, but could not be agreed.[17] Thus, the negotiators decided to include a "marker" (a clause that demands action from the treaty bodies and keeps the subject on the agenda)[18] into the text of the Convention. According to Article 13 FCCC the COP "shall, at its first session, consider the establishment of a multilateral consultative process, available to Parties on their request, for the resolution of questions regarding the implementation of the Convention". This provision, however, neither requires the Parties to actually establish a multilateral consultative process,[19] nor does it mandate a certain design.

At COP 1, the Parties to the FCCC in Berlin established an *ad hoc* group of technical and legal experts to "study all issues" and to report to COP 2.[20] This "*Ad Hoc* Group on Article 13" (AG 13) met five times between COP 1 and COP 3, after its mandate had been extended.[21] The initial open discussions in the group were characterised by a great divergence of views concerning the "advisory" of "supervisory" character of the process.[22] China, in particular, was opposed to any procedure that would have entailed a scrutiny of a Party's individual compliance,

16 See "Protection of the Atmosphere: Statement of the International Meeting of Legal and Policy Experts, 20–22 February 1988, Ottawa"; reprinted in: The American University Journal of International Law and Policy, No. 5 (1990), pp.539–542.

17 See UN Doc. A/AC.237/Misc.13, pp. 30 et seq.

18 See Széll 1995, p. 99.

19 See also the Note by the interim Secretariat, UN Doc. A/AC.237/59, p. 10.

20 Decision 20/CP.1 in FCCC/CP/1995/7/Add.1; this approach had been recommended by the interim Secretariat after a careful analysis of the issue, see UN Doc. A/AC.237/59, p. 11.

21 Decision 4/CP.2 in Report of COP 2, FCCC/CP/1996/15/Add.1.

22 See Ott 1996b.

whereas most industrialised Parties, such as the EC, Japan or Canada appeared to take the Non-compliance Procedure of the Montreal Protocol on Substances that Deplete the Ozone Layer (1987) as a model for departure (see Box 16.1).

An agreement between the EU and China at the beginning of the fourth session of AG 13 in 1997 set the process firmly on a purely advisory track – an approach that was further refined afterwards.[23] The "process" would thus not entail any compliance control, but serve mainly in the function of a "helpdesk", as it was frequently referred to. The Parties at COP 3 extended the mandate of the group until the fourth Conference of the Parties,[24] where a final decision was supposed to be taken regarding the design of the multilateral consultative process in the FCCC.

At its sixth session in June 1998, the group finally adopted the terms of reference for a multilateral consultative process, but failed to agree on the terms for the composition of its standing Multilateral Consultative Committee that was to be at the core of the procedure.[25] The G-77 and China favoured a composition according to the principle of "equitable geographical distribution", which would provide them with a majority. This, for example, is the basis for the composition of the Implementation Committee under the non-compliance procedure for the Montreal Protocol (see Box 16.1). Many industrialised countries, by contrast, opted for a composition that would designate half of the seats to developing and the other half to industrialised countries. It is not evident why agreement could not be reached on this point, since the envisaged body would play a purely advisory role and neither developing nor industrialised countries would thus be in a position to force anything upon the other group of countries against their will. The reason for the disagreement might rather be seen in the precedent that would be set by such a decision for the composition of future committees in the context of the climate regime, not least under a non-compliance procedure for the Kyoto Protocol (see below).

As there was no agreement, COP 4 approved the text of the Multilateral Consultative Process, with the exception of the composition of the Multilateral Consultative Committee. The Argentinian President of COP 4 was called upon to conduct consultations on this issue with a view to adopting the terms of reference for the process and to establishing the committee at the fifth session of the COP.[26]

16.2.2 The Multilateral Consultative Process Under Article 13 of the FCCC

The Multilateral Consultative Process recommended by AG 13 and approved by the fourth Conference of the Parties[27] has the following objectives

- to provide advice to Parties with difficulties in implementing the Convention,

23 See the reports of the fourth and fifth sessions of AG 13, FCCC/AG13/1997/2 and 4.
24 Decision 14/CP.3 in FCCC/CP/1997/7/Add.1.
25 FCCC/AG13/1998/2 and Annex II.
26 Decision 10/CP.4 in FCCC/CP/1998/16/Add.1.
27 Annex II of the report of the sixth session of AG13 in FCCC/AG13/1998/2.

- to promote understanding of the Convention, and
- to prevent disputes from arising (para. 2).

In line with the agreement on its advisory character at the fourth session between the EU and China, the process "shall be conducted in a facilitative, co-operative, non-confrontational, transparent and timely manner, and be non-judicial" (para. 3). In order to emphasise the independent nature of this process, para. 4 mandates that it shall be separate from the dispute settlement provision of Article 14 FCCC.

The process can be initiated or "triggered" by:

- a Party or by a group of Parties if they experience difficulties with respect to their own implementation;
- a Party or a group of Parties with respect to another Party or a group of Parties; and
- the Conference of the Parties itself (para. 5).

Despite some attempts to provide the Secretariat with a triggering function, as is the case in the non-compliance procedure under the Montreal Protocol (see Box 16.1), it was not possible to entrust the FCCC Secretariat with a stronger role regarding implementation supervision. It will play its part, however, in providing information on the performance of individual Parties (see Section 16.1 above).

Members of the Multilateral Consultative Committee will be experts in relevant scientific, socio-economic and environmental fields that are nominated by governments (para. 8). As referred to above, however, neither AG 13 nor the Conference of the Parties has so far been able to agree on the terms of its composition. The draft still contains brackets as regards the number of experts sitting on this panel (10, 15 or 25 members) and whether these experts would be drawn according to "equitable geographical distribution" or whether one-half would be nominated by developing countries and the other half by industrialised countries. This question should be resolved by COP 5.[28]

The mandate of the Committee is confined to clarifying and resolving questions arising in the implementation of the Convention, providing advice on technical and financial resources and compiling and communicating information (para. 6). In order to meet some legitimate concerns about unnecessary duplication of work, the Committee shall refrain from activities performed by other bodies of the Convention (para. 7). Its primary function is not to report to the Conference of the Parties, but to provide recommendations to the Party or Parties concerned (para. 12). The Committee shall, however, forward its conclusions or recommendations and any written comment to the COP (para. 13). The last paragraph allows for the amendment of these terms of reference by the Conference of the Parties (para. 14).

[28] Decision 10/CP.4 in FCCC/CP/1998/16/Add.1, para. 2.

16.2.3 *Application of the Multilateral Consultative Process to the Kyoto Protocol*

Negotiations on the Multilateral Consultative Process developed in parallel with the negotiations on the Kyoto Protocol. Since it was not clear whether the Parties to the Kyoto Protocol would elaborate a non-compliance procedure at all, the Parties at COP 3 decided to include at least a marker to the Multilateral Consultative Process under the FCCC into the Kyoto Protocol. Article 16 of the Kyoto Protocol therefore requests the COP/MOP to "consider" its application to the Protocol. This provision highlights the independence of the COP/MOP from the decisions taken within the framework of the Convention. It was also due the fact, however, that not all Parties were convinced of the merits of this procedure for the Kyoto Protocol and, furthermore, that AG 13 had not yet delivered a final draft at the time of COP 3. This latter point, on the other hand, was partly due to the fact that AG 13 was waiting for the results of Kyoto before adopting a final procedure.

According to Article 16, the Parties to the Kyoto Protocol, with its different membership, may thus modify the Multilateral Consultative Process in light of the special needs of the Protocol. It is safe to predict, however, that because of its rather uncontroversial character any modification will be considered unnecessary.

16.3 The Procedure on Non-compliance (Article 18)

In addition to Article 16 on the possible application of the Multilateral Consultative Process, with its advisory function elaborated in the Convention, the Kyoto Protocol requires the COP/MOP to "approve appropriate and effective procedures and mechanisms to determine and to address cases of non-compliance with the provisions of this Protocol" (Article 18). This provision leaves much room for further negotiations. It does specify, however, that the procedure should include the "development of an indicative list of consequences, taking into account the cause, type, degree and frequency of non-compliance".

Phrase two of Article 18 sets a limit on the powers conferred upon such a mechanism and stipulates that "any procedures and mechanisms under this Article entailing binding consequences shall be adopted by means of an amendment to this Protocol". Paragraphs 5 and 6 of Article 8 on the review of communications furthermore contain some rudimentary provisions for the treatment of Annex B Parties experiencing difficulties with the implementation of their obligations (see also Section 16.1 above).

16.3.1 Negotiating History

The negotiations on this article were closely related to those that took place on a multilateral consultative process preceding and following the adoption of the FCCC (see Section 16.2).[29] The advisory procedure agreed in that context may be considered appropriate for the Convention, which is characterised by the absence of legally binding obligations to reduce GHG emissions. The implementation of Article 3 of the Protocol, however, represents a formidable task for all the Annex B countries and thus suggests the need for adopting procedures that allow for the supervision of individual performances.

In the Kyoto process, the first suggestion of a non-compliance procedure for the Protocol was contained in a Secretariat's paper on institutional issues submitted in early 1996 prior to AGBM 3.[30] Most subsequent submissions by Parties on this point were not very elaborate, except for the EU's proposal submitted to AGBM 5 the same year.[31] It was modelled after the Non-compliance Procedure of the Montreal Protocol and proposed the establishment of an "Implementation Committee".

These early proposals were contained in the Chairman's synthesis of November 1996 and in the framework compilation of proposals of February 1997.[32] At AGBM 6 in March 1997, however, the US proposed a procedure for non-compliance that was part of its larger proposal on a framework for a protocol.[33] Based in large part on its "flexible" approach arising from the extensive use of economic instruments, such as Emissions Trading and Joint Implementation, the proposal contained a modified in-depth review procedure by "expert review teams" and the possibility of sanctions, should the allowable emissions be overdrawn. These rather drastic proposals for enforcement were considered indispensable by the US administration for a credible regime of emissions trading

Another landmark proposal was submitted to AGBM 7 in July/August 1997 by Brazil.[34] Should an industrialised country Party fail to keep its emissions below its allowed ceiling, it was supposed to compensate by paying a fixed amount per ton of excess carbon into a "Clean Development Fund". 90% of the resources of the fund should be spent on climate change mitigation in developing countries, with 10% earmarked for adaptation projects of these countries. This Brazilian proposal was later supported by the G-77 but it failed to win the support of industrialised countries. However, it built the basis of what was later to become the Clean Development Mechanism under Article 12 of the Kyoto Protocol (see Chapter 14).

---

29 For the negotiations see also Ehrmann 1999; Wang 1998.

30 FCCC/AGBM/1996/4, "Possible Elements of a Protocol or Another Legal Instrument – Institutional Issues".

31 FCCC/AGBM/1996/Misc.2/Add.2, "Elaboration of the EU Draft Protocol Structure".

32 For the synthesis and the framework compilation see FCCC/AGBM/1996/10 and FCCC/AGBM/1997/2 and Add.1.

33 US Draft Protocol Framework, Article 4; FCCC/AGBM/1997/Misc.1, p. 78.

34 FCCC/AGBM/1997/Misc.1/Add.3.

Box 16.1: The Non-compliance Procedure for the Montreal Protocol[35]

Based upon Article 8 of the Montreal Protocol that calls for the adoption of a non-compliance procedure, the Meeting of the Parties to the Montreal Protocol first adopted an interim non-compliance procedure in 1990 and then established the final non-compliance procedure in 1992. At the heart of the procedure is an Implementation Committee that consists of representatives of 10 parties elected by the Meeting of the Parties for two years, based on equitable geographical distribution (two countries each: Western Europe and other states, CEITs, Asia, Africa, Latin America).

The Implementation Committee usually meets twice a year to perform the following functions: consider any information regarding possible cases of non-compliance, request (through the Secretariat) further information on relevant matters, undertake information gathering in the territory of the Party concerned, and maintain an exchange of information with the Executive Committee of the Multilateral Fund for the Implementation of the Montreal Protocol in developing countries.

The non-compliance procedure can be triggered either by written complaints by other parties, by the non-compliant party itself or by the Secretariat. In considering the relevant cases, the Implementation Committee is to try to find an amicable solution. It reports to the Meeting of the Parties, including any recommendations for further action it might deem appropriate. The Meeting of the Parties retains the authority to take relevant decisions. The provisions for the non-compliance procedure include an indicative list of measures that might be taken by the Meeting of the Parties in respect of non-compliance with the Protocol. Accordingly, the Parties may offer assistance, issue cautions or suspend specific rights and privileges under the Protocol.

In the course of the 1990s, the Implementation Committee focused on two main areas of activity. First, it has regularly addressed cases of failure to report data as required under the Protocol. Second, the Committee has dealt with various cases of non-compliance with the Montreal Protocol schedules for phasing out ozone depleting substances by CEITs, most importantly Russia. The prevailing response to these cases of non-compliance has been the offering of financial and technical assistance through the GEF, conditional on a commitment by the recipient to accelerate its phase-out schedules. In this process, the Implementation Committee co-operates closely with the GEF, the Executive Committee of the Multilateral Fund and the agencies through which these provide assistance to CEITs and developing countries (UNDP, UNEP, UNIDO and the World Bank).

---

35 See in general Ott 1998a, pp. 227 et seq.; Victor 1998.

Given the diverging views of Parties on the issue, the consolidated negotiating text by the Chairman prior to AGBM 8 in October 1997 merely contained a marker requesting the Parties to elaborate procedures and mechanisms to determine and address cases of non-compliance.[36] In the revised text under negotiation of November 1999 this provision was already very much in its final form.[37] The main differences between the EU on the one hand, and many JUSSCANNZ and developing countries, on the other hand, concerned the application of "binding penalties" under the procedure. Whereas most JUSSCANNZ countries objected to binding penalties, the US was prepared to go along with them as long as they were clearly defined so that the US Senate could know exactly what it was ratifying.

In the last phase of the negotiations in Kyoto, the Netherlands (acting for the EU) introduced language that required the procedure or mechanism to be adopted by means of an amendment of the Protocol, if it entailed "binding consequences". This won the agreement of JUSSCANNZ and other Parties because it ensured that their parliaments would be consulted before the introduction of any specific penalties. This compromise, although creating certain problems (see below), was eventually adopted without major discussion in the last night of Kyoto.

The resulting Article 18 has the character of a marker and further preparatory work is thus needed to establish a non-compliance procedure under the Kyoto Protocol. To this end, the Parties at COP 4 established a "joint working group on compliance" under the SBI and SBSTA, which is to report through those bodies to the fifth session of the COP.[38] The Parties are invited to submit their views and the Secretariat has been requested to facilitate a consultation among the Parties on this issue. The joint working group is to consider various elements of a compliance regime:

- identify compliance-related elements in the Kyoto Protocol,
- follow the development of such elements in various negotiating circles and identify gaps,
- develop ideas to the extent that they are not being addressed by other groups, and
- ensure a coherent approach to developing a comprehensive compliance system.

According to the decision taken by COP 4, the fifth session of the Conference of the Parties is to take further steps including, if necessary, the establishment of an *ad hoc* working group on compliance (or other procedure), with a view to adopting a decision by COP 6. The envisaged *ad hoc* working group might then formulate the draft terms of reference of a concrete non-compliance procedure.

36 FCCC/AGBM/1997/7, Article 9.
37 FCCC/CP/1997/2, Article 18.
38 Decision 8/CP.4, Annex II in FCCC/CP/1998/16/Add.1.

16.3.2 Possible Rules Concerning Non-compliance

It is not easy to predict the final character of the "procedures and mechanisms to determine and address cases of non-compliance" that will be elaborated and adopted according to Article 18 Kyoto Protocol. The law of nations is characterised by the absence of a formal hierarchical structure and it has never been easy to establish means of ensuring compliance with and enforcement of its obligations.[39] Treaty regimes thus frequently provide for the peaceful settlement of disputes and establish a so-called "self-contained régime", which generally excludes resorting to unilateral reprisals.[40]

Modern environmental treaties increasingly use political and sometimes quasi-judicial supervisory procedures, although generally all still contain traditional dispute settlement provisions.[41] Prominent examples of the latter are the non-compliance procedure established under the Montreal Protocol on Substances that Deplete the Ozone Layer (1987),[42] and the procedure established under the Second Sulphur Protocol (1994) to the Geneva Convention on Long-Range Transboundary Air Pollution of 1979.[43] These procedures are part of the implementation mechanism of these environmental treaties, built on multilateral communication and review procedures and generally non-confrontational, non-punitive and rather facilitative in character (although limited "sticks" are usually within their scope).[44]

This approach reflects the fact that the reasons for non-compliance are very often not a wilful disregard for legal obligations, but rather the incapacity of governments, the ambiguity of norms and changing circumstances between the time a treaty is adopted and its entry into force. In general, any effective mechanism for compliance control and enforcement of obligations should have three elements: review, correction and creativity.[45] The review provides the basis for a determination of the factual situation with regard to implementation. The correction function is aimed at preventing or eliminating non-compliant behaviour of a Party through negotiations, financial or other assistance, internal or public shaming or, as a last resort and with all due care, by way of sanctions. The third function, that of creativity, allows the rules of a regime to be adjusted, if changed circumstances demand and justify such adjustment.

---

39 On enforcement non-compliance in international environmental law in general see Ehrmann 1999.

40 The terminology of the International Court of Justice (ICJ) in the Case Concerning United States Diplomatic and Consular Staff in Tehran (USA v. Iran), ICJ Rep. 1980, pp. 3, 40, 43.

41 See Ott 1998a; Gehring 1990; Koskenniemi 1992.

42 For its history see Ott 1991; Barratt-Brown 1991.

43 See Széll 1995, pp. 104 et seq. In other cases no formal procedure has been elaborated, but practice has developed like in the framework of the Convention on International Trade in Endangered Species of Wild Fauna and Flora of 1973, see Sand 1997.

44 See e.g. Chayes/Handler Chayes 1993 and Handler Chayes et al. 1995; for an application to the climate regime see Werksman 1998b.

45 See Hoof et al. 1984, pp. 11 et seq.

One clause in Article 8 on the expert review might already be considered as part of a non-compliance procedure, at least with regard to industrialised countries' obligations. Article 8.5 requests the COP/MOP to consider all information submitted by Parties, the expert review teams and the Secretariat (review function). In accordance with its function as supreme decision-making body under the Protocol, it is furthermore required to "take decisions on any matter required for the implementation of this Protocol" (Article 8.6). The COP/MOP is thus already in a position, without having an elaborated procedure at hand, to deal with cases of non-compliance (correction/creativity functions).

The issue that must be dealt with under Article 18 in particular is the "correction function", which aims to determine whether the obligations of the Protocol have been fulfilled by individual Parties and to decide on the adequate response. Article 8.6 mentioned above remains rudimentary in this respect. The formulation of Article 18 strongly points to the precedent of the non-compliance procedure of the Montreal Protocol (see Box 16.1). This is not only suggested by the term "non-compliance", which had been carefully avoided in the last sessions of the AG 13 because it was considered too supervisory in nature, but also by the "indicative list of consequences" to be developed.

The Montreal Protocol approach was favoured by the EC and Japan and it comes very close to the "review and compliance process" suggested by the US in the Kyoto process (see above). The merits of this procedure for the multilateral resolution of conflicts has furthermore been expressed by most NGOs and scientists that responded to a questionnaire on the design of a "multilateral consultative process" drawn up by the Chairman of AG 13 and issued prior to COP 2.[46] However, it would have to be adjusted to the different circumstances of the climate regime, as was done in the case of the procedure developed under the Second Sulphur Protocol (1994).[47]

Should the precedence of the ozone regime be followed, the ultimate decision-making power would lie with the COP/MOP as the supreme body of the Protocol. Furthermore, a smaller, non-universal body would have to be established in order to deal with questions of non-compliance in an effective and expeditious manner. As such, an "Implementation Committee" was contained in the proposals of both the EC and the US. Like the Implementation Committee of the Montreal Protocol, a committee consisting of 10 members with equal geographical distribution would probably be a good size and structure of such a body. The negotiations on the composition of the Multilateral Consultative Committee under Article 13 of the Convention suggest, however, that equitable geographical distribution of the membership would meet with resistance from industrialised countries (see above).

As desirable as it might be to give non-state actors the right to sit on the Implementation Committee, considerations of sovereignty and political expediency

46 FCCC/AG13/1996/Misc.2 and Add.1.

47 See Széll 1995, pp. 104 et seq.

speak strongly in favour of a body composed only of state representatives. Going beyond the model of the Montreal Protocol, it could, however, be useful to allow qualified non-state actors to observe (part of) the proceedings and make oral interventions as well as written submissions. The Implementation Committee should try to reach an amicable solution to the problem, report to the COP/MOP and recommend a suitable course of action. Responses might consist of assistance to the afflicted Party, the issuance of cautions, withdrawal of voting rights, the loss of the right to participate in the Emissions Trading regime under Article 17 of the Kyoto Protocol or the compulsory devaluation of a Party's "assigned amount" of emissions or emission reduction units/certified emission reductions acquired through JI under Article 6 or the CDM under Article 12 of the Kyoto Protocol (see the respective Chapters on these provisions).

The precedent of the ozone regime further suggests that the initiation (the "triggering") of this non-compliance procedure should not be restricted to the Parties. Experience with other procedures, such as the enforcement of EC law and human rights conventions, indicate that the right to initiate proceedings should also be given to a body independent of the Parties. The Secretariat is in a good position to do this since it oversees the implementation of the Protocol and has the required skilled personnel. To some extent, Article 8.3 of the Protocol already provides the legal basis for this function of the Secretariat, since it requires the Secretariat to "list the questions" with regard to the national implementation identified by the reports of the expert review teams for "further consideration by the COP/MOP". Under the non-compliance procedures of the Montreal Protocol and the Second Sulphur Protocol the Secretariat is allowed to report to the Implementation Committee on possible non-compliance no matter how it becomes aware of it. The Secretariat thus has the first opportunity – and duty – to notice and possibly report implementation problems of a particular Party (even if such problems are detected and communicated to the Secretariat by non-state actors).

The establishment of such a non-compliance procedure would therefore not only provide a structure for detecting problems in implementation by individual Parties and for resolving conflicts regarding non-compliance, but would also act as a "conflict avoidance mechanism".[48] Since the ultimate decision lies with the COP/MOP, such a procedure furthermore has the potential to adjust rules of the regime through decision-making if they are considered to be inadequate (creativity function). Through a phased procedure, taking the expert review or individual submissions as a starting point, difficult cases would be subject to a co-operative process of communication and (possibly) coercion that could facilitate the smooth and effective implementation of the Convention's commitments.

The most severe constraint on the elaboration of such a procedure is posed by the last phrase of Article 18. It requires that "any procedures and mechanisms under this Article entailing binding consequences shall be adopted by means of an

48 Similarly, see the the submission of the UK in FCCC/AG13/1996/Misc.2.

amendment to this Protocol". Inserted after a proposal by the EC, this provision either takes the teeth out of any procedure or will prevent its universal application. According to Article 20.4 of the Protocol, amendments shall enter into force only after ratification by at least three-fourths of the Parties, which is likely to take several years. Also, the amendment would subsequently only bind the ratifying Parties.

Judging from the experience with other multilateral environmental agreements, this might not be considered a serious shortcoming, since concrete sanctions (e.g. financial penalties) are very rarely applied in practice in the context of existing non-compliance mechanisms and, if applied at all, are very limited in scope.[49] The demand for "binding consequences" is, however, bound to be far higher under the Kyoto Protocol because of the far-reaching economic and societal implications of the treaty (which may be more directly measurable than action taken under other environmental treaty regimes). Compliance with the Kyoto Protocol might thus need to be ensured by way of binding sanctions, which would require an amendment.

One possible solution to this dilemma might be the adoption of such an amendment by consensus or, as a last resort, with a three-fourths majority (Article 20.3), and its subsequent provisional application by a decision of the COP/MOP. If adopted by consensus, this approach would allow the procedure to be applied by a simple decision of the COP/MOP until the entry into force of the amendment. Parties joining the Protocol after its adoption would be bound by this decision.

16.4 Dispute Settlement (Article 19)

As regards the settlement of disputes, the Protocol simply incorporates the provisions of the Convention and provides that these shall be applicable *mutatis mutandis* (Article 19). This approach of applying provisions elaborated in the "mother treaty" is common in international environmental regimes and contributes to the efficiency in international law making. The provision of Article 19 mirrors Article 14.8 FCCC stipulating that its provisions are applicable to any "related legal instrument which the Conference of the Parties may adopt". Because of this clear provision, Article 19 of the Kyoto Protocol was indeed one of the rare provisions that were undisputed in the negotiating process.

Like most multilateral environmental treaties, the FCCC contains a provision on judicial dispute settlement modelled after Article 33 of the United Nations Charter. Article 14 of the Convention mandates a step-by-step process for the settlement of disputes: the Parties shall first attempt to settle their disputes through negotiations or "other peaceful means" (Article 14.1 FCCC).[50] An amicable solution of conflicts between the Parties to the Convention or the Protocol is facilitated by the

49 See Széll 1995; Ott 1998a, pp. 215 et seq.; Victor et al. 1998.

50 This may be the most effective means of conflict resolution, see Sohn 1983, p. 1121–1146, p. 1122.

institutionalisation of both treaties, i.e. the existence of several multilateral treaty organs like the COP (COP/MOP) and the subsidiary bodies. These bodies allow disagreements to surface early and provide a communicative structure for finding co-operative solutions, even in the absence of formal procedures.[51]

As a second step, the dispute may be submitted to the International Court of Justice (ICJ) or to arbitration. For the latter, Parties are required under Article 14.2 of the FCCC to adopt an annex specifying the procedures. The COP has not yet acted on this matter. If both Parties to the dispute have made a declaration to accept compulsory jurisdiction in accordance with Article 14.2 of the Convention, either of them may unilaterally bring the matter before the ICJ or to arbitration. However, the Solomon Islands are the only Party to the FCCC to have deposited such a declaration.[52] Since the compulsory jurisdiction only exists with regard to those Parties that have made a declaration, it is currently not a viable option. Although not explicitly mentioned in the Convention, disputing Parties may also jointly submit conflicts for compulsary jurisdiction to the ICJ or to arbitration on an ad hoc basis.

If the dispute has not been settled after 12 months by negotiations or other peaceful means, and has not been submitted to compulsory jurisdiction, a third option may be invoked. Either of the Parties may then submit the issue to a conciliation commission (Article 14.5 and 6 FCCC). This commission is to be composed of an equal number of members appointed by either Party and by a chair chosen jointly by those members; its decisions are not binding and have a recommendatory character (Article 14.6 FCCC).[53] Any decision of the International Court of Justice, however, is automatically legally binding in accordance with Article 59 of its Statute. In contrast, the question of whether arbitral awards are binding has not yet been resolved since the Parties to the Convention have so far not adopted the arbitration procedure.. There are at present no activities underway to elaborate this procedure, although precedents would point to these awards being binding.

So far, there has never been a case where, within the framework of an environmental treaty, parties to a dispute have resorted to these dispute settlement procedures. First, only a limited number of treaties provide for the unilateral submission of a dispute to an Arbitration Tribunal or the ICJ.[54] Second, there are only a few declarations by Parties that they are willing to accept the compulsory jurisdiction of the ICJ or an *ad hoc* tribunal. Thus, it is most unlikely that *both* Parties to a dispute would have made the required declaration.

51 For an elaboration of this function see Ott 1998a, pp. 273 et seq.

52 Whereas the Republic of Cuba expressly declared that, in relation to Article14 FCCC, disputes should be settled by way of diplomatic negotiations, see FCCC/CP/1998/Inf.5, Declarations 9 and 11.

53 The compulsory or mandatory character of the dispute resolution was a contentious issue in the negotiations of the Convention, see Bodansky 1993, p. 549.

54 For examples see Ott 1998a, p. 219.

---

33 [illegible] recommendatory character of the dispute [illegible] Convention see Sachariew 1990, p. 343.

17. Developing Country Participation (Articles 10, 11)

As the most vulnerable countries, developing countries have a profound interest in mitigating and adapting to climate change. At the same time, the dominant development pattern established by industrialised countries and followed by most developing countries depends to a large extent on the availability and the use of fossil fuels. Hence, many developing countries perceive action to limit and reduce GHG emissions as a potential threat to their economic future. Even action by industrialised countries is worrisome for a number of developing countries because of the possible negative repercussions on their own economic development.

Although there are signs that economic growth in some developing countries is not absolutely proportional to higher emission trends,[1] the GHG emissions of developing countries will continue to rise with the growth of their populations and economies. Within a few decades, the emissions of developing countries are expected to exceed those of the industrialised countries (see Figure 17.1). It is obvious and generally accepted, that, in the long run, these trends will have to decrease in order to prevent further global warming. However, a number of industrialised countries, particularly the US, requested already during the negotiations on the Kyoto Protocol that developing countries should take on binding obligations to limit their GHG emissions. Since, historically, these industrialised countries were the major contributors to the problem and have themselves failed to take decisive action to limit and reduce their own emissions, this demand was considered ethically questionable and was successfully rejected by developing countries.

As pointed out earlier (see Chapter 2), developing countries can hardly be said to represent a monolithic block, although they are to some extent united in the G-77. In the climate change context, at least three different groupings need to be distinguished. AOSIS countries, which are particularly vulnerable to the negative impacts of climate change, stressed the urgent need to act and demanded support for adaptation. OPEC members, in contrast, have tried to slow the pace of international co-operation on climate change and demanded compensation for any economic losses they may incur through the mitigation of climate change by industrialised countries. The majority of developing countries, including China, India, Brazil and others, have pushed industrialised countries to accept binding targets and attempted to protect their interest in economic development.

Each of these groupings articulated its special demands in the negotiations on the Kyoto Protocol, but all of them were faced with the demand from the US and others to take on binding obligations that were contrary to the Berlin Mandate.

---

1 According to a study undertaken by UNDP, India experienced an annual economic growth of 6.8% during the 1990s, whereas GHG emissions only rose by 4.9%; see Goldemberg/Reid, pp. 6–8.

This chapter is divided as follows: first, it sketches out the negotiating history of the main proposals related to developing countries, beginning with the provisions of the Convention, since they represent the starting point for the Kyoto negotiations; second, the chapter analyses the provisions of the Kyoto Protocol pertaining to developing countries, most importantly Articles 10 (obligations of all Parties) and 11 (financial mechanism); and third, as the participation of developing countries remains a highly controversial topic, this chapter concludes with a brief outlook towards the future.

17.1 Negotiating History

The provisions of the Framework Convention on Climate Change are key to understanding the negotiating process on the Kyoto Protocol. Indeed, the Convention incorporates the demands of developing countries to a large extent. The inclusion of developing countries' crucial concerns into the Convention received its strongest expression in the statement that it will be taken "fully into account that economic and social development and poverty eradication are the first and overriding priorities" of developing country Parties (Article 4.7 FCCC). The Convention obliges the Parties to protect the climate "on the basis of equity" and according to their respective capabilities. Moreover, it acknowledges the specific problems of developing countries (Article 3.1 and 3.2 FCCC). The Convention also explicitly recognises the historical responsibility of the industrialised countries[2] regarding the emission of GHGs and, in relation to developing countries, demands that they

- "take the lead in combating climate change" (Article 3.1 FCCC),
- "provide new and additional financial resources to meet the agreed full costs incurred by developing country Parties" for their obligations under the Convention (Article 4.3 FCCC), the most important of which consists in submitting national communications (cf. Articles 4.1 and 12.1 FCCC),
- "provide such financial resources (...) needed by the developing country Parties to meet the agreed full incremental costs of implementing measures" (Article 4.3 FCCC),
- "assist the developing country Parties that are particularly vulnerable (...) in meeting the costs of adaptation" (Article 4.4 FCCC), and
- "take steps to promote, facilitate, finance the transfer of (...) technologies and know-how (...) particularly (to) developing country Parties" (Article 4.5 FCCC).

These provisions calling for substantial assistance are given additional emphasis by Article 4.7 FCCC, which makes the fulfilment of developing country Party commitments conditional on the realisation of this assistance. This article has

2 These are usually referred to as "developed country Parties" or "developed country Parties and other Parties included in Annex I".

therefore been described as "the carrot that lured the developing countries to the negotiating table".[3] Developing countries were, however, dissatisfied with the Global Environment Facility (GEF), which was entrusted with the task of operating as the Conventions' financial mechanism. Apart from it being underfunded, developing countries complained about the GEF's lack of transparency (see Chapter 3).

Furthermore, the Convention explicitly names categories of countries, including small island countries (Article 4.8 FCCC) and least developed countries (Articles 4.8, 4.9 FCCC), most likely to be effected by global warming. The specific situation of OPEC members as countries that are highly dependent on the income generated from the production, processing and export of fossil fuels is recognised as well and is to be taken into consideration in the implementation of the Convention (Article 4.10 FCCC).

Subsequent to the climate negotiations, one of the structural deficiencies inherited from the Convention (and the wider UN practice) is the blanket division between industrialised countries (Annex I Parties) and developing countries (non-Annex I Parties). This does not sufficiently reflect the "differentiated responsibilities and respective capabilities" (Article 3.1 FCCC) of various developing countries. Non-Annex I Parties such as Singapore, South Korea, or Israel are among the highest GHG emitters and enjoy GDP per capita outputs that are higher than, for example, those of Spain or Portugal. That Mexico (1994) and South Korea (1996) have joined the OECD is perhaps one of the most obvious signs of the differentiation among developing countries. As a result of this differentiation, developing countries have been drawn between the twin objectives of sticking to the G-77 solidarity and pursuing their national interest. This has led to complications and delays in the negotiating process as the G-77 and China have tried, at times unsuccessfully, to find common positions.

In light of their great interest in climate protection, it was not surprising that the first draft of a protocol was proposed by AOSIS prior to COP 1 in Berlin.[4] This proposal provided for a "basic commitment" of all Parties, "taking into account their common but differentiated responsibilities and their specific national and regional development priorities". In particular, it provided that Annex I Parties "shall reduce (their) 1990 level of anthropogenic emissions of carbon dioxide by at least 20% by the year 2005" and that Annex I Parties transfer "the best available technologies, practices and processes" under "fair and most favourable conditions".

With regard to the commitments of developing countries, the draft protocol contained a procedure, similar to Article 4.2(g) of the Convention, which would have allowed a Party not included in Annex I to notify its intention to be bound by the obligations of industrialised countries. Article 4.2(g) FCCC, for example,

3 Gupta 1997, p. 88.

4 UN Doc. A/AC.237/L.23.

allowed countries such as Monaco and Luxembourg to join the Annex I "club" without revising the Convention's annexes. Similarly, the procedure proposed by AOSIS would have allowed non-Annex I Parties to take over Annex I commitments concerning, in particular, the quantified targets. Thus, the AOSIS draft protocol included a rather simple procedure for developing countries to assume, with certain restrictions, Annex I status when they felt it to be appropriate.

COP 1 adopted the Berlin Mandate, which called for a strengthening of industrialised countries' commitments (see Chapter 4). Despite some attempts to put the commitments of developing countries on the negotiating table, the Berlin Mandate determined "not to introduce any new commitments for the Parties not included in Annex I", a result of the strong position of the developing countries and the support of the European Union. Consequently, during the first rounds of negotiations after COP 1, the debate was not very concerned with developing country participation.

At AGBM 5 in December 1996, however, the EU submitted its first proposal for a protocol.[5] One element of it was received with a lot of scepticism by developing countries: the development of a new "Annex X" to the protocol, which would list those countries with quantified targets. Furthermore, similar to the AOSIS draft protocol mentioned above, non-Annex X Parties could voluntarily decide to take on the same commitments as Annex X Parties. This proposal was rejected by the G-77, which strongly opposed the elaboration of any new Annexes, because they would, contrary to the Berlin Mandate, introduce new commitments for developing countries and have the potential to split the group.

The EU initially failed to name the Parties to be listed in Annex X, causing some confusion. It took the Union until AGBM 7 in July/August 1997 to specify that it had the new OECD members Mexico and South Korea in mind. Also at AGBM 7, the European Union recognised "that in the long-term emissions of greenhouse gases from countries not included in Annex I must also be regulated", addressing the difficult issue of timing or "evolution" of commitments, as it was referred to.[6]

Apart from the EU and developing countries, the third major player on this issue was the JUSSCANNZ group. At AGBM 6 in March 1997, the US submitted a draft protocol framework[7] that outraged many developing countries. It suggested a two speed approach, i.e. the development of two Annexes A and B. The only difference between Annex A- and Annex B Parties would be the scale of the quantified emission limitation and reduction objectives. In addition, it contained a concept of evolution: (all) "Parties shall adopt, by (2005), binding provisions so that all Parties have quantitative greenhouse gas emission obligations." Both the two

---

5 See FCCC/AGBM/1996/MISC.2/Add.2.

6 See AGBM/1997/MISC.2/Add.1, p. 14.

7 See FCCC/AGBM/1997/MISC.1.

speed reduction approach and the evolution element were rejected outright by developing countries.

A third element of the US proposal further upset developing countries. Article 5, titled "Advancement of the Implementation of Article 4.1 of the Convention" (which sets forth the basic obligation of all Parties to the Convention), contained a bundle of extra requirements for Parties that appear neither in Annex A nor in Annex B, namely the remaining developing countries. First, it urged those Parties to "identify and implement "no regrets" measures" (i.e. measures that could be taken at no or low cost). The identification of such "no regrets" measures would be part of a "process for reviewing communications". Under this process, a developing country would be obliged to "report to the Secretariat (...) on the measures it has implemented, plans to implement, and the barriers to adopting potential measures". Developing countries perceived this proposal as yet another attempt to establish double standards: although the IPCC had identified a potential for GHG-emission reductions of up to 30% at no or very low costs in industrialised countries,[8] the "no regrets" paragraph should not apply to Annex I Parties. Further, developing countries were expected to submit "on an annual basis, its inventory of GHG emissions".[9] The burden of meeting these costs, however, was not allocated in the draft protocol and would thus, by implication, have to be met by the developing countries themselves.

In order to calm the dispute, Estrada used his full authority to rule out evolution as proposed by JUSSCANNZ at AGBM 7 in the summer of 1997. He decided that evolution did not fall under the Berlin Mandate and that JUSSCANNZ could put the item on the agenda of COP 3.

Even though Estrada had disapproved of evolution, he was aware of the necessity to establish long-term goals that explicitly included developing countries. A new idea was needed to reconcile the strong resistance to "evolution" and the urgency of global GHG emission reductions in the future. The most suitable vehicle appeared to be an idea that was already contained in the AOSIS proposal: non-Annex I Parties might voluntarily adopt quantified targets. Thus, according to Article 10 of Estrada's draft protocol, a party that wished to voluntarily commit itself could notify the Depositary of GHG-emission limitations or reduction targets and a base year at any time. The Secretariat would then put such a notification on the agenda of the next Meeting of the Parties, which would decide on the acceptance of the notification.[10]

The proposal contained potential advantages for those Parties that opted to be bound by Article 10 of Estrada's proposal, since they could participate in Emissions Trading and Joint Implementation. Furthermore, the proposal matched equity concerns, as non-Annex I Parties could choose their emission reduction

---

8 See FCCC/CP.3/1997/CRP.5.; IPCC 1996.

9 See FCCC/AGBM/1997/MISC.1.

10 See FCCC/CP/1997/2.

level and the base year (subject to acceptance by the Meeting of the Parties). Argentina, AOSIS and a number of other developing countries consequently lent their support to "Article 10", as the provision on voluntary commitments was referred to throughout the negotiations in Kyoto. However, major developing countries, such as India, China and Brazil, rejected the idea. They feared being pressured into accepting voluntary commitments by the US and other industrialised countries using levers such as bilateral development aid and trade. Thus, the internal divergence of interests and the split between developing countries was clearly revealed in the discussions on Article 10.

Disputes over how to involve developing countries in the protocol continued until COP 3 in Kyoto. Positions hardened when US President Bill Clinton announced the US position for Kyoto in October 1997 and called for "meaningful participation" of developing countries. This was due not least to the resistance in the US Senate to the protocol.[11] At the end of the first week in Kyoto, the discussion became even less fruitful when New Zealand proposed a concrete procedure and timetable for "evolution". The reaction of a large number of developing countries culminated in one final word of the Chair of G-77: "no".[12]

In the last night of COP 3, Article 10, concerning voluntary commitments, was still a matter of hot debate. The supporters – Annex I Parties, Israel, South Korea, the Philippines, AOSIS and Argentina – met fierce resistance from the big GHG emitting developing countries like India, Brazil and China. They were joined by OPEC countries, who had been advised by the fossil fuel industry that voluntary commitments on the part of the developing countries would significantly raise the likelihood of ratification of the protocol in the US Senate. Attempts by the US, South Korea and Mexico to save Article 10 by introducing compromise language shared the same destiny. Finally, Chairman Estrada, visibly disappointed, withdrew his draft Article 10. Argentina, in particular, regretted the failure to establish the possibility of voluntary commitments in the protocol and announced that it would request to have the matter put on the agenda of COP 4.

Other major issues surrounding developing country participation concerned the "advancement" of commitments in Article 4.1 of the Convention (commitments of all Parties) and financial resources. Ambassador Bo Kjellén, who had chaired the informal meetings on these matters, informed Chairman Estrada on the last night of negotiations in Kyoto that the Parties were sharply divided around two formulations.[13] The US and most industrialised countries favoured Alternative A, which obliged all Parties to "formulate, implement, publish and regularly update national programs containing measures to mitigate climate change". Moreover, Alternative A contained a comparatively detailed list of policies and measures (PAMs), similar

---

11 See the Byrd Hagel Resolution of the American Senate, 25 June 97, Report No. 105–54, S.Res. 98; see Chapter 4.

12 See Earth Negotiations Bulletin, Vol. 12, No. 76, 13 December 1997, p. 13.

13 This was Article 11(b) of the draft, see FCCC/AGBM/1997/CRP.1/REV.1, which later became Article 10(b).

to that contained in Article 2 of the Protocol which binds Annex I Parties only, that all Parties would have to undertake. In contrast, G-77 supported Alternative B, which was less strict: "Each developing country Party shall seek to include in its national communications, as appropriate, information on programmes which contain measures that the Party believes contribute to addressing Climate Change".[14]

Obviously, Alternative A placed a much higher burden on developing countries. Hence, a compromise was put forth by Ambassador Bo Kjellén. In his draft, he deleted the list of PAMs and various detailed provisions concerning capacity building, awareness raising, sustainable development and resources management. Instead, he inserted a general commitment that national mitigation programmes "would, inter alia, concern the energy, transport and industry sectors as well as agriculture, forestry, and waste management" (Article 10(b)(i)).

The Kjellén draft was not much liked by the US and other industrialised countries. On the other side, Tanzania, on behalf of G-77 and China, rejected the compromise as well. Instead of talking it through, the spokesman of G-77 asked Estrada "to use his gavel". Although this request was ambiguous, Estrada did not hesitate: the Kjellén compromise was adopted within seconds.

This secured the unequivocal adoption of Article 11 on the Financial Mechanism, which had not been a major issue in the Kyoto negotiations. Following strong arguments regarding "institutional economy", it was agreed relatively early that the mechanism would operate in conjunction with the Financial Mechanism of the Convention (see Chapter 3). Agreement on this arrangement was, however, politically linked by the US to agreement on how to advance the commitments of developing countries. Thus, once agreement on this was reached, acceptance of Article 11 on the Financial Mechanism was also secured.

Two further issues relating to the involvement of developing countries acquired prominence during the Kyoto negotiations: adaptation and compensation. Securing assistance for adaptation had been a major concern particularly for AOSIS. This was eventually reflected in various places in the Kyoto Protocol, perhaps most visibly in the provisions concerning the Clean Development Mechanism (see Section 14.2) and policies and measures (see Section 10.2).

OPEC countries, for their part, requested compensation for any economic losses they might incur as a result of mitigation measures taken by industrialised countries (e.g. reduced revenues from oil exports). As such, they proposed the establishment of a compensation fund.[15] Although this proposal was formally included in the G-77 position unveiled in 1997 (Section 4.6), thus securing OPEC endorsement of the position, compensation never received widespread support. As a result, the issue was not even on the agenda of a contact group in Kyoto. However, OPEC was successful in keeping the issue on the agenda of the post-Kyoto process.

---

14 See FCCC/CP/1997/2.

15 See the Negotiating Text by the Chairman, FCCC/AGBM/1997/3/Add.1, para. 152 (proposal 2). Attribution by the authors.

17.2 The Outcome: Articles 10 and 11 of the Kyoto Protocol

The final result of the negotiations on the participation and commitments of developing countries can be found in Articles 10 (commitments of all Parties) and 11 (financial mechanism) of the Protocol.[16] These provisions hardly go beyond what had already been agreed upon in the Convention. In particular, no new (voluntary) commitments for developing countries are included in the Protocol at all: Article 10 states explicitly that its requirements shall be applied "without introducing any new commitments for Parties not included in Annex I, but reaffirming existing commitments under Article 4, paragraph 1, of the Convention".

It is important to note that the commitments found in Article 10 of the Protocol, like in Article 4.1 of the Convention, apply not only to developing countries but to "all Parties, taking into account their common but differentiated responsibilities and their specific national and regional development priorities, objectives and circumstances" (Article 4.1 FCCC and Article 10 Kyoto Protocol). As in this case, most of Article 10 corresponds almost exactly to Article 4.1 of the Convention.

Article 10(a) requires all Parties to formulate national or regional programmes in order to improve their emission inventories and national communications required by Article 4.1(a) and Article 12.1 of the Convention. Article 10(b) mirrors Article 4.1(b) of the Convention in referring to the formulation and implementation of national and regional programmes for climate change mitigation and adaptation. It goes slightly beyond the provisions of the Convention, however, since it specifies that such programmes "would" concern the energy, transport and industry sectors as well as agriculture, forestry and waste management. Furthermore, reference is made to adaptation technologies and the improvement of spatial planning (Article 10(b)(i)). Article 10(b)(ii) asks Parties not included in Annex I to submit information on programmes containing measures "that the Party believes contribute to addressing climate change and its adverse impacts" and refers specifically to capacity building and adaptation measures.

The subject of Article 10(c) is the transfer of environmentally sound technologies and know-how, particularly to developing countries. The corresponding Article 4.1(c) of the Convention is enlarged upon by referring to policies and programmes for effective transfer of such technologies that are publicly owned. It also goes beyond the provisions in the Convention by calling for the creation of an enabling environment for the transfer of relevant technology by the private sector. Article 10(d) calls for the improvement of co-operation in scientific and technical research as mirrored in Article 4.1(g) of the Convention. In addition, it requests the promotion of the development and strengthening of related endogenous capacities.

---

16 The CDM under Article 12 is of course also of crucial importance for developing countries and their participation in the Protocol; see Chapter 14.

Following from Article 4.1(i) of the Convention, Article 10(e) seeks to establish education and training programmes that facilitate public awareness of and access to information on climate change at the national level. According to paragraph (f), information on programmes and activities pursuant to Article 10 shall be included in the national communications (cf. Article 4.1(j) of the Convention). In the implementation of Article 10, Article 10(g) states that Parties should give full consideration to Article 4.8 of the Convention, which lists various categories of developing countries particularly vulnerable to climate change or to the implementation of response measures.

Article 11 specifies the obligations of industrialised countries concerning the transfer of financial resources and technology needed by developing countries to meet their commitments under Article 10. This extends slightly the existing commitments of industrialised countries under Article 4.3 of the Convention and the financial mechanism as regulated in Article 11 of the Convention. After making reference to the pertinent provisions in Article 4 of the Convention (Article 11.1; see also above), Article 11.2 obliges Parties included in Annex II to the Convention to provide "new and additional financial resources" to developing countries.

This additional obligation refers in particular to the transfer of funds under the existing official development assistance (ODA). These new financial resources, including the transfer of technology, shall meet the "agreed full costs" of national communications according to Article 10.1(a), and the "agreed full incremental costs" incurred by the implementation of commitments under Article 10. Paragraph 2 further states that the guidance provided by the COP to the GEF, as the entity entrusted with the operation of the financial mechanism under Article 11 of the Convention, shall apply "mutatis mutandis" to the Financial Mechanism under the Protocol.

In general, the formulations contained in Articles 10 and 11 are rather imprecise and leave much room for interpretation. Expressions like "where appropriate" or "to the extent possible" hardly provide for exact standards. It is probably fair to state that these provisions, while ensuring the funding of related activities under the Protocol, can at best build the basis for future moves concerning domestic implementation and the further development of international regulations. Therefore, these Articles, as well as many others, nurture the impression of an unfinished Protocol.

17.3 Assessment and Outlook

Discussions on developing country participation did not stop with the adoption of the Kyoto Protocol. On the contrary, the US call for "meaningful" participation of these countries remained prominent in the post-Kyoto negotiations. There were renewed attempts by the US, supported by Japan and Canada, to place voluntary commitments on the agenda of COP 4. At COP 4, the US and Argentina engaged in what looked like a concerted action: President Menem of Argentina, who failed to make voluntary commitments an item on the official agenda, announced his willingness to adopt quantified emission reduction/limitation obligations by the next Conference of the Parties.[17] Kazakhstan signalled its willingness to take a similar step. Within 24 hours the US signed the Kyoto Protocol.

Overshadowed by these high politics, the decisions taken at Buenos Aires are to a large extent favourable to developing countries. First, the GEF, as the operating entity of the Financial Mechanism, was requested to provide funding for an enhanced array of adaptation activities.[18] With regards to technology transfer, developing countries failed to find sufficient support for a "technology transfer mechanism" but secured the establishment of a consultative mechanism under SBSTA. This will possibly include regional workshops and a SBSTA workshop.[19] OPEC countries, on the other hand, were unsuccessful in their attempt to link the issue of compensation with PAMs. They did, however, manage to insert the issue of "implementation of response measures under the Convention" into a decision.[20] This was another attempt to keep compensation for declining revenues from fossil fuel exports on the agenda, making it a sleeping "time bomb" that is pushed carefully from one round of negotiations into the next.

In any case, developing country participation remains a key to the entry into force of the Kyoto Protocol. The call for "meaningful participation" on the part of the US and other industrialised countries is effectively stifling the negotiating process. First, the call for "meaningful participation" serves as a useful negotiation instrument to put pressure on the negotiating partners and it can be used occasionally to disrupt the process. Second, it effectively divides the block of G-77 countries. The US Administration considers the CDM to be an important cornerstone of "meaningful participation". However, while India and China are still taking a cautious approach, various Latin American and African countries, disappointed with the benefits of the pilot phase on AIJ, have welcomed the CDM as a second chance. And third, the call for "meaningful participation" has served, and still serves, as a vehicle to avoid taking on domestic implementation measures.

Finding an adequate reaction to these demands is difficult. Non-governmental organisations are also split on this issue. For some NGOs, a ratification of the Protocol by the US Senate would outweigh the concerns of the developing countries. There is, however, no guarantee that the US Senate will ratify the Protocol, even after substantial commitments are made by developing countries. It thus seems appropriate for developing countries and a leadership group of industrialised countries to engage in constructive confidence-building and enter into a truly co-operative effort on climate change (see Chapter 25).

---

17 This was linked to the participation in the Kyoto mechanisms: "We shall continue working to ensure that countries like the Republic of Argentina, which wish to commit themselves to their own targets, are able to participate in all the mechanisms of the system"; see the report of COP 4, FCCC/CP/1998/16, p. 35.

18 Decision 2/CP.4 in FCCC/CP/1998/16/Add.1.

19 Decision 4/CP.4 in FCCC/CP/1998/16/Add.1.

20 Decision 5/CP.4 in FCCC/CP/1998/16/Add.1.

Figure 17.1: Projected CO_2 Emissions of Industrialised and Developing Countries

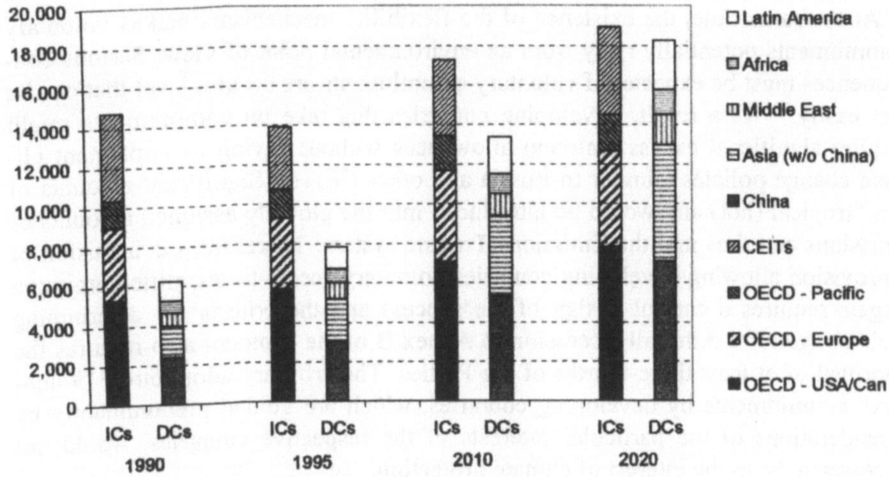

Source: IEA 1998.

Apart from these political considerations, there is no doubt that the rising GHG emissions in developing countries actually do pose a challenge to effective climate protection. Eventually, this will need to be addressed. All projections predict rising emissions in the developing countries. The various studies undertaken differ only in the actual point of time when the emissions of these countries will exceed those of industrialised countries. For example, the International Energy Agency (IEA) projects that CO_2 emissions in the southern countries will exceed those of Annex I countries after the year 2020 (see Figure 17.1).[21] Therefore, in the mid and long term perspective, developing countries will be expected to adopt binding GHG emission reduction or limitation obligations in order to intensify global efforts to mitigate climate change.

A potentially successful approach might be the adoption of voluntary commitments as proposed in the draft protocol submitted by AOSIS countries. After all, voluntary commitments offer an opportunity to effectively differentiate between developing countries (even the US did not demand that all developing countries

21 Taking a historical perspective on cumulative emissions, the Brazilian proposal from AGBM 7 calculates, however, that non-Annex I emissions will equal Annex I emissions in 2147; see FCCC/AGBM/1997/Misc.1/Add.3, pp. 19–23.

should take on commitments). The inclusion of flexibility instruments into the Protocol provides some developing countries (as exemplified by Argentina and Kazakhstan) with a real incentive of being able to participate in Emissions Trading and to sell any excess emission allowances. This excess would ideally be achieved by implementing domestic policies to reduce GHG emissions.

At the same time, the existence of the flexibility mechanisms makes voluntary commitments potentially risky from an environmental point of view. Serious consequences must be expected if voluntary commitments are set at a level that can be met easily.[22] As a result, developing countries that take on commitments might acquire significant excess emission allowances without having to implement climate change policies (similar to Russia and other CEITs). Significant amounts of this "tropical (hot) air" would be introduced into the globally assigned amounts of emissions and thus into the Emissions Trading system. Therefore, the inclusion of a provision allowing developing countries voluntary access to quantified emission targets requires a careful design of the process and the criteria for determining their obligations. After all, accession to Annex B of the Protocol also requires the approval of at least three-fourths of the Parties. The arbitrary adoption of "voluntary" commitments by developing countries, which are guided predominantly by considerations of the particular interests of the respective countries, would not necessarily be in the interest of climate protection.

In order to make developing countries take on concrete obligations, a different approach will most likely be required. This is because of the risks inherent in a strategy based on voluntary commitments, and, perhaps most importantly, because of the strong rejection of this approach by the major developing country GHG emitters (e.g. China, India and Brazil). This acceptance will hinge decisively upon whether the post-Kyoto process and any resulting obligations on developing countries will be considered as equitable. Equity is not simply a concern of the developing countries,[23] but is of prime importance.[24] In this respect, a number of factors need to be taken into account. First, historic responsibility for the disturbance of the climate system lies with industrialised countries. This is acknowledged in Article 3.1 of the Convention.[25] Second, developing countries will suffer most from climate change because of their geographical and climatic conditions and their lack

22 According to President Menem, the annual growth rate of the Argentinian economy was 6% in the last decade, whereas GHG emissions only increased by 1%; see FCCC/CP/1998/16, p. 35.

23 For example, equity considerations influenced the distribution of commitments within the EU "bubble", see Chapter 12.1.

24 Gupta 1997, pp. 74 et seq.; Grant 1995.

25 "(T)he Parties should protect the climate system for the benefit of present and future generations of humankind, on the basis of equity and in accordance with their common but differentiated responsibilities and respective capabilities. Accordingly, the developed country Parties should take the lead in combating climate change and the adverse effects thereof" (Article 3.1 FCCC).

of capacities to adapt to a changing climate.[26] And, finally, mitigating climate change will be a major challenge for most developing countries, not least because of their limited financial and technological resources.

Even with a system of Emissions Trading that ideally should lead to a levelling of mitigation costs across countries, important equity concerns remain.[27] These are mainly related to the fair and equitable initial distribution of emission allowances. The present rather arbitrary allocation of "assigned amounts" to industrialised countries under the Kyoto Protocol follows the so-called "grandfathering rule", i.e. emission allowances are derived from historic emission levels (generally the base year 1990). As highlighted by China, India and other developing countries that resisted emissions trading during the negotiations at Kyoto, starting from the status quo can hardly be seen as equitable in this case.[28]

A number of alternatives for the fair and equitable allocation of obligations among industrialised and developing countries have been developed over the past few years.[29] One of the possible and likely the most prominent approaches is the process of "contraction and convergence" towards equal per capita emissions.[30] This is based upon the assumption that the world's population seeks to stabilise CO_2 concentrations at 450 ppmv (a level that might prevent dangerous impacts). Annual emissions of about 2 Gt carbon by the end of next century and not more than 600 Gt cumulative carbon emissions in the period from 1990 to 2100 would be the upper limit to the worlds carbon dioxide emissions.[31] Under the convergence approach, equal per capita emissions would guide the allocation procedure over the long-term, i.e. per capita emissions of the various countries would converge to an amount considered to be sustainable. Obviously, this scenario would demand that industrialised countries curtail emissions significantly. Yet imposing this limit would also necessitate caps on developing countries in the near future.

Most developing countries perceive equal per capita emissions in the long-term as an acknowledgement of the "equity" concerns of the Convention. However, the difficulty of determining what is "fair" remains. The long-standing debate on the fair differentiation of commitments between industrialised countries, which was most intense in the pre-Kyoto process, has evidenced that there are a number of factors in addition to population size that need to be taken into account in order to allocate GHG emissions under a global limit. These include geographical as well

26 See IPCC 1996b.

27 See also Roson/Bosello 1999.

28 Since this is seen as the creation of new property rights for the atmosphere, see Politics in the Post-Kyoto World, CSE Briefings Paper 2, Centre for Science and Environment, New Delhi, p. 2.

29 Different notions of equity are discussed in IPCC 1996c, pp. 80–124 ; Müller 1998; see also Ringius et al 1998; Berk et al. 1999; Rose et al. 1999.

30 See "Contraction and Convergence: A Global Solution to a Global Problem", Global Commons Institute, 18/07/1997 <http://www.gci.org.uk/contconv/cc.html> as of 9 June 1999; Kinzig et al. 1998; see also Grubb 1995b.

31 IPCC 1996a.

as climatic conditions, strength of the economy, size of the country and others. Judging from the discussion on differentiation between industrialised countries, there is scope for more refined approaches. The small relative historical contribution of developing countries to the current GHG concentrations in the atmosphere deserves special consideration. According to a pre-Kyoto Brazilian calculation, the cumulative emissions of developing countries will not equal those of the developed nations until 2147. This would need to be taken into account when assigning emission allowances.[32] In 1999, the World Resources Institute suggested that developing country commitments could be defined in terms of GHG "intensity".[33] The GHG intensity indicator to be used would be the absolute amount of GHG emissions per GDP. The use of this measure is, however, impeded by the lack of accurate and comprehensive data.

Under these circumstances, there is demand for an intensified North-South dialogue on the equitable allocation of future emission allowances. It might be possible to conduct such a dialogue if industrialised countries accept their obligation to demonstrate leadership by implementing domestic measures before pushing developing countries to take obligations upon themselves. Many developing countries have publicly accepted that they will have to act once industrialised countries take decisive steps. Apart from that, many developing countries have made much progress in de-linking economic growth from GHG emissions.[34] On this basis, a political dialogue could and should begin soon, as the issues are complex and will need time to be resolved (see Chapter 25).

32 FCCC/AGBM/1997/Misc.1/Add.3, pp. 19–23.

33 Baumert et al. 1999.

34 See Reid/Goldemberg 1997; 1998.

18. Institutions (Articles 13, 14, 15)

As in the case of the FCCC[1] and most other recent multilateral environmental agreements[2], negotiators of the Kyoto Protocol faced a dilemma with respect to the institutional arrangement. Principles of "institutional economy" (i.e. using existing bodies) stood against the advantages of a fresh start that could be achieved with the help of new institutions better suited to the requirements of this particular treaty. Most debates in the negotiations leading up to Kyoto concerned the question of whether the Protocol should use the existing institutions of the Convention and, if so, whether these institutions should be identical (i.e. whether the Conventions' bodies should simply take over tasks under the Protocol) or whether they should have their own identity due to the potentially different membership of the respective treaties. The negotiating history is dealt with in Section 18.1 below)

In the end, the Parties to the Convention chose a mixed approach. One completely new permanent institution under the Protocol has been created: the Executive Board of the Clean Development Mechanism (Article 12.4).[3] On the other hand, the Kyoto Protocol makes use of the same institutional arrangements for the provision of financial assistance to developing countries as the FCCC.[4] Furthermore, Parties decided to use the existing Secretariat of the Convention for the purposes of the Protocol without making any specific arrangements regarding the allocation of funds etc. (Article 14; see Section 18.3 below).

Although the language on the main institutions seems at first sight to indicate that the Protocol's bodies are identical to those of the Convention, the Protocol in fact establishes three new main organs: a Conference of the Parties serving as the meeting of the Parties (COP/MOP, Article 13; see Section 18.2), the Subsidiary Body for Scientific and Technological Advice (SBSTA, Article 15) and the Subsidiary Body for Implementation (SBI, Article 15) (Section 18.4). In addition, the COP/MOP may, according to Article 13.4(h), establish subsidiary bodies, including ad-hoc groups, to consider specific questions. Pending the entry into force of the Kyoto Protocol, the COP of the Convention established in 1998 a "joint working group on compliance" under the subsidiary bodies to discuss the non-compliance procedure under the Protocol (Section 18.5).

The independence of the main bodies of the Protocol from those of the Convention and their rather hybrid nature follows from the rules for voting and the

1 See Bodansky 1993, pp. 532 et seq.

2 See generally Ott 1998a.

3 See Chapter 14 for details. Note, however, that some Parties would like to entrust an existing institution like the GEF with this task.

4 See Chapter 17. COP 4 established the GEF as an entity entrusted with the operation of the financial mechanism, Decision 3/CP.4 in FCCC/CP/1998/16/ADD.1; see Chapter 3.

composition of the Bureau, on which representatives of the different regional groupings sit to decide on administrative issues between the meetings of the respective bodies. Article 13.2 of the Kyoto Protocol reiterates the principle contained in Article 17.5 of the FCCC that decisions under the Protocol shall only be taken by those that are Parties to it. Non-Parties to the Protocol may participate in the proceedings as observers. Furthermore, those members of the Bureau that are not Party to the Protocol shall be substituted by another member (Article 13.3). The same rules for voting and membership of the Bureau apply to the Subsidiary Body for Scientific and Technological Advice and the Subsidiary Body for Implementation under the Protocol (Article 15.2/3). Despite the language of Articles 13.1 and 15.1 stipulating that the COP/SBSTA/SBI of the Convention "shall serve as" the respective bodies of the Protocol, the institutions established under the Kyoto Protocol are therefore separate bodies from those serving the Convention, since composition is different and decision-making is independent. The ambiguous language will create problems for the Secretariat, however, when it comes to budgeting and allocating costs.

18.1 Negotiating History

A majority of Parties did not want to establish new institutions, partly for reasons of "institutional economy" and partly to keep as much power as possible in the hands of the Convention's bodies. The first reason was advanced by industrialised countries, which would have to pay the main part of the administration and operational costs of any new institution. In addition to these financial reasons, however, there are also reasons relating to practicality and efficiency that mitigate against a proliferation of institutions on the international level.

Many important developing countries opposed the creation of independent new institutions in a protocol for another reason.[5] Their primary aim was to keep as much political control within the Convention' bodies as possible, since at the time of the negotiations, they could not be sure that they would be willing or able to join a protocol. Ratification of the Kyoto Protocol by major developing countries would have been very unlikely, for example, had the Kyoto Protocol included binding commitments for developing countries or some graduation mechanism that would eventually have subjected them to commitments. Developing countries thus preferred that the affairs of the Protocol be regulated mainly by the Convention's bodies so as to prevent an independent evolution of the Protocol.

The case of the ozone regime served as a warning to those developing countries, because the Montreal Protocol on Substances that Deplete the Ozone Layer of 1987 had quickly replaced the Vienna Convention on the Protection of the Ozone Layer of 1985 as the more important legal instrument. The Montreal Proto-

---

5 See the statement of Costa Rica on behalf of the Group of 77 and China in FCCC/AGBM/1997/MISC.1, p. 15; see also the proposal contained in FCCC/AGBM/1997/MISC.1/Add.2, pp. 71, 74.

col, which was dominated in its early days by industrialised countries, contained restrictions on trade with non-Parties that would have prevented access by non-member developing countries to controlled substances. This virtually forced a number of developing countries to join the Protocol with its binding phase-out obligations, even though this may not have been their priority.[6]

The first proposal for the main institutions was submitted by AOSIS in their protocol proposal, which advocated a mixture of existing and separate institutions.[7] Whereas the Secretariat and the subsidiary bodies of the convention were supposed to also serve the protocol, a new "Meeting of the Parties" was to be established, very much modelled after the approach of the Montreal Protocol. The US proposal, which took the wording of the Montreal Protocol as a blueprint ("The Parties shall hold meetings at regular intervals"), would have amounted to creating a new institution as well, without expressly saying so.[8]

It was the Union's proposal, however, that anticipated the agreement that prevailed in the end. Under the EU proposal, the COP of the FCCC should also serve as the COP of the protocol.[9] However, this proposal did not imply identical organs, since voting in this body was restricted to those Parties that were Parties to the protocol. Similarly, those members of the Bureau that were not Party to the protocol would have to be replaced by another Party. Except for the name of this new body, this proposal finally was incorporated in the Kyoto Protocol.

The consolidated negotiating text of 13 October 1997 by the Chairman still included both the AOSIS and the EU-option.[10] The final formula in the Kyoto Protocol, however, appeared already in the revised text under negotiation prepared by Chairman Estrada prior to COP 3. According to the proposed Article 14, "The Conference of the Parties, the supreme body of the Convention, shall serve as the Meeting of the Parties to this Protocol".[11]

Institutional questions did not represent a major issue in the negotiations in Kyoto. Both the provisions on the Secretariat and the Subsidiary Bodies were essentially agreed upon prior to COP 3. Differences in the "non-group" on institutions during Kyoto concerned the question of whether the "meeting of the Parties" should be written with a capital "M" or not. Disputes also arose over the question whether the COP should serve the Parties to the Protocol "as" the supreme body of the Convention, thus indicating a higher degree of authority. The basic principle, however, already incorporated in Article 17.5 of the FCCC, remained unchanged: that "decisions under any protocol shall be taken only by the Parties to the protocol concerned" and that the institutions of the Protocol thus could create their own

6 In this respect, the trade restrictions worked well to achieve near universal participation; see Brack 1996; Ott 1998a, pp. 67 et seq.

7 UN Doc. A/AC.237/L.23 and UN Doc. FCCC/AGBM/1997/3/Add.1, p. 77 (proposal 2).

8 UN Doc. FCCC/AGBM/1997/3/Add.1, p. 79 (proposal 3).

9 UN Doc. FCCC/AGBM/1997/3/Add.1, p. 77 (proposal 1).

10 UN Doc. FCCC/AGBM/1997/7, Article 14.

11 UN Doc. FCCC/CP/1997/2, Article 14.

"will" independent from the bodies of the FCCC. In the end, the wording as contained in the revised text under negotiation dating from November 1997 was kept and the issue effectively settled well before the end of COP 3. Consequently, the provisions on the institutions were accepted without discussion in the final night of Kyoto (see Chapter 7).

Figure 18.1: The Institutions of the Kyoto Protocol

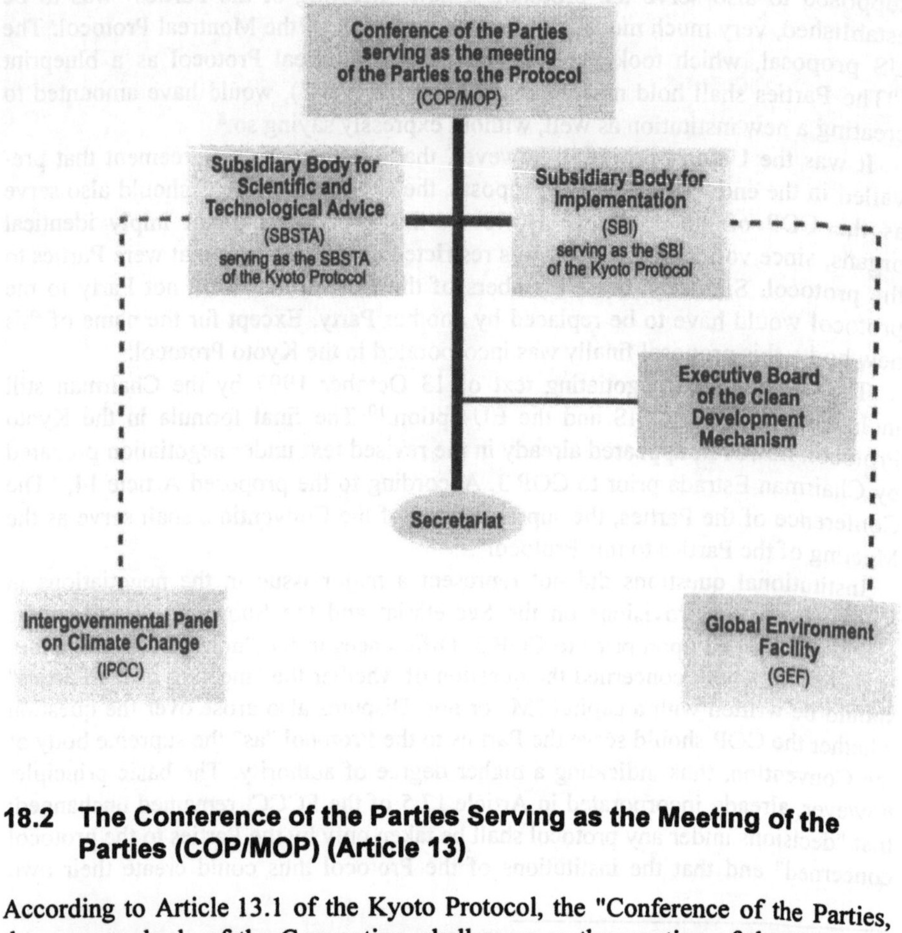

18.2 The Conference of the Parties Serving as the Meeting of the Parties (COP/MOP) (Article 13)

According to Article 13.1 of the Kyoto Protocol, the "Conference of the Parties, the supreme body of the Convention, shall serve as the meeting of the Parties to this Protocol". Although not expressly stated as in the Convention, the "Conference of the Parties serving as the meeting of the Parties" (COP/MOP), as this body is referred to throughout, is the supreme governing body of the Protocol. This is evidenced by the broad range of functions allocated to the COP/MOP and corresponds with the tendency in modern international environmental law to use plenary organs of this type as the highest decision-making body of a treaty.[12]

The COP/MOP of the Kyoto Protocol has been assigned administrative and legislative functions.[13] Like the COP of the Convention,[14] the primary function of the COP/MOP is to "keep under regular review" the implementation of the Protocol and to take, within its mandate, the "decisions necessary to promote its effective implementation" (Article 13.4). To this end, the COP/MOP shall assess the overall effects of the measures taken pursuant to the Protocol and the extent to which progress towards the objective of the Convention is being achieved (Article 13.4(a)). It shall, furthermore, periodically examine the obligations of the Parties in light of the Convention's objective and adopt regular reports on the implementation of the Protocol (Article 13.4(b) and Article 9).

The Conference of the Parties serving as the meeting of the Parties shall furthermore promote and facilitate the exchange of information on measures adopted by Parties for the implementation of their obligations (Article 13.4(c)) and facilitate the co-ordination of measures between two or more Parties (Article 13.4(d) and Article 2.4). Further functions of the COP/MOP include the development and periodic refinement of methodologies for the effective implementation of the Protocol (Article 13.4(e) and Article 5) and seeking to mobilise additional financial resources in accordance with Article 11.2 (Article 13.4(g)).

Additional tasks of the COP/MOP are spread throughout the Protocol's articles:

- the COP/MOP may promote the implementation of policies and measures in such a way as to minimise adverse effects (Article 2.3),
- the COP/MOP may consider ways and means to elaborate the co-ordination of policies and measures (Article 2.4),
- the COP/MOP shall take decisions on additional categories of sinks that might be included in the inventories at its first session or as soon as practicable thereafter (Article 3.4),
- the COP/MOP shall decide on the flexibility granted to CEIT countries, especially as regards the settling of the baseline for those countries that had not yet submitted their first national communications under Article 12 of the Convention at the time the Kyoto Protocol was adopted (Article 3.5, 3.6),[15]

12 See Ott 1998a, pp. 101 et seq.

13 There are not yet any explicit quasi-judicial functions, as exercised by the MOP of the Montreal Protocol, but Article 18 provides for a non-compliance procedure to be adopted at the first session of the COP/MOP and Article 8 paragraphs 5 and 6 contain rudimentary provisions for the exercise of quasi-judicial functions; see Chapter 16.

14 See Bodansky 1993, pp. 533 et seq.

15 Thus, formally a decision of the *COP/MOP* (not the COP) will be required, once the Kyoto Protocol has entered into force, regarding the acceptance of a "base year or base period other than 1990 for the implementation" of the commitment of those CEITs that want to use such a different baseline and submitted their first national communications after COP 3. As of mid-

- the COP/MOP shall initiate the consideration of commitments for subsequent commitment periods at least seven years before the end of the first period (Article 3.9),
- the COP/MOP shall consider what actions are necessary to minimise the adverse effects of climate change and/or the impacts of response measures on developing countries at its first session (Article 3.14),[16]
- the COP/MOP may further elaborate guidelines for the implementation of the article on Joint Implementation at its first session or as soon as practicable thereafter (Article 6.2),
- the COP/MOP shall determine the frequency of national communications under Article 7 and, at its first session, adopt guidelines for their preparation (Articles 7.3, 7.4),
- the COP/MOP shall establish guidelines for the in-depth review of national communications at its first session, consider the information received and take decisions on any matter required for the implementation of the Protocol (Article 8),
- as regards the Clean Development Mechanism, the COP/MOP shall provide guidance to its Executive Board, designate operational entities for the certification of emission reductions, elaborate, at its first session, modalities and procedures to ensure transparency, efficiency and accountability, and it shall ensure that a share of the proceeds is used to cover administrative expenses as well as to assist particularly vulnerable developing countries to meet the costs of adaptation to climate change (Article 12),
- the COP/MOP shall consider the application of the multilateral consultative process under Article 13 of the FCCC (Article 16), and
- the COP/MOP shall, at its first session, approve a non-compliance procedure (Article 18).

In addition, Article 13.4(j) contains a blanket clause requesting the COP/MOP to exercise such other functions as may be required for the implementation of the Protocol. These are standard in multilateral environmental agreements, so as to prevent challenges to the legal authority of supreme decision-making bodies.[17] If any evidence was needed that the COP/MOP indeed is the supreme decision-making body of the Protocol, this empowering clause would dissolve any remaining doubts.

The formulation "as may be required for the implementation of the Protocol" differs from that of Article 7.2(m) FCCC, which requests the COP to "exercise

---

1999, Slovenia was the only relevant case, since COP 4 granted it the right to use 1986 as its baseline under the Convention (Decision 11/CP.4). According to Article 3.5, the COP/MOP would have to accept this base year separately for the purposes of the Protocol.

16 On the controversial issue of "impacts of response measures" see Chapters 10 and 17.

17 Bodansky 1993, p. 534.

such other functions as are required for the achievement of the objective of the Convention...". This difference in wording is not a difference in function, but points to the fact that the Protocol itself has been adopted "in pursuit of the ultimate objective of the Convention as stated in Article 2".[18] To "exercise such other functions as may be required for the implementation of the Protocol" therefore requires the COP/MOP to perform those functions required for the achievement of the FCCC objective.

Article 13.4(j) of the Protocol furthermore requests the COP/MOP to "consider any assignment resulting from a decision of the Conference of the Parties". This clause does not force the COP/MOP to implement those recommendations or requests by the COP, but merely to "consider" its assignments. Any other arrangement would have been contrary to the position of the COP/MOP as an independent body that is not subject to the authority of the COP of the FCCC. However, a duty to consider the decisions of the COP, as the supreme body of its "mother treaty", is an appropriate task of the COP/MOP. Because of the overlap of membership it is, however, hardly conceivable that the COP/MOP would not act in line with the intentions of the COP in most cases.

Some of the functions mentioned above will in fact have to be tackled first by the COP of the Convention. The Parties at COP 3 did not establish any new body for the interim period until the entry into force of the Protocol. Instead, the COP of the FCCC (and its Subsidiary Bodies) will perform the functions of an *ad hoc*-body for the preparation of the first session of the COP/MOP of the Kyoto Protocol. In this capacity, the COP has already been requested by COP 3 to perform certain tasks in order to allow a "prompt start".[19] This is especially needed for those issues that should be decided at the first session of the COP/MOP. In 1998, the fourth Conference of the Parties thus established a workplan for the COP and the subsidiary bodies on preparations for the first COP/MOP of the Protocol.[20]

The COP furthermore adopted a workplan for the Kyoto Mechanisms with a view to taking decisions on all mechanisms by COP 6 in the year 2000 (see Chapter 24).[21] This refers most importantly to the Clean Development Mechanism (see Chapter 14). Article 12 on the CDM reflects the haste and exhaustion of negotiators in Kyoto. Whereas certified emission reductions obtained from 2000 to 2008 may be banked for subsequent use in the first commitment period (Article 12.10), the COP/MOP of the Protocol has been assigned the task to elaborate the specific

---

18 Preamble, second paragraph. On the "ultimate objective" of the FCCC see Chapter 3 and Sands 1992; Ott 1996a.

19 Decision 1/CP.3 in FCCC/CP/1997/7/Add.1. On the "prompt start" of the FCCC see Chayes/Skolnikoff 1992.

20 Decision 8/CP.4 in FCCC/CP/1998/16/Add.1. The requirement that the first COP/MOP take a decision on an issue does not necessarily mean that the issue will be removed from the agenda of the negotiations afterwards. Many of the issues require further or even constant (re-)consideration.

21 Decision 7/CP.4 in FCCC/CP/1998/16/Add.1.

rules required for the operation of the CDM (Article 12 paragraphs 4, 5, 7). Since the Protocol is unlikely to enter into force before the year 2000, however, waiting for the COP/MOP's decisions would result in the necessity to retroactively assign and validate the emission reductions – with all the political and technical uncertainties connected with this task. Correcting the mistake, COP 4 determined that the relevant rules will be elaborated by the Conference of the Parties to the FCCC. They will, however, subsequently have to be approved by the first session of the COP/MOP.

The instruments available to the COP/MOP for carrying out its functions mirror those of the COP and are similar to those available to supreme decision-making bodies in other treaties. The COP/MOP may make recommendations on any matters necessary for the implementation of the Protocol (Article 13.4 (f)), it may adopt amendments to the Kyoto Protocol according to Article 20 and it may adopt and amend annexes to the Protocol in accordance with Article 21. Ordinary working instruments of the COP/MOP are the "decisions", which have been employed to achieve far-reaching regime developments in various other modern environmental treaty systems.[22]

It was not possible to resolve at COP 3 the questions concerning the rules of procedure (most importantly the rule on voting) that have plagued the Convention's proceedings since COP 1. On the contrary, this stalemate has been effectively incorporated into the Protocol. According to Article 13.5, the rules of procedure of the COP[23] that are applied under the Convention shall be applied *mutatis mutandis* under the Protocol, except as may otherwise be decided by consensus by the COP/MOP. Until the COP resolves the differences regarding the rule on voting, the consensus requirement under the Convention will thus also govern the proceedings of the COP/MOP. There is a possibility, however, that the COP/MOP, with its different membership, will be in the position to take a consensual decision on the adoption of rules of procedure including some majority voting rule even without agreement in the COP, as allowed by Article 13.5.

Voting rights in the COP/MOP have been established in accordance with Article 17.5 FCCC (see Section 18.1): whereas the Parties to the Convention that are not Parties to the Protocol may participate as observers, decisions of the COP/MOP under the Protocol shall only be taken by those that are also Party to the Protocol (Article 13.2). Similarly, any member of the Bureau of the COP that is not Party to the Kyoto Protocol shall be replaced by another member of the same geographical group if the Bureau meets for the purposes of the COP/MOP (Article 13.3).

The rules of the Protocol concerning the frequency of ordinary meetings follow the well-established precedents of other multilateral environmental agreements and

---

22 Ott 1998a.

23 See FCCC/CP/1996/2 for the draft rules of procedure.

are modelled after the FCCC.[24] The COP/MOP will meet annually and in conjunction with a session of the COP under the FCCC, unless otherwise decided by the COP/MOP (Article 13.6). The first session of the COP/MOP will take place in conjunction with the first session of the COP following the entry into force of the Protocol.[25] Extraordinary sessions of the COP/MOP shall be held if the COP/MOP so decides or at the written request of any Party, provided that it is supported by at least one-third of the Parties within six months (Article 13.7).

The clause on participation of observers follows the example of the FCCC as well.[26] As it has been standard practice in the climate regime, not only "the UN, its specialized agencies, the International Atomic Energy Agency as well as any State member thereof or observer thereto" may participate in the proceedings of the COP/MOP as observers, but also all governmental and non-governmental organisations that are "qualified in matters covered by the Kyoto Protocol", unless at least one-third of the Parties objects (Article 13.8).[27] Those organisations that have been admitted as observers to the COP or the COP/MOP automatically have the right to observe also the proceedings of subsidiary bodies.

The practice of admitting observers to the bodies of the FCCC has been rather liberal and it is not to be expected that the COP/MOP will deviate from this. However, this rule regarding participation is only applicable to the proceedings of the COP/MOP and its subsidiary bodies. Substantial negotiations frequently take place in contact groups and so-called "non-groups", where access is more restricted. An important decision in this respect was adopted by COP 4, that observers could in principle participate in contact group meetings, unless at least one-third of the Parties present in the group object to such participation.[28]

18.3 The Secretariat (Article 14)

According to Article 14, the Secretariat of the Convention shall also serve as the Secretariat of the Protocol.[29] There were none of those disputes and differences over its scope and authority that had characterised the negotiations on the Secretariat of the FCCC.[30] The arrangements made for the Secretariat had proven to be workable and did not require any revision in principle. However, the functions of the Secretariat have evolved over the years and are no longer limited to those assigned to it under the text of the treaty. The Secretariat plays, for example, a central role in the in-depth review procedure under the Convention (see Section 16.1).

---

24 See Article 7.4 and 5 FCCC.
25 See Section 20.1 for the details of the entry into force requirements.
26 See Bodansky 1993, p.534.
27 On the role of NGOs in the FCCC process see Talaab 1998.
28 FCCC/CP/1998/16/Add.1.
29 As already indicated by Article 8.2 (g) FCCC.
30 See Bodansky 1993, p. 534.

Article 14.2 determines that the arrangements made for the functioning of the Secretariat under the Convention shall apply to the Protocol. Prior to the adoption of the FCCC, differences in view existed with regard to the question whether the Convention should be located with UNEP or some other organisation. The compromise finally reached deferred the ultimate decision on the Secretariat to the first session of the Conference of the Parties (Article 8.3 FCCC).

At COP 1, the Parties decided to link the Secretariat institutionally with the United Nations and in effect designated the INC Secretariat as the permanent Secretariat.[31] This arrangement was to be reviewed at COP 5, but not later than 31 December 1999. However, a major revision of the arrangement is unlikely to take place. Also at COP 1, the Parties decided to locate the permanent Secretariat in Bonn at the offer of the Government of Germany.[32] By the end of 1998, the Secretariat employed about 100 staff, including part-time members and consultants.[33]

According to Article 14.2 of the Kyoto Protocol, the functions of the Secretariat comprise those stipulated in Article 8.2 FCCC and other functions assigned to it under the Kyoto Protocol. The basic administrative functions therefore are the same for both the Convention and the Protocol, i.e. that the Secretariat will:

- make arrangements for sessions of the COP/MOP and its subsidiary bodies,
- compile and transmit reports submitted to it,
- facilitate assistance to Parties, especially developing country Parties, in the preparation of reports,
- prepare reports on its activities for the COP/MOP, and
- ensure the necessary co-ordination with the secretariats of other relevant international bodies (Article 8.2 FCCC).

Other functions of the Secretariat have been assigned to it by the Protocol. In particular, the Secretariat is called upon:

- to receive any notification by Parties that they have agreed to fulfil their commitments under Article 3 jointly and to inform the Parties of the terms of the agreement (Article 4.2), and
- to co-ordinate the expert review teams for the review of information submitted under Article 7 and to list those questions of implementation indicated in the reports of the review teams (Article 8.2 and 3).

This last function is one of the most important, since it provides the Secretariat with a relatively independent role as regards the future evolution of the Protocol. The provision effectively codifies the practice that had been developed under the Convention in previous years. Despite the opposition of some Parties, the in-depth

31 Decision 14/CP.1 in FCCC/CP/1995/7/Add.1.

32 Decision 16/CP.1 in FCCC/CP/1995/7/Add.1.

33 The Secretariat's mailing address: Climate Change Secretariat, P.O. Box 260 124, D-53153 Bonn, Germany; website: <http://www.unfccc.de>.

review procedure under the Convention *de facto* gave the Secretariat a dominant position in evaluating and analysing Parties' national communications and national climate policies (see Section 16.1). This function has now been formally incorporated into the Protocol and will further enhance and strengthen the position of the Secretariat.

18.4 The Subsidiary Bodies (Article 15)

According to Article 15.1, the subsidiary bodies under the Convention shall also serve as subsidiary bodies of the Protocol and the same provisions relating to their functioning shall apply. The Kyoto Protocol thus contains provisions for a "Subsidiary Body for Scientific and Technological Advice" (SBSTA) and a "Subsidiary Body for Implementation" (SBI). Sessions of those bodies shall be arranged in conjunction with the subsidiary bodies of the Convention.

The rules for voting and the composition of the Bureau follow the principles established for the COP/MOP (see Section 18.2 above): Parties to the Convention that are not Parties to the Protocol may participate in the proceedings as observers, but decision-making on matters concerning the Protocol is restricted to those that are Parties to the Protocol (Article 15.2). Members of the Bureau of the Convention's bodies that are not Party to the Protocol will be replaced by other members of the same regional group (Article 15.3).

The Kyoto Protocol does not enumerate the functions of the subsidiary bodies as the Convention does (Articles 9 and 10 FCCC). Nor does it refer explicitly to the respective provisions of the Convention as in the case of the Secretariat's functions (see Section 18.3 above). One possible explanation may have been the objective to avoid touching on the complicated issue of the division of work and competence between the Convention's bodies. From their inception, the work of the two subsidiary bodies has been characterised by a certain degree of overlap and by a lack of co-ordination, that was not anticipated by the provisions of the Convention (see Section 3.2). A fragile balance has developed in practice. Any attempt to formally incorporate the evolving division of tasks of the subsidiary bodies to the Kyoto Protocol would thus probably have triggered a complicated and time-consuming debate in the negotiations. As a result, the "muddling through" practised since COP 1 continued. The Parties at COP 3 requested the Chairmen of the subsidiary bodies to make a joint proposal to the 8th sessions of those bodies, that took place in summer 1998, on the allocation of preparatory work to enable the first session of the COP/MOP to accomplish the tasks assigned to it by the Protocol.[34] COP 4 then took corresponding decisions (see below).

Under the Convention, SBSTA (Article 9) and SBI (Article 10) are plenary organs like the COP, open for all Parties and observers from other states, international organisations and NGOs. Their main functions are outlined in the Conven-

---

34 Decision 1/CP.3, para. 6 in FCCC/CP/1997/7/Add.1.

tion and have been elaborated on by later decisions of the COP.[35] The role of the SBSTA under the Convention is to provide the scientific and technological input into the COP's decision-making process. In this capacity, it has summarised scientific findings by the IPCC, considered scientific, technical and socio-economic aspects of the in-depth review of national communications and dealt with methodological issues. The SBI has an advisory role with regard to the review of national communications, the effects of measures agreed on by the Parties, the Financial Mechanism, technology transfer and adequacy of commitments. One of its main functions is to assist the COP in making arrangements concerning the Financial Mechanism (Article 11 FCCC) and the institutional linkages with the United Nations.

Some specific tasks have been assigned to the subsidiary bodies by the substantive provisions of the Protocol. The SBI is mandated to assist the COP/MOP in the consideration of the information submitted by Parties under Article 7 of the Protocol on reporting and national communications and the reports of the expert review teams under the review procedure (see Section 16.1). It is furthermore called upon to consider the questions of implementation listed by the Secretariat or raised by Parties (Article 8.5). The SBSTA may be invited to assist in this task "as appropriate", if questions of a scientific or technological character should arise.

In addition, SBSTA has been assigned the following tasks:

- to consider the data submitted by Parties prior to the first COP/MOP to establish their level of carbon stocks in 1990 and to make an estimate of their changes in carbon stocks in subsequent years (Article 3.4),
- to provide advice to the COP/MOP on modalities, rules and guidelines that should govern the inclusion of sinks in the Parties' inventories (Article 3.4),
- to provide advice to the COP/MOP on the revision of methodologies and adjustments for estimating anthropogenic emissions of GHGs, based on the work of the IPCC (Article 5.2), and
- to provide advice to the COP/MOP on the revision of global warming potentials of GHGs (Article 5.3).

As a sign of the enhanced role of the IPCC in the climate regime, the SBSTA shall act as an intermediary between the IPCC and the COP/MOP for most of these functions. In the Convention, the IPCC is only mentioned in the "interim arrangements" provisions in Article 21 due to the fierce resistance of some Parties. Despite the strengthened role of the IPCC accepted by the Kyoto Protocol, the challenge for SBSTA will remain to live up to its role as a true liaison between the primarily scientific IPCC and the political COP/MOP.[36]

---

35 On the negotiating history see Bodansky 1993, pp. 535 et seq.
36 See Bodansky 1993, p. 536.

COP 4 allocated further work to both subsidiary bodies. Most importantly, the COP adopted the work programme for the Kyoto Mechanisms[37] and decided on the allocation of preparatory work for the first session of the Protocol's COP/MOP.[38] Taking account of their different character, the SBI is mainly responsible for procedural and institutional questions, whereas SBSTA will deal with methodological and technical matters. Reflecting the continuing difficulty of dividing up work between those bodies, both will be involved in the consideration of the more general and basic issues. To avoid subordination of one body to the other, joint sessions have been held regularly post-Kyoto. Since the AGBM has now concluded its work, these joint sessions have become a plenary body for the Parties to consider general political issues between the sessions of the COP.

18.5 *Ad Hoc* Bodies

At Kyoto, the Parties did not establish any temporary subsidiary organs. However, Article 7.2(i) of the Convention and Article 13.4(h) of the Kyoto Protocol empower the COP and the COP/MOP respectively to "establish such subsidiary bodies as are deemed necessary for the implementation" of the Convention/Protocol. On this basis, COP 4 established a "joint working group on compliance" under the SBI and SBSTA. The establishment of an independent working group under the direct authority of the COP was resisted by the G-77 and China. As a compromise, the Parties agreed on a joint working group under the two subsidiary bodies, which is to report through those bodies to the fifth session of the COP.

The joint working group established by COP 4 is confined to considering various elements of a compliance regime:

- identifying compliance-related elements in the Kyoto Protocol,
- following the development of such elements in various negotiating circles and identifying gaps,
- developing ideas to the extent that they are not being addressed by other groups, and
- ensuring a coherent approach to developing a comprehensive compliance system.

37 Decision 7/CP.4 and Annex in FCCC/CP/1998/16/Add.1.
38 Decision 8/CP.4 and Annex I in FCCC/CP/1998/16/Add.1.

Upon the report of the joint working group on these issues, COP 5 might establish an *ad hoc* working group on compliance. This group would be in a position to formulate the draft terms of reference of such a mechanism, with a view to COP 6 adopting a decision on it.[39]

---

[39] See Decision 8/CP.4, Annex II in FCCC/CP/1998/16/Add.1; see Section 16.3.

19. Review, Development and Amendment of the Kyoto Protocol (Articles 3.9, 9, 20, 21)

Provisions on regular review and assessment are amongst the most important for the dynamic evolution of international treaties.[1] These programme norms[2] therefore are increasingly used in international environmental agreements[3] in order to regularly adjust the rules to new scientific findings, technological and economic developments or enhanced concern about environmental degradation. They furthermore can provide windows of opportunity for adjusting treaty rules to changing political circumstances and serve to keep up the regulatory pressure on laggard states. This increasing importance is reflected by the fact that these clauses are frequently only agreed on at the end of the negotiating processes as part of a package deal.

The provisions on review and assessment have to be analysed in connection with the rules governing the amendment of the treaty, since any adjustment of the treaty rules following from the review is bound to make use of these. It is the combination of the review process with the rules on decision-making and the procedures for amendment that eventually determines the ability of a treaty to develop dynamically and flexibly in line with changing circumstances.

The Kyoto Protocol contains explicit rules for both the future review of its substantive regulations and its amendment. On both accounts, it is not as progressive as other modern environmental treaties. Before and in Kyoto, most attention of delegations and observers focused on what became the review clause of the Protocol in Article 9. The negotiating history presented in the following section therefore focuses on this Article (Section 19.1). Subsequently, the Articles governing review and development (Section 19.2) and amendment of the Protocol and its Annexes are addressed (Section 19.3). The chapter concludes with an assessment of these rules of the Protocol that will impact decisively on its future development.

19.1 Negotiating History

The main provision on reviewing the Protocol's adequacy remained contentious nearly until the end of the negotiations, although the final formulation was already contained in the draft Protocol text issued by Ambassador Estrada as Chairman of the Committee of the Whole on 7 December.[4] AOSIS, in particular, assisted by

---

1 See in more detail Ott 1998a, p. 270; see also Gehring 1994.

2 Széll refers to these clauses as "markers", see Széll 1995, p. 99.

3 See Article 6 of the Montreal Protocol on Substances that Deplete the Ozone Layer of 1987 and Article 15.7 of the Basel Convention on the Transboundary Movement of Hazardous Waste and its Disposal of 1989.

4 FCCC/CP/1997/CRP.2.

environmental NGOs, continued to push for a clause that would allow an early review of the commitments.

The formulation of Article 4.2 (d) of the Convention was widely seen as providing for a strong review procedure for the Convention, because it contained concrete deadlines for such reviews to take place. Building upon this experience, the European Union, Switzerland, and Canada proposed a "review of the adequacy of commitments" also for the emerging Protocol.[5] These proposals, and that of AOSIS, based the review on the "best available scientific information and assessment of climate change", an implied reference to the IPCC as the most authoritative source of information.

In contrast, the US proposal placed strong emphasis on the "evolution" of commitments, which, for the US, meant involving developing countries as part of the "global solution".[6] As regards the general review, the US proposed a "periodical" unspecified review of the Protocol in light of the "evolving scientific knowledge". Japan did not submit a formal proposal, but merely asked for "a mechanism to regularly review" the Protocol. The Australian proposal on "review of commitments", on the other hand, was very detailed, with the emphasis placed firmly on a revision of individual Parties' obligations due to difficulties in implementation and economic factors. This proposal thus aimed at a possible relaxation of existing commitments, rather than strengthening them in the future, arising from Australia's fear that it would face special difficulties in meeting any quantified target because of its economic situation.

Mirroring the insistence of the US and other JUSSCANNZ-members on a process for the "evolution" of commitments, the G-77 proposal avoided the term "adequacy" and "commitments" altogether. Instead, the review was supposed to be restricted to the "implementation of the Convention and any related legal instrument". This formula aimed at averting the discussion of developing country commitments and placed the emphasis on implementation of existing obligations of industrialised countries. In line with its general objective to avoid the emergence of an institutional structure under the Protocol that would become separate from that of the Convention (see Chapter 18), the G-77 furthermore insisted on the review being carried out by the Conference of the Parties to the Convention under Articles 7.2 and 4.2 (d) FCCC.

The "consolidated negotiating text by the Chairman" of October 1997 followed the general approach of the EU and called for a periodical review of the adequacy of commitments of Annex I Parties, without specifying at which session of the COP/MOP such review was to take place.[7] The final "revised text under negotiation" of November 1997 contained a formula that combined the review of the

---

5 These and the following proposals are contained in the Negotiating Text of the Chairman of 22 April 1997, FCCC/AGBM/1997/3/Add.1. Attribution of the proposals to individual Parties or groups of Parties by the authors.

6 See the US non-paper of December 1996, FCCC/AGBM/1996/MISC.2/Add.4, p. 35.

7 FCCC/AGBM/1997/7.

implementation of the Protocols obligation – as requested by the G-77 – and that of the adequacy of commitments.[8] The text that the Chairman presented in Kyoto, however, avoided both terms ("implementation" and "adequacy of commitments") altogether and provided for an unspecified "review" only.[9] This text was eventually adopted in the final plenary at Kyoto without discussion. It was part of the compromise package that included provisions regarding the consideration of commitments in future commitment periods.

As regards amendments of the Protocol that might follow from its review, there was little hope that Parties would be able to agree on a procedure that would facilitate and support a dynamic and flexible development of the treaty. The European Union and AOSIS submitted a proposal for amending the treaty by two-thirds majority decision-making, as has, for example, been laid down in the Vienna Convention for the Protection of the Ozone Layer (Article 9.4) and in the Basel Convention on Hazardous Wastes (Article 17.4) with regard to amendments to its Protocols.[10] However, none of the submissions provided for circumventing the lengthy ratification procedure for entry into force of any agreed changes, as it has been successfully employed by the Montreal Protocol.[11] Given the experience with the negotiations of the FCCC, where similar attempts also failed, it was quite obvious that these proposals would be rejected by Parties that were either opposed to the treaty's dynamic development or feared that decisions impacting on their economic well-being might bind them against their will.[12] As a result, the amendment procedure of the Protocol was not a major point of discussion in the final negotiations. The rules eventually agreed follow a majority rule employed by most environmental treaties and require that an amendment of the Protocol be ratified by three-fourths of the Parties before it enters into force (for those Parties that have ratified it).

19.2 The Rules Applying to Review and Development

Although the term "adequacy of commitments" does not appear in the Kyoto Protocol in its final form, it contains several provisions that provide for the general review of commitments. Article 9, whose development has been described above, provides for a general review process with a rudimentary schedule. Furthermore, Article 3.9 requests the Parties to "consider" commitments for subsequent periods. Finally, Article 13 requests the COP/MOP to keep the implementation of the Protocol under regular review and, in Article 13.4(b), to "examine" the obligations of Parties. These provisions are dealt with in the following paragraphs.

8 FCCC/CP/1997/2.

9 FCCC/CP/1997/CRP.2.

10 FCCC/AGBM/1997/3/Add.1, p. 86 (proposal 2, attribution by the authors).

11 See Ott 1998a, pp. 155 et seq.

12 See for example the extremely tight rules proposed by OPEC countries in FCCC/AGBM/1997/3/Add.1, pp. 86/87 (proposal 3, attribution by the authors).

According to Article 9, the COP/MOP is called upon to "periodically review" the Kyoto Protocol and to "take appropriate action" if so required. The first review shall take place at the second session and further reviews "at regular intervals and in a timely manner". The timing of further reviews will thus be in the discretion of the COP/MOP. Article 9 in its final form falls short of the demand of the EU and many environmental NGOs; it is, however, considerably stronger than the text submitted by the US. Furthermore, although this provision does not expressly refer to the "adequacy of commitments", it is clearly not meant to be a pure "implementation review" as demanded by G-77 countries (see above, Chapter 19.1). The requests of the G-77 are to some extent reflected, however, in the call for coordination of reviews of the Protocol with "pertinent reviews under the Convention".

The requirement that the first review shall take place not earlier than at the second session of the COP/MOP was the outcome of a political compromise. This review could well have been conducted by the first session of the Protocol's supreme organ, if the Convention's bodies had been assigned the task to prepare it. However, such "prompt-start" arrangements were only agreed with respect to the "Kyoto Mechanisms", sinks and the pending non-compliance procedure. At least the first review will take place according to a fixed schedule. Unlike similar provisions in other treaties, the Kyoto Protocol does not require the COP/MOP to decide on the intervals for any reviews thereafter but leaves it to the Parties to adopt a schedule.

Since the review shall be based on the "best available scientific information and assessments on climate change and its impacts", the IPCC, the most important and authoritative source of such information, is set to retain its prominent role as the provider of pertinent information and the engine for the regime's progress. In addition, the review shall be based on "relevant technical, social and economic information". The call for "relevant" rather than "best available" information (as in the case of science) information implies that other actors beyond the IPCC may provide related input on such matters as economics and social sciences.

A second review procedure is prescribed in Article 13.4(b) of the Kyoto Protocol, on the functions of the COP/MOP. With almost the same wording as Article 7.2 of the Convention, the Protocol requests the Parties to "periodically examine the obligations of the Parties", taking into account the objective of the FCCC, the experience gained in its implementation and the evolution of scientific and technological knowledge. The provision does not request the review of the "institutional arrangements", as the Convention does, largely because the Protocol, in contrast to the Convention, does not contain any interim arrangements (cf. Article 21 FCCC). Finally, as in the case of Article 9 dealt with above, the COP/MOP shall give due consideration to the reviews executed under the Convention.

A procedure for establishing further obligations for a second commitment period is contained in Article 3.9 of the Protocol. This clause originally started out as an amendment procedure for the inclusion of greenhouse gases not yet controlled.[13] Such a provision was no longer required after the adoption of the six gases approach, since the most pertinent gases are now covered. Towards the end of the negotiations, it became apparent that agreement beyond the first commitment period could not be expected and a provision was required to account for this failure. Article 3.9 thus mandates the COP/MOP to start consideration of further obligations for Annex I Parties at least seven years before the end of the first commitment period, which is 2005. Depending on when the Kyoto Protocol enters into force, it may well be that the first review of the Protocol under Article 9 and the consideration of further obligations under Article 3.9 will be conducted concurrently.

19.3 Amendments to the Protocol and its Annexes

The Kyoto Protocol follows the UNFCCC in its rules for adopting amendments as well as adopting new annexes and amendments to new annexes (cf. Articles 15 and 16 FCCC). Before an amendment to the Protocol can be considered, the proposed text has to be communicated to the Parties through the Secretariat at least six months prior to the meeting at which it is to be adopted (Article 20.1). The language on the decision-making procedure for adopting amendments follows the traditional formula that the "Parties shall make every effort to reach agreement on any proposed amendment ... by consensus" and "that, if all efforts at consensus have been exhausted, and no agreement reached, the amendment shall as a last resort be adopted by a three-fourths majority vote...".

In practice there have only been a few cases where Parties to an international environmental agreement have actually voted. On the contrary, in almost all cases Parties prefer to call a decision "consensual", even if a minority of Parties has expressed its opposition.[14] In the context of the FCCC, where in the absence of any express provision decisions have to be taken by consensus, this "consensus" aproach happened once when the Parties at COP 2 "took note" of the Geneva Declaration despite the protest of 14 OPEC countries with Russia (see Section 4.5). After the adoption of an amendment, the Kyoto Protocol requires ratification by three-fourths of the Parties for it to enter into force. This is often the slowest part of international law-making, since it can take many years until the required number of states completes the cumbersome process of getting a proposed ratification through their own legislative bodies.[15]

Annexes are an integral part of the protocol (Article 21.1) and *new* Annexes are confined to "lists, forms and any other material of a descriptive nature that is of a scientific, technical, procedural or administrative character". The procedure for their adoption and subsequent amendment follows the rules for the adoption of

13 FCCC/AGBM/1997/7, Article 3.15.

14 For an example of "consensus minus one" in the ozone regime see Werksman 1996.

15 On this problem of the "slowest boat rule" and possible remedies see especially Sand 1990.

amendments of the Protocol itself according to Article 20; proposals have to be submitted at least six months prior to any meeting and they can be adopted by a three-fourths majority. Afterwards, the procedure deviates from this route. The new annexes or amendments to them enter into force automatically for all Parties six months after adoption, except for those Parties that register a formal objection with the Secretary-General of the United Nations as the Depositary of the Protocol (see Article 21.5). The adoption and amendment of annexes thus follows the traditional "opting-out" procedure employed by a great number of treaties and organisations.[16] This procedure reverses the "burden of activity" and thus provides an effective instrument for accelerated rule-setting in technical and administrative matters.

These rules are, however, not applicable to the existing Annexes of the Kyoto Protocol, both of whom were adopted in Kyoto: Annex A lists the GHGs that are subject to regulation under the Protocol, whereas Annex B specifies the quantified emission limitation and reduction commitments of industrialised countries under the Protocol. The procedure for amending these Annexes follows the general procedure for amendments of the *Protocol* set out in Article 20: after adoption, ratification by three-fourths of the Parties is required for the amendment to enter into force for those Parties that have ratified. In addition, due to the special features of Annex B, Article 21.7 stipulates that any amendment to it shall only be adopted with the written consent of the Party concerned. This ensures that new or modified targets in Annex B will have received the express governmental approval of the Party concerned before submitted for ratification to the national parliaments.

This in fact is a return to "unanimity" in voting and is a deviation from the now generally accepted consensus rule – a further indication of the importance attached to these obligations by all Parties. At the same time, this procedure represents an admission control for potential new members of the "club". Decision-making by a three-fourths majority ensures that any Party not yet listed in Annex B of the Protocol that wishes to take on a binding quantified target (e.g. in order to be able to participate in Emissions Trading) needs the consent of a large majority of the Parties to do so.

19.4 Assessment and Outlook

Since international environmental law – and certainly the issue of climate change – is characterised by fast changing scientific and political parameters,[17] numerous devices for a flexible and dynamic law-making process have been elaborated within a variety of international treaties.[18] The review of the adequacy of commitments and the rules for amending the treaty provisions are two of the most

16 See in general Ott 1998a, pp. 160 et seq.; Sand 1990.

17 See Contini/Sand 1972, p. 38; Bodansky 1993, p. 549 for the climate regime.

18 For a general treatment of this issue see Ott 1998a, pp. 145 et seq.; see also Sand 1990, pp. 14 et seq.; Palmer 1992, pp. 270 et seq.

important aspects for ensuring such a flexible and dynamic law-making process. Various examples of modern dynamic international environmental agreements have proven the importance of prescriptions for review that are specified in terms of date and subject as well as of rules that facilitate adjustment of a treaties rules to changing circumstances.

For example, the specified provisions contained in the FCCC regarding the first review of adequacy of commitments greatly helped launch the Kyoto process: according to Article 4.2(d) FCCC the first Conference of the Parties had to review whether the commitments under Article 4.2(a) and (b) were adequate to reach the Convention's objective.[19] This review highlighted the inadequacy of those measures that had been taken by industrialised countries: nine Parties projected increased CO_2 emissions in 2000 compared to 1990 levels if no additional measures were taken.[20] The review thus served as the basis for the Berlin Mandate that led to the adoption of the Kyoto Protocol two and a half years later.

The most advanced blueprint for both a specified review and a flexible amendment procedure has probably been elaborated by the negotiators of the Montreal Protocol on Substances that Deplete the Ozone Layer. Its Article 6 determines that the control measures shall be assessed "at least every four years" with input from appropriate expert panels under the Protocol. As a result, *existing* control measures may be strengthened by a two-thirds majority representing majorities of both developing and industrialised countries. Such adjusted control measures become automatically binding on all Parties. By contrast, the introduction of new control measures (i.e. controls of new substances) requires ratification (see Article 2.9 and 10 of the Montreal Protocol). On this basis, the Montreal Protocol has been strengthened flexibly four times since 1987.[21]

In comparison, the design of the relevant provisions of the Kyoto Protocol is far less conducive to dynamic development. Possibly because of its successful operation in the ozone regime, procedures similar to the ones of the Montreal Protocol were not seriously considered in the climate regime. Instead, rather unspecific language was used as regards the content, goal and (apart from the first review) timing of future reviews. Furthermore, the entry into force of new binding obligations under the Kyoto Protocol will require a lengthy and cumbersome ratification process.

Thus, the dynamic development of the Kyoto Protocol will largely depend on the existence of suitable political framework conditions. Whether the COP/MOP of the Protocol will review the adequacy of commitments in regular intervals will depend on strong advocacy of leading groups of Parties. Furthermore, the flexible

---

19 This objective is spelled out in Article 2, i.e. the stabilisation of greenhouse gas concentrations in the atmosphere at a level which would prevent dangerous anthropogenic interference with the climate system, see Chapter 3.

20 UN Doc. A/AC.237/81; see also the review by environmental NGOs: CNE/USCAN 1995; on the impact of the review see Oberthür/Ott 1995.

21 See Benedick 1998a; Oberthür 1997; Ott 1998a.

development of the regime's rules will only be possible if political conditions in major countries allow for quick ratification of amendments. Alternatively, Parties may resort to a development of the rules by way of normal decisions of the COP/MOP, including concerning the interim application of adopted amendments – a route that has been pursued successfully in a number of other modern international environmental agreements.[22]

22 See Ott 1998a.

20. Final Provisions of the Kyoto Protocol (Articles 22–28)

The final provisions of the Kyoto Protocol on signature, ratification and entry into force generally follow the established practice under international law. Therefore, no major disputes arose on the relevant articles of the Protocol with the exception of Article 25 on the entry into force of the Protocol. This is dealt with in more detail in Section 20.1. Subsequently, the provisions on voting rights, the Depositary of the treaty, reservations and withdrawal are presented (see Section 20.2).

20.1 Signature, Ratification and Entry into Force (Articles 24, 25)

The provisions on signature, ratification and entry into force of the Kyoto Protocol give the impression of a carefully drafted procedure that is designed to stretch out the process of entering into force. The reason behind this may be the unfinished nature of the Kyoto agreement,[1] which will take some time to be fleshed out. Stretching out the process, however, may also serve to distract from slow substantive progress. Article 24 of the Protocol thus provides for a signature period of one year, beginning on 16 March 1998 through to 15 March 1999. 84 Parties to the Convention signed the Protocol during that year, among them all major GHG emitters. This act already imposes legal obligations: under general international law, the signatories are obliged to at least not obstruct the spirit and the entry into force of the Kyoto Protocol; according to Article 18 of the Vienna Convention on the Law of Treaties (1969), a State is obliged to refrain from acts that would defeat the object and purpose of the treaty after it has signed it.

Since 15 March 1999, the Protocol has been open for ratification by Parties to the Convention. The ratification process is more cumbersome and has been designed so as to require participation by all major polluters. Therefore, Article 25 of the Protocol establishes a rather high threshold and employs a "double-trigger". It not only requires ratification by 55 Parties to the Convention, but sufficient Annex I Parties must ratify so as to represent at least 55% of the total CO_2 emissions of Annex I Parties in 1990. In order to avoid uncertainty as to the exact numbers, the authoritative data is that which has been communicated by the Parties in their first national communications according to Article 12 of the FCCC.[2]

1 This character is reflected by the great number of articles carrying "unfinished" in their titles, see Ott 1998b; Jacoby et al. 1998 and Schneider 1998.

2 According to some calculations, actual emissions in 1990 were much higher; see Greenpeace: Analysis of the Kyoto Protocol, Greenpeace 1998, p. 14.

Table 20.1: Emission Percentages of Parties or Groups of Parties

Party / Group of Parties	% of Annex I emissions in 1990
EU	24.2
CEITs (without Russia)	7.4
EU+Russia+Japan+CEITs	57.5
Russia	17.4
JUSSCANNZ	50.8
USA	36.1
Japan	8.5
Canada	3.3

Source: FCCC/CP/1997/7/Add.1, Annex. Note that since Ukraine had not submitted a first communication, its emissions in 1990 of roughly 5% of Annex I Parties are not considered for ratification purposes.

As can be seen in Table 20.1, the US as the major emitter represented roughly 36% of global CO_2-emissions in 1990, the EU Member States cumulatively emitted some 24% and Russia accounted for 17.4%. The next biggest emitters are Japan with 8.5% and the Central and Eastern European countries together representing 7.4%. Thus, the US on its own is just 9% short of blocking the entry into force of the Kyoto Protocol, it can effectively bring the ratification process to a standstill if it allies itself either with either Russia (53.5%) or the rest of JUSSCANNZ (nearly 51%). If the EU wanted to bring into force the Protocol without the US, it would effectively need to win the ratification by both Russia and Japan, in addition to central and eastern European countries (57.5%).

The "double-trigger" ensures that the Protocol only enters into force once a certain "critical mass" is reached to support its effectiveness. Since mitigation of climate change is considered by almost all Parties as costly and creating a comparative disadvantage on global markets, the "double-trigger" serves the purpose of preventing or at least limiting the opportunities for such "free riding".

While there is logic to the "double-trigger" requirement, there was nothing that forced negotiators to opt for the threshold of 55%. Indeed, there were various options of the entry into force provision on the negotiating table until the final night of the negotiations. The Chairman's draft contained the formula of 55/55. A last minute proposal by the US suggested to lower the threshold to 50% of Annex I emissions. Although this US proposal also included a slightly higher number of required ratifications for the entry into force (65–70), it would have diminished the US's position of possessing near-veto power. The European Union was apparently unwilling to assume its obligations without the US signing on. In fact, the UK did not support the US proposal and no other member of the Union grasped this opportunity to increase the chances for an early entry into force. Canada, on the other hand, proposed a threshold of 65%, whereas Japan and Australia offered a compromise at 60%. Finally, Chairman Estrada followed his own rules and reverted to his original proposal since there was no consensus to change his text.[3]

The 55/55 formula finally agreed confers substantial power upon the US. This poses problems for the entry into force of the Protocol because of the internal political battles between the (Democratic) President and the (Republican dominated) Congress. Ratification by the US-Senate requires a two-thirds majority. This has already proved an insurmountable barrier to US ratification of several international environmental treaties, including the UN Convention on Biological Diversity of 1992 and the Basel Convention on the Transboundary Movement of Hazardous Waste and Its Disposal of 1989. Ratification of the Kyoto Protocol by the US, therefore, cannot be expected without major changes in the current composition of the Senate. However, there are political options for an entry into force of the Kyoto Protocol without participation of the US (see Chapter 25).

20.2 Voting Rights, Depositary, Reservations and Withdrawal (Articles 22, 23, 26, 27 and 28)

Voting rights under the Kyoto Protocol follow the well established rules of international law and Article 22 repeats the language of the Convention. According to its paragraph one, each Party has one vote, except in the case of regional economic integration organisations. This provision so far concerns only the European Community, which may ratify the Protocol, but may exercise its right to vote only if none of the member states exercises its right. If the Community decides to vote, it can do so with a number of votes equal to the number of their Member States that are Party to the Protocol (Article 22.2).

Article 23 determines that the Secretary-General of the United Nations is the Depositary of the Kyoto Protocol, as in the case of the Convention. He or she receives notifications of ratification, withdrawal and non-acceptance of new annexes or amendments of annexes. The Depositary also receives declarations on the extent of competence of regional economic integration organisations with respect to the matters governed by the Protocol under Article 24.3, and subsequently has to inform the Parties. Article 28 determines, in line with the general practice, that the text of the Kyoto Protocol is "equally authentic" in any of the six official languages of the United Nations.[4]

As is common practice in modern international environmental agreements, the Kyoto Protocol does not allow Parties to enter reservations (Article 26).[5] The

---

3 Authors' observations.

4 Arabic, Chinese, English, French, Russian and Spanish.

5 See also Article 18 Montreal Protocol (1987); Article 26 Basel Convention (1989) Article 37 Biodiversity Convention (1992). The definition of a "reservation" is contained in Article 2.1(d) of the Vienna Convention on the Law of Treaties (1969).

objective of this prohibition is usually to provide for uniform obligations among the Parties and to prevent "free riding".[6] However, the Protocol does not prevent Parties from elaborating "interpretations" annexed to the ratification document, a usage increasingly common and already extensively used by Parties when they signed and ratified the UNFCCC. These declarations do not have a legal effect and therefore do not change the legal relationships of Parties.

Withdrawal from the Protocol according to Article 27 is allowed after three years from the date it entered into force for a Party. The withdrawal will take effect one year after notification of this request to the Depositary. This time lag is meant to provide for a certain stability of expectations among the Parties, since the seriousness of ratification is underscored by the fact that immediate withdrawal from a treaty is prohibited. In comparison to the length of the commitment period (five years), the effectiveness of a withdrawal one year after its notification appears to be rather short to support much stability of expectations. The right of withdrawal is, however, generally seldom used in international environmental relations. Whereas a Party is free to withdraw from the Protocol but stay within the FCCC, a withdrawal from the Convention at the same time constitutes withdrawal from the Kyoto Protocol (Article 27.3). This mirrors Article 24.1 which allows only Parties to the Convention to sign and ratify the Protocol. Since the Protocol in many ways refers to the "mother treaty", being a Party to the Kyoto Protocol alone would create serious difficulties.

6 Bodansky 1993, p. 552.

PART III
Conclusions and Outlook

Regarding its genesis as well as its design, the Kyoto Protocol is quite unique in international environmental affairs. It has set the stage for the international climate policy of the next century. The dynamics of the Kyoto process have been fascinating for everyone involved. In many respects, the Kyoto Protocol remains a work in progress: it represents the end of a process as well as the beginning of a new one. It contains a number of innovative provisions (e.g. on Emissions Trading and the Clean Development Mechanism) that will have to be elaborated and tested in the coming years before any final assessment will be possible. Adding to this uncertainty, the Protocol mirrors the many diverse interests that needed to be reflected in the final agreement. Therefore, large parts of the text of the Kyoto Protocol remain ambiguous.

Nevertheless, the last part of this book makes an attempt to assess the Protocol, explore the challenges of the future and cautiously discuss some steps that might be helpful for the progressive evolution of the still nascent climate regime. Chapters 21 and 22 evaluate the Kyoto Protocol and the political process that led to its adoption. Chapter 23 explores the potential of international linkages and synergies, i.e. the relationship between the evolving international regime on climate change and some of the other international institutions in the field of the environment and beyond. With the growth of the climate regime, these linkages have increased considerably in importance. This is destined to continue in the future. Chapter 23 thus digs into one of the emerging fields of future climate policy and international environmental policy in general. Chapter 24 then describes the political landscape on the international as well as the national and regional levels after Kyoto and COP 4 in Buenos Aires. And finally, Chapter 25 takes a more political viewpoint and, drawing on the experience of the authors as experts and practitioners, outlines some elements of a Leadership Initiative that might help to create new momentum for the international process in the first years of the next century.

21. Lessons from the Kyoto Process

Apart from the fascinating details of the Kyoto process, a number of more general lessons can be derived from the negotiating process. These are related to the importance of leadership, including the ability of the European Union to exert such leadership in international environmental affairs (Section 1.1), and the outstanding role of certain situational factors in bringing about the final agreement (Section 1.2). Furthermore, the Kyoto process provides clear signs of the evolving role of climate policy in world politics on the threshold to the 21st century as well as of the intergovernmental and non-governmental character of politics in this field (Section 1.3).

21.1 The Importance of Leadership

A number of scholars dealing with the formation and evolution of international environmental regimes have stressed the importance of leadership for successfully concluding the respective negotiating processes.[1] Leadership is particularly needed in the early stages of the development of environmental regimes, when a stable institutional framework for developing the rules and procedures of the regime is absent. The Framework Convention on Climate Change of 1992 did establish a basic institutional framework. This, however, was not mature enough to provide clear guidance to the development of climate policy, not least because of the failure to set clear targets for climate protection or the development of climate policy. At the institutional level, the Convention lacked a procedure for the adoption of these substantive rules. Therefore, the Kyoto process still represented an early phase of the international regime on climate change very much in need of decisive leadership.

Whereas academic research has identified a number of different leadership categories, the existence of a strong actor who is determined to build an effective and strong regime appears to be most important. This precondition is especially important in highly politicised contexts, as is the case with international climate policy. Only few countries are in a position to exert such "structural leadership".[2] Given the general political circumstances, the US and the EU are the main candidates for such leadership on international environmental and climate policy-making. Under exceptional circumstances, a coalition of developing countries may take on such a role as well, as was illustrated by the "green group"-initiative of India at COP 1 in Berlin (see Section 4.2).[3] In recent years, the US has not sub-

---

1 E.g. Young 1991; see also Gupta/Grubb 1999.

2 On structural leadership see Young 1991.

3 One of those exceptional circumstances was the existence of a strong personality as Head of the Indian delegation.

stantiated its claim for leadership in global environmental affairs and in international climate politics in particular, despite its leadership rhetoric.

For some years, this has left the EU as the sole potential leader in international climate policy. The Union attempted to take advantage of this situation by claiming a leadership position in the Kyoto process. Several of its Member States consistently pushed for progressive targets and timetables and other stringent provisions in the international negotiations. As a result, the EU managed to remain ahead of most other industrialised countries, JUSSCANNZ in particular, for most of the Kyoto process. This was made evident by the early willingness of the EU to engage in serious negotiations on a binding agreement and by the relatively far-reaching proposals on targets and timetables by the EU (see generally Part I).

However, EU leadership in the Kyoto process virtually collapsed during the final days of COP 3. EU delegates became stuck in what was known in Kyoto as the "EU bunker": as EU Member States cumbersomely tried to bridge their internal differences on various proposals on the table, they were hardly able to react to outside developments, let alone take the initiative to steer events. The increasing demand for creating innovative solutions and building winning coalitions for stringent rules was not met by EU leadership during the rather chaotic days of Kyoto (see Chapter 7).

The Kyoto conference itself therefore revealed what had long been visible to observers: the apparent EU leadership on climate change was built on shaky foundations. The weakness of the EU leadership on climate change has its roots both on the domestic and international level. On the domestic level, Union climate policy has not advanced much. Since EU decision-making on various legislative initiatives relevant to climate policy requires consensus, national differences concerning policies as well as broader fears of a shift of competences to the European level have frequently prevented their adoption. As a result, common and co-ordinated policies and measures of the EU on climate change have only been implemented to a very limited degree.[4] This, in turn, has not only weakened the credibility of EU leadership at the international level. It has also rendered the internal EU co-ordination of foreign climate policy positions more difficult, as European legislation did not provide a common denominator for Member States' positions.[5]

Furthermore, the elaboration of effective EU negotiating positions at the international negotiations has proven to be particularly difficult. Reaching agreement between 15 Member States has been complicated and lengthy, despite some attempts to improve co-ordination through, for example, the establishment of an EU Ad Hoc Group on Climate Change. Changing the Presidency twice a year has not provided a stable basis, whereas the role of the Commission has remained

---

4 See Oberthür/Bär 1998; Michaelowa 1998.

5 See on the "inadequacy" of EU leadership in the negotiations on climate change also Sjöstedt 1998; contributions in Gupta/Grubb 1999; on the contrasting situation in the framework of the Montreal Protocol Oberthür 1999a.

sharply limited. As a result, most EU Member States have focused their attention on the internal affairs while failing to link up with other countries.

To be sure, EU co-ordination is indispensable to the establishment of the EU as an actor in international environmental policy. As regards international climate policy, the EU and its Member States deserve credit for their determination to push serious negotiations throughout most of the Kyoto process, for taking the initiative with respect to targets and timetables in spring 1997 and for standing firm on those targets until Kyoto. Many observers believe that, without the EU, targets and timetables might not have been agreed upon at all at Kyoto. Nevertheless, the weaknesses of the EU leadership were more than obvious. As a result, many if not most of the other features of the Kyoto Protocol, most importantly the Kyoto Mechanisms, were to a large extent devised and pushed by the US and its allies in JUSSCANNZ.

The leadership of a powerful actor on international climate policy will be needed for some time in the future. The climate change regime has not yet developed the self-supporting dynamics necessary to drive the process further in the absence of strong leadership. The Kyoto Protocol could be the starting point for such dynamics but needs to be developed carefully for this purpose. Nevertheless, the EU is the only potential leader in international climate policy. If the Kyoto Protocol is to become an effective instrument for climate protection, the conditions of EU leadership need to be improved by alleviating the constraints mentioned above (see also Chapter 25).

21.2 The Importance of Situational Factors

Different schools of international relations have highlighted variables that help explain countries' positions in negotiations on international (environmental) regimes and the outcome of these negotiations. The most prominent of these factors are (1) the countries' varying economic, political, cultural and other interests, (2) knowledge and ideas, and (3) the distribution of power and influence in the international system and in respect to a particular policy field.[6]

Without doubt, these factors also played a major role in the Kyoto process. The different players' interests (see Chapter 2) certainly help to make their negotiating positions and the dynamics that developed because of these positions understandable. The influence of knowledge and ideas is perhaps made most evident by the impact of the IPCC and its Second Assessment Report on the negotiating process. Lastly, the overall distribution of influence in the field of climate change, as expressed not least by the varying contributions to global GHG emissions of the different players, affected the weight of the actors in the negotiations.

In addition, recent literature has stressed the importance of negotiations as communication processes. Players thus experience a need to bargain *and* argue –

6 See e.g. Haas 1993; Levy et al. 1995.

and to argue consistently in order to retain influence.[7] Evidence for the importance of this factor can be found in the Kyoto process: the weakness of the EU during the final stages of the process (see above) was not least the result of its limited involvement in communication. The impact of the host country Japan also appears to have been restricted by its limited engagement in the ongoing communication process.

And yet, a close analysis of the Kyoto process reveals that neither the three "master variables" of international relations analysis[8] nor the importance of "communicative action" suffice to paint the full picture. Perhaps more predominant than in other cases, a number of situational factors (i.e. factors that appear to be special to the particular process or are at least not part of any well-established concept of negotiation analysis) made a decisive difference in the negotiating process. Three of them in particular deserve closer attention: (1) the role of modern communication technology, (2) the role of the Chairman, and (3) the importance of exhaustion (see Chapter 7).

Of these, the role of modern communication technology will have to be integrated into established concepts of negotiation analysis in the future. The "virtual diplomacy", characterised by the use of e-mail for exchanges prior to Kyoto and mobile phones for conversations at the conference as well as by the use of video broadcasting for providing transparency (see Chapter 7), will become a more common feature of international (environmental) negotiations in the 21st century. Mobile phones enabled delegations and NGO participants to stay in close contact, even during the final days and hours of Kyoto, when expert advice was in high demand. Furthermore, communication technology exposed the proceedings in Kyoto to public scrutiny and thus enhanced the responsiveness of negotiators to public sentiment. Arguably, the Kyoto process would have taken quite a different (and less dynamic) course and would have produced different results without the devices of modern communication technology.

Much the same can probably be said about the other two situational factors mentioned. Although international negotiations frequently last until the last minute and benefit from the exhaustion of the delegates (the Climate Convention was agreed upon in the early morning of 8 May 1992), the "negotiation by exhaustion" procedure applied during the final hours of Kyoto was special. It appears that exhaustion hindered several negotiators in their ability to continue arguing. The fact that the US alleviated the problem by replacing its chief negotiators every few hours indicates that relieve was available (but not to countries with little capacity to adapt in such a manner). Because of the sensitivity of the issue and the varying specific interests of countries, the Kyoto Protocol might not have been agreed upon without the exhaustion factor.

---

7 See Gehring 1996.

8 Haas 1993.

The influence of the Chairman, lastly, is a factor that can hardly be anticipated. Chairman Estrada turned out to be a strong leader, who pushed the negotiations in the direction of compromise. It is nearly impossible to imagine what would have happened without Estrada as Chairman, as he shaped and directed the process not only at Kyoto, but also during the pre-Kyoto process. Any agreement would have taken on quite a different form without the leadership of the "Hero of Kyoto", who in the end used his gavel to force a compromise between the countries. It was this innovative procedure that secured agreement at Kyoto.

In sum, the analysis of the Kyoto process points out the importance and potential of taking situational factors into account when analysing international negotiations on the environment. The importance of these situational factors stresses the benefit of personal experience with and exposition to the negotiating process. This does not mean that the established variables of international relations analysis should generally be neglected. They need to be complemented by other less common context-specific factors of influence in order to understand international environmental negotiations and their outcome.

21.3 Climate Policy 21: High Politics in Global Society

Environmental policy on both the national and international levels has traditionally been considered "low politics", i.e. it has been considered subordinate to the "hard" interests of nation states like security and economic welfare ("high politics"). This is evident on the international level, where, in the past, Foreign Ministers and Heads of State or Government have generally restricted their involvement to posturing at international conferences. Despite attempts to define a "foreign environmental policy",[9] they had not actually become involved in the substance of international environmental negotiations to any considerable extent prior to the Kyoto process.

With the Kyoto process, a turning point appears to have been reached in this respect. Even before Kyoto, world leaders like US President Clinton (as well as Secretary of State Albright), UK Prime Minister Blair and other Heads of State or Government and Foreign Ministers got involved personally in the international climate negotiations. There was virtually no high-level governmental meeting in 1997 in which participants did not "talk climate" (see Chapter 5). This engagement of high-level officials intensified as Kyoto approached. Although only the Environment Ministers were represented at the Kyoto conference, various Heads of State and Government, including the ones mentioned above, Japanese Prime Minister Hashimoto, German Chancellor Helmut Kohl and many more, were closely involved from their capitals in hammering out the final deal in Kyoto during the last days and hours. This was largely made possible because of the extensive use of modern communication technologies (Chapter 7).

---

9 See Prittwitz 1984.

Kyoto may therefore have marked the entrance of the environment into the sphere of high politics. Further evidence for increasing importance of environmental policy is provided by an increasing trend to link the environment (and in particular global climate change) with the high-politics field of foreign and security policy.[10] Overall, environmental issues thus appear to be growing in importance in general politics.

This trend may also partly explain the comparatively great constraints that have been placed on the participation and input of representatives of various non-governmental interests.[11] Governments generally tend to allow NGOs less open access to decision-making in areas of high politics. Compared to policy fields like security and economics, the influence of NGOs on the international climate negotiations has nevertheless been sizeable. Using modern communication technologies, building upon and offering their respected expert knowledge and making effective use of existing participatory rights and their established links to the media and the public at large, environmental non-governmental forces were able to give decisive input to the Kyoto process.

To be sure, many other environmental institutions have granted non-governmental players more and more extensive rights to participate in and provide input for intergovernmental negotiations. Also, extensive and extended non-governmental participation in the climate talks is warranted to support further progress. Considering that climate change has come close to becoming high politics, however, puts the non-governmental input in the Kyoto process in another perspective. In this view, the impact of NGOs on the Kyoto Protocol represents a major advance towards an emerging "global civil society" uniting a variety of governmental, societal, economic and other actors.[12]

---

10 See, for example, United States Department of State 1997; on the issue of environment and security see generally Carius/Lietzmann 1998; Nato/CCMS 1999.

11 Yamin 1997.

12 See Wapner 1997.

22. Evaluation of the Kyoto Protocol

The Kyoto Protocol, despite its apparent flaws, can be regarded as a milestone in the history of climate protection. For the first time in history, the vast majority of countries, among them all the world's major players, have recognised the fact that economic and social prosperity is not necessarily linked to the unfettered emission of GHGs. Since the Protocol was adopted unanimously by all Parties to the Convention present at COP 3, this recognition is shared even by oil-producing states. After this statement, all future battles will be battles of retreat and battles over the allocation of burden.

At the same time, the Kyoto Protocol provides an institutional basis on which to proceed into the next century. Nevertheless, the correct accounting of absorption of GHGs by sinks, the careful elaboration of the rules for the Kyoto Mechanisms and the establishment of a truly effective non-compliance procedure will remain major challenges. Much will depend on the decisions that will be taken by the next Conferences of the Parties, most notably COP 6, scheduled for late 2000 or early 2001, which has been assigned an enormous workload by COP 4 (see Chapter 24). The uncertainty regarding the ultimate design of the Kyoto agreement does, however, turn the assessment and evaluation of the Protocol into a gamble.[1] Nevertheless, certain features shall be addressed and evaluated in the following paragraphs.

A number of different criteria can be applied for evaluating international environmental institutions such as the climate change regime and its Kyoto Protocol.[2] The following analysis is confined to three central criteria. (1) Will the Kyoto Protocol change the behaviour of governments, businesses and individuals and thus lead to less GHG emissions than would have otherwise occurred (behavioural effectiveness)? (2) Will it succeed in being sufficient for combating climate change (environmental effectiveness)? (3) Does the Kyoto Protocol provide the means for building a comprehensive regime that can be strengthened and broadened in the future, as appropriate (institutional effectiveness)?

From an ecological point of view, the short-term *environmental effectiveness* of the Protocol is rather limited. Compared with the reduction by 4.6% of total emissions by Annex I countries from 1990 to 1995 – mainly due to economic decline in the Countries with Economies in Transition (CEITs) – the 5.2% overall reduction target agreed upon in Kyoto represents a mere stabilisation of emissions at levels of 1995 (see Section 11.4). Even if this target was achieved and the obligations of Article 3 of the Kyoto Protocol fully implemented by industrialised countries, global emissions would still rise considerably because of the rise in developing countries' emissions. Nevertheless, the implementation of the Kyoto Protocol

---

1 See Ott 1998c; Müller-Kraenner 1998; see also Jacoby et al. 1998; Schneider 1998; Grubb et al. 1999.

2 See generally Young 1994; Bernauer 1995; Oberthür 1997; Victor et al. 1998.

would clearly cause a deviation from current emission trends. Moreover, experience with other environmental agreements suggests that the process set in motion with the adoption of the FCCC and the Kyoto Protocol may well begin to take on a life of its own and lead to steeper reductions in the medium and long-term.[3]

Figure 22.1: Modelling Scenarios to Stabilise the Climate System

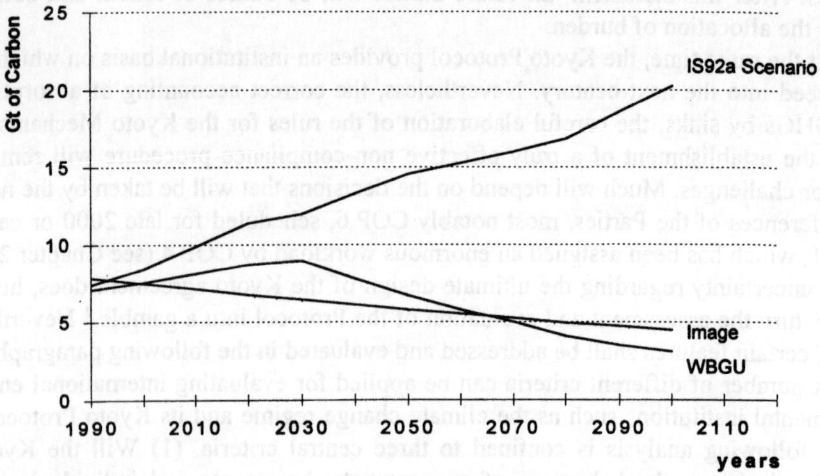

Sources: IS92a: IPCC 1996a; Image: Alcamo/Kreileman 1998; WBGU: Petschel-Held/Schellnhuber 1998. The normative scenarios (WBGU, Image) have been contrasted with the IS92a scenario of the IPCC, that makes assumptions about the future development based on key projections for population growth and economic development.

Modelling scenarios indicate that much more drastic changes in emission patterns would be required to "prevent dangerous anthropogenic interference with the climate system" (Article 2 of the FCCC).[4] According to the scenarios of the Potsdam Institute for Climate Impact Research, only a decrease in global emissions of CO_2, CH_4 and N_2O of at least 1% per year in the next century would keep the climate system within a "tolerable window".[5] Taking account of an increase in

---

3 One of the best examples is presented by the development of the Montreal Protocol (1987), see in particular Benedick 1998a; Ott 1998a; Oberthür 1997.

4 An overview of different scenarios is presented on a website that is part of the "open process" of the Special Report on Emission Scenarios (SRES) of Working Group III of the IPCC: <http://cres.ciesin.org/index.html>.

5 Petschel-Held/Schellnhuber 1998; see also the assessment by Schellnhuber/Fuentes 1997, pp. 445 et seq.

developing country emissions until 2010 would thus require annual reductions in the order of 3% for industrialised countries. This would demand reduction obligations for Annex I countries in the range of minus 23% by the year 2010. The "Image"-scenario likewise illustrates the emission path that would have to be followed in order to achieve a "safe landing" of emission concentrations at or below 450 ppm by the year 2100 (see Figure 22.1).

The *behavioural effectiveness* of the Kyoto Protocol may be somewhat more pronounced, since the reductions are much steeper when compared with a business-as-usual path. According to projections of the International Energy Agency,[6] emissions of CO_2 from the combustion of fossil fuels in the United States and Canada will rise by 32% relative to 1990 by 2010 if no additional measures are taken. For Pacific OECD countries the figure is almost the same (plus 31%) and for OECD Europe slightly lower (plus 26%). Taking these figures into account, major policy changes are required in all industrialised countries in order to achieve the Kyoto targets.

The exact dimensions of these changes will, however, depend on the future evolution of the provisions on sinks and the design of the Kyoto Mechanisms. The limited gross/net approach through the inclusion of three activities related to sinks has already considerably weakened the targets for a number of countries (see Section 11.3). Developments after Kyoto have revealed an increasing tendency to widen this approach through the inclusion of more activities and sectors related to carbon sequestration. Due to the large scientific uncertainties and the difficulties to establish reliable baselines for sink projects, this might undermine the very basis on which the Kyoto Protocol rests: without fairly reliable data and inventories, the Protocol will lack any credibility.

Emissions Trading, Joint Implementation and the Clean Development Mechanism (see Chapters 13–15) certainly have the potential to become the most important cornerstones of the emerging climate regime of the next century. Their political importance can hardly be overestimated. These instruments will transform the climate regime from a predominantly environmental agreement into a "hard" economic treaty. The CDM might one day rival multilateral development assistance and the establishment of a system for the trading of emissions (and derivatives like futures) could create an enormous new market involving large scale financial transactions.

However, there are many issues that need clarification and elaboration in the next rounds of negotiations. Post-Kyoto, numerous workshops and conferences have evidenced the vast complexity of the matter. The possible postponement of COP 6 until early 2001 is also partly due to the difficulties of coming to terms with the intricate questions posed by each of the mechanisms individually and by the interlinkages between them. Since all of them will be elaborated in parallel, there is a good chance to address these linkages, for example with regard to monitoring,

---

6 IEA 1998.

verification and certification. Both the "fungibility" of those credits generated under the respective instruments and the establishment of a common procedure for the enforcement of their rules and regulations play an important role. The Parties will also have to clarify whether common guidelines shall apply to the participation of the private sector in each of the mechanisms.[7]

Furthermore, the design of each of the Kyoto Mechanisms will have direct consequences for the other two. Restrictions on the use of Emissions Trading, for example, will increase the potential of JI and the CDM. The latter two are competing with regard to the transaction costs, which should ideally not differ too much. The CDM, however, is already slightly disadvantaged because of the "share of the proceeds" that shall be used for its administration and for adaptation purposes. This is one of the reasons which speaks in favour of a transaction fee on all three mechanisms (see Chapters 15 and 25). This might also fulfil the demand for supplementarity of the EU, because raising transaction costs would improve prospects for domestic implementation.

There is a real danger of a "race to the bottom" between the mechanisms. If each of them was designed so as to maximise its use, this might lead first to an inflation and, consequently, a devaluation of emission allowances, as the market adjusts. As a result, the mechanisms would not only fail to generate additional revenues, but the system as such would be endangered with profound implications for the efforts to protect the climate system.[8]

At the beginning of the Kyoto Process, transparency and the building of trust are the bases on which the systems will be established. Above all, therefore, the rules to be elaborated must correspond to the need for accurate, transparent and verifiable information on all activities under the mechanisms. This will contribute to alleviating fears about hugely inflated emission allowances in the system beyond the already included hot air. Above all, the Kyoto Mechanisms should lead to *more* emissions being avoided or reduced than without them.

And last, the extent to which JI and the other mechanisms will be used depends very much on the success of domestic implementation of the Kyoto obligations. If this was tackled vigorously, climate policy induced technological, behavioural and structural change might well render domestic measures cost-effective and thus significantly reduce the potential market for the Kyoto Mechanisms.[9]

The evaluation of the Protocol's *institutional effectiveness* presents a mixed picture. This concept refers to the appropriateness of the institutional provisions (organs and decision-making procedures) as a basis for the strengthening and expansion of a treaty's obligations in line with advancing scientific and technologi-

---

7 On the role of the private sector see Campbell 1998.

8 See Grubb et al. 1999, p. 193.

9 Even under current conditions the European Commission estimates that about two-thirds of the EU commitment of 8% reduction may be achieved at a price below 5 Euro per ton of CO_2 equivalent, see European Commission 1999, p. 10.

cal knowledge as well as changing political circumstances. It refers furthermore to the instruments available for the enforcement of a treaty's provisions. On the one hand, the institutions that serve the FCCC have simply been duplicated (see Chapter 18). The COP, the Secretariat and the subsidiary bodies have worked reasonably well, however, and the same may be expected from the Protocol's organs in the future. As regards the Kyoto Mechanisms, the institutional set-up is yet unclear, excepting the establishment of a largely undefined executive board under the CDM (see Chapter 14). These institutions, especially those administering the CDM and Emissions Trading, might well develop their own dynamics and provide an effective structure for the climate regime in the next century.

Nevertheless, the Parties to the Kyoto Protocol will have to resolve one of the biggest problems that also plagues the Convention, namely the voting procedure. Due to fierce resistance on part of oil-producing countries, the Protocol, like the Convention, lacks any provision on voting except for the adoption of amendments to the treaty and its annexes, which require a three-fourths majority (Article 20.3 and 21.4). For the Protocol to be truly effective, however, some kind of majority voting is essential.

As regards the enforcement mechanism, Article 18 explicitly calls for a non-compliance procedure to be elaborated. Furthermore, the Parties at COP 4 included a programme to consider and eventually adopt such a procedure.[10] There seems to be a growing recognition that the Kyoto Protocol with its legally binding obligations, demands the possibility to strictly enforce these commitments (see Section 18.3).

Due to the unfinished nature of the Kyoto Protocol, it will only enter into force in a form amended either formally or by way of extensive decision-making. Formal amendments might, for example, be necessary for the adoption of a non-compliance procedure involving binding consequences. The trading system, Joint Implementation and the Clean Development Mechanism, on the other hand, might only need elaboration through decisions of the Conference of the Parties.

Despite this critical appraisal, the Protocol can become a success and should form the basis for further action on climate change. There are enough reasons to be dissatisfied with the snail's pace of international climate policy in the context of the existing climate regime.[11] But despite calls for a completely different approach,[12] there is no credible alternative to the FCCC and its Kyoto Protocol, which represent the result of 10 years of climate diplomacy. In addition, there is ample room to strengthen the existing regime, e.g. by launching a Leadership Initiative within its framework (see Chapter 25). Abandoning the Protocol without a credible and politically feasible alternative would lead to a weakening of interna-

---

10 Decision 8/CP.4 Annex II in FCCC/CP/1998/16/Add.1

11 Flavin 1998.

12 See Benedick 1998b; Cooper 1998; see also the cogent reply by the chief US negotiator, Eizenstat 1998a.

tional climate policy by playing into the hands of those who have an interest in its failure.

The Kyoto Protocol, therefore, does not represent a final solution to the manifold problems confronting the international community as regards co-operation on climate change. It does, however, provide a firm basis for international climate policy to embark on the urgently required transformation of industrial society into a fossil-free economy in the next century. Finally, and perhaps most importantly, it does send the clear signal to policy-makers, industry and other relevant domestic actors that climate change has been firmly placed on the world's agenda (see Chapter 24). Because of its far-reaching implications for the way we produce and consume, the Kyoto Protocol is likely to affect the life of every person living on this planet in the next century more than any other international agreement.

23. Synergies and Conflicts with Other International Institutions

International institutions and treaties on the environment have proliferated over the last few decades, as they have in other policy areas as well.[1] As a result, the potential for linkage and overlap among these institutions has increased steadily, providing both the opportunity for creating synergies and the danger of conflicts between those international institutions – be it international organisations or, more importantly, international treaty regimes.[2] It is hardly surprising then, that numerous linkages exist between the climate change regime and other international institutions, which have been touched upon throughout this book. Examples include the IMO and ICAO in the issue of GHG emissions from bunker fuels (Chapter 10) and the GEF in the context of being the Financial Mechanism of the FCCC (Chapter 3). Without aiming at exhaustiveness, this Chapter spells out some of the issues involved in the most pertinent relationships between the climate change regime and other instruments. These are the World Trade Organization (WTO) (see Section 23.1), the Montreal Protocol on Substances that Deplete the Ozone Layer (see Section 23.2), as well as the Convention on Biological Diversity (CBD) and the Convention to Combat Desertification (CCD) (see Section 23.3).

23.1 GATT/WTO and Multilateral Investment Rules

The relationship between multilateral environmental agreements and the international trade rules (GATT/WTO) has been a subject of debate for some time.[3] In the climate change context, negotiators paid attention to the issue from the beginning, and both the FCCC and the Kyoto Protocol contain references to international trade and economics. Thus, Article 3 of the FCCC requires that "Parties should co-operate to promote a supportive and open international economic system" and that any measures "should not constitute a means of arbitrary or unjustifiable discrimination or a disguised restriction on international trade". In softer language, Article 2.3 of the Kyoto Protocol obliges Parties to implement PAMs "in such a way as to minimise ... effects on international trade".

GATT/WTO rules prohibit members from taking measures that go against the following key principles:

- no discrimination as between "like products" originating from any member ("most favoured nation" principle; Article I GATT),

---

1 This Chapter has received substantial input by Richard G. Tarasofsky.

2 See generally Young 1996 and 1999.

3 E.g. Tarasofsky 1998; Moltke 1997; Housman et al. 1995; on WTO rules and climate change see especially Sampson 1998.

- no discrimination as between imported and domestic "like" products (Article II GATT), and
- no quantitative restrictions on imports or exports (Article XI GATT).

It remains uncertain to what extent these principles affect trade-related environmental measures, since the GATT/WTO also contains exceptions for measures aimed at animal, plant or human health, and the conservation of natural resources (Article XX GATT). For example, a key judicial ruling is that "like products" refers to similarities relating to a product's characteristics, rather than how the product is produced. However, until the concept of "likeness" in relation to products is defined as being like in the sustainability of the process of production, tensions between environment and trade policies will persist. Environmental issues are continuously dealt with in the WTO Committee on Trade and Environment, and possibly in the forthcoming Millennium Round of trade negotiations.

Neither the FCCC nor the Kyoto Protocol have yet introduced any trade restrictions. Indeed, so far, no conflicts between MEAs and the GATT/WTO have become the subject of international litigation, although the potential for conflict exists, e.g. in relation to the trade measures in the Montreal Protocol, Basel Convention, and CITES. The near-universal membership of all treaties involved has no doubt helped conflicts from escalating into full-blown disputes, as has the strong signals from the WTO itself that it prefers multilaterally agreed trade restrictions to ones taken unilaterally. However, from the legal point of view, the precise relationship between the GATT/WTO and environmental treaties is unclear and thus very much on the international agenda.

Conflicts are more likely to arise in the case of unilateral trade measures, in their direct (prohibitions) as well as indirect form (voluntary standards, subsidies, certification and labelling), although it is unclear what will happen if the measure is taken in the implementation of an environmental treaty. Problems may also arise in such cases where Parties, for competitiveness reasons, apply border tax adjustments to imported products to offset any disadvantages caused by domestic carbon/energy taxes.[4] Once Parties start to implement the Kyoto Protocol more effectively, issues like this may come to the fore. If such complaints are filed with WTO, Parties may be called upon to prove the compatibility of their PAMs with existing trade rules, which may lead to the undesirable effect of being ruled WTO-illegal. As this would be good neither for the WTO nor for the climate regime, there are good reasons to find an international accommodation between the trade and climate regimes. However, until the FCCC and its Kyoto Protocol become more focused on trade issues, the overall process will be somewhat unbalanced.

Further consideration needs to be devoted to the relationship between the Kyoto Mechanisms and international trading rules. It is, for example, not certain whether countries could be excluded from Emissions Trading under WTO rules. Prima

---

4 See, e.g., UNU 1998; Chambers 1998b.

facie, the WTO rules should not affect the trading of emission rights, since those rules apply only to goods, services and intellectual property rights. However, some observers have raised concerns.[5] Another issue to be resolved is the WTO-compatibility of trade restrictions imposed in cases of non-compliance with the Kyoto Protocol.

But there is potential not only for conflict but also for synergy and learning. As such, the 50 years evolution of the GATT from a provisional arrangement to a fully-fledged international organisation may well be seen as an important and encouraging example for the climate regime.[6] The GATT/WTO system thus has informed thinking about the design of the international regime on climate change – and will continue to do so in the future. This relates in particular to the area of implementation review, non-compliance and dispute settlement.[7] In principle, the climate regime could develop an effective mechanism for the resolution of disputes in order to provide an alternative to the WTO procedure in relevant cases.

As regards the more technical provisions, the WTO Subsidies Agreement aiming at prohibiting government assistance to industries to improve their competitive position may well provide some useful lessons. This is very much in line with the Kyoto Protocol's approach to subsidies to the fossil fuel sector (see Article 2.1 and Chapter 10) and might be used to enhance pressure on Parties to reduce subsidies that lead to increased GHG emissions. On the other hand, the Subsidies Agreement also permits members, under specific circumstances, to grant subsidies to comply with environmental standards.

In addition to trade, investment flows are an important aspect of the international economic order. While no international agreement has been concluded yet, the Multilateral Agreement on Investment (MAI) was under negotiation under the auspices of the OECD during the second half of the 1990s. The Most Favoured Nation clause and the National Treatment standards have been at the heart of the draft MAI. These state that all investors should be treated alike and no discrimination be applied irrespective of the country of origin of the investor. Foreign and domestic investors would thus have to be treated alike.

The future of the Multilateral Agreement on Investment is very much uncertain, since negotiations within the OECD framework were effectively halted in 1998. However, there are calls for the negotiations on international investment rules to be conducted within the WTO framework. Any such agreement based on the standards mentioned above might easily turn out to be at odds with the CDM under Article 12 of the Kyoto Protocol (and possibly also with JI under Article 6). The specific rules of the CDM are yet to be determined. If they encourage or require host countries to discriminate between investors according to their status under the

---

5 See UNU 1998; Chambers 1998b, pp. 48–49.

6 See Jacoby et al. 1998. See also Ott 1998a for a legal comparison of the status of GATT and environmental regimes.

7 See, for example, FCCC/AG13/1996/MISC.2 and Add.1; Victor 1995; Ott 1996b.

Kyoto Protocol (Annex I Party, Party in non-compliance), this might easily lead to tensions with future multilateral investment rules.[8]

Overall, these and other issues call for intensified exchange and co-operation between WTO and the broader economic arena, on the one side, and the climate change regime, on the other. The "Millennium Round" of trade talks may provide an opportunity to make progress towards reconciling both areas.

23.2 The Montreal Protocol on Substances that Deplete the Ozone Layer

From the very beginning, there has been a special relationship between international climate politics and the international co-operation for the protection of the ozone layer.[9] This is not least because the Montreal Protocol has served – and continues to serve – as a useful precedent and model for the FCCC and its Kyoto Protocol. Over the years, it has informed various aspects of the evolving regime on climate change, including the elaboration of rules on implementation review (see Chapter 16). Insights from the Montreal Protocol furthermore have led the opponents of effective action to prevent the adoption of provisions that have proven successful in the ozone regime, as shown by the inconclusive discussions on the establishment of technical and economic assessment panels (which exist under the Montreal Protocol).

The Montreal Protocol will certainly continue to function as a useful precedent. New opportunities for learning from the Montreal Protocol have appeared by the inclusion of HFCs, PFCs and SF_6 in the Kyoto Protocol (see Section 11.2). Because of the similar characteristics of these GHGs and ozone depleting substances controlled under the Montreal Protocol, policy-making on these fluorinated GHGs may easily draw lessons from the experience under the Montreal Protocol, in particular with regard to data collection,[10] monitoring of emissions and effective emission abatement policies. For example, there is a long history of efforts in the framework of the Montreal Protocol to assign common customs codes to ozone depleting substances in order to enable and facilitate data collection on imports and exports of these chemicals. This has involved co-operation with the World Customs Organization and has proven to be a rather complicated and long-term task.[11] Parties to the Kyoto Protocol may face similar challenges with respect to the fluorinated GHGs.

---

8 Chambers 1998a; Werksman/Santoro 1998.

9 Based on the Vienna Convention for the Protection of the Ozone Layer of 1985 and its Montreal Protocol on Substances that Deplete the Ozone Layer of 1987; for the development of these see Oberthür 1997; Ott 1998a; Benedick 1998a; the issues raised in this section are dealt with in more detail in Oberthür 1999b.

10 For data collection and reporting under the Montreal Protocol see UNEP 1999.

11 See various related decisions of the Meeting of the Parties to the Montreal Protocol as contained in Ozone Secretariat 1996; 1998.

In addition, closer co-operation between the two treaty regimes may be needed because of the close relationship between the causes and effects of climate change and ozone depletion (e.g. ozone depleting substances are also GHGs: Chapter 1).[12] To some extent, this has already been taken care of by the two treaty regimes. To avoid overlap the FCCC explicitly refers only to GHGs "not controlled by the Montreal Protocol".[13] However, the fluorinated GHGs, and HFCs in particular, have been promoted as substitutes for CFCs and other ozone depleting substances in the ozone regime. Thus, while they have been part of the solution under the Montreal Protocol, they are part of the problem under the Kyoto Protocol. Figures included in the Scientific Assessment of Ozone Depletion 1998 have furthermore indicated that the Global Warming Potential (GWP) of HFCs and PFCs might be greater than previously thought (see Table 23.1).

Table 23.1: GWPs of Selected Halocarbons (100-year Time Horizon)

Substance	GWP (IPCC 1996a)	GWP (UNEP 1998)
HFC-32	650	880
HFC-43-10	1300	1700
HFC-125	2800	3800
HFC-134a	1300	1600
HFC-143a	3800	5400
HFC-152a	140	190
HFC-227(ea)	2900	3800
CF_4 (a PFC)	6500	5700
C_2F_6 (a PFC)	9200	11400

Source: IPCC 1996a; UNEP 1998.

Parties to both treaties acknowledged and started to confront the problem in 1998. The Tenth Meeting of the Parties to the Montreal Protocol and COP 4 of the FCCC requested/encouraged the convening of a workshop by the IPCC and the Technology and Economic Assessment Panel of the Montreal Protocol in 1999 on available and potential means of limiting HFC and PFC emissions. On the basis of information compiled by the Secretariat from this workshop and submissions by Parties, SBSTA is to report to COP 5, while the Technology and Assessment Panel will report to the Eleventh Meeting of the Parties to the Montreal Protocol in

12 UNEP 1998.

13 This excluding proviso was the main reason why a Japanese initiative in Kyoto to give credit to the destruction of ozone depleting substances under the Kyoto Protocol has never been included into the official agenda of the climate negotiations.

1999.[14] The joint IPCC/TEAP workshop took place in May 1999 and identified a number of alternative technologies to the use of HFCs and PFCs as well as numerous options for reducing their emissions from applications.[15]

There are a number of policy implications of the HFC/PFC issue. The first one relates to the future policies of the Multilateral Fund for the Implementation of the Montreal Protocol that provides the incremental costs of projects for phasing out ozone depleting substances in developing countries.[16] Until 1995, the Fund gave preference to the most cost-effective solution in these projects, which in many cases led to funding conversion to HFC technology. Since 1995, hydrocarbon technology has been granted a special 35% discount on the applicable cost limits to provide for an equal footing of this alternative technology in the approval process.[17] Nevertheless, HFCs have continued to be used as the ODS alternative in more than 80% of projects approved by the Executive Committee of the Multilateral Fund for the refrigeration sector.[18] To minimise use of HFCs and PFCs, alternatives could be prioritised wherever technically feasible and economically acceptable – as has been done in the case of GEF support for the phase-out of ozone depleting substances in CEIT countries.[19] If this involves higher costs than HFC technology, such projects might be eligible under the Clean Development Mechanism of the Kyoto Protocol in that GHG emissions in developing countries would be thereby avoided.

Furthermore, the similarity in applications suggests that countries might apply similar political concepts and instruments for controlling emissions of HFCs as have proven to be effective for ozone depleting substances. These instruments might include targeted financial incentives, voluntary agreements, regulatory limits to production and/or consumption, prohibitions of all except essential uses, taxes, trade controls, prohibitions of specific uses etc.[20] One particular related problem that is rooted in the history of the Montreal Protocol process concerns possible effects on developing countries that policies to limit HFC and PFC emissions might have. In the Montreal Protocol framework these countries have, as a result of pressure from industrialised countries, converted their CFC industry partly to HFC technology. Thus, industrialised countries will face political and ethical (if not legal) difficulties in restricting imports of relevant products (that contain HFCs like, e.g., refrigerators) from developing countries.

---

14 Decision 13/CP.4 in FCCC/CP/1998/16/Add.1 and Decision X/16 in UNEP/OzL.Pro.10/9.

15 IPCC/TEAP 1999.

16 See DeSombre/Kauffman 1996; Biermann 1997.

17 See Decision 17/14 in UN Doc. UNEP/OzL.Pro/ExCom/17/60; also Multilateral Fund 1998, p. 84.

18 See Pinto et al. 1999.

19 See GEF 1996, p. 64.

20 See, e.g., Cook 1996; UNEP/SEI 1996.

23.3 The Conventions on Biodiversity and Desertification

There are obvious linkages between the climate regime and those related to biodiversity and desertification. These arise because the conventions on biodiversity and desertification both adopt holistic and integrated approaches for achieving their objectives. As such, both instruments deal with climate change, without actually mentioning this expressly.

For example, Article 7(c) of the Biodiversity Convention requires Parties to "identify processes and categories of activities which have or are likely to have significant adverse impacts on the conservation and sustainable use of biological diversity ...". Article 8(l) of that Convention goes on to require Parties to regulate such processes and activities. The Subsidiary Body for Scientific, Technical and Technological Advice of the Biodiversity Convention, in 1996, approved a framework for identifying these processes and activities which specifically included climate change.[21]

By the same token, the Convention to Combat Desertification requires Parties to "adopt an integrated approach addressing the physical, biological, and socio-economic aspects of the processes of desertification and drought" (Article 4(a) CCD). This too would encompass climate change, to the extent that it affects desertification and drought. In fact, the Convention to Combat Desertification provides in Article 8(1) that:

> "The Parties shall encourage the coordination of activities carried out under this Convention and, if they are Parties to them, under other relevant international agreements, particularly the United Nations Framework Convention on Climate Change and the Convention on Biological Diversity, in order to derive maximum benefit from activities under each agreement while avoiding duplication of effort."

Nevertheless, the challenge to make sure that these conventions and the FCCC, which aim at achieving sustainable development, all are implemented in a complementary manner remains. A major challenge arises from the treatment of forests under the FCCC and its Kyoto Protocol, which views them mainly in terms of their "carbon sequestration" potential (see Chapter 11.3), although Article 4.1(d) of the FCCC requires the promotion of the sustainable management and conservation and enhancement of sinks, including biomass and forests. In addition, the Kyoto Protocol requires each industrialised country to implement policies and measures related to the protection and enhancement of sinks and reservoirs "taking into account its commitments under relevant international environmental agreements; promotion of sustainable forest management practices, afforestation and reforestation" (Article 2.1(a)(ii)). Despite this provision, these issues were not high on the agenda in Kyoto and post-Kyoto.

Indeed, Article 3 of the Kyoto Protocol can be an incentive to maximise carbon absorption of forests (and possibly other ecosystems in the future) without taking

---

21 UNEP/CBD/SBSTTA/2/3.

full account of other environmental values. For example, the maximisation of carbon absorption might lead to transformation of biologically diverse forest ecosystems into "high yielding mono-culture tree-plantations" that "resemble fields of crops as opposed to natural forest".[22] Such effects would also place the Kyoto Protocol at odds with the objectives of the UN Intergovernmental Forum on Forests (IFF). Unsustainable forestry practices could also have detrimental effects on sensitive soils (eventually contributing to soil erosion and desertification).[23]

Many developing countries are rich in biological diversity and possess ecologically sensitive and valuable forests, as well as other ecosystems. The need for harmonisation of measures under the Kyoto Protocol and international instruments aiming at biological diversity conservation, sustainable forestry and combating desertification will increase if sinks projects become eligible under the CDM (see Chapter 14).[24] The same holds if developing countries take on commitments under the Kyoto Protocol, which they may wish to fulfil by enhancing their sinks capacity (see Chapter 17).

Synergy issues have received attention since Kyoto, which is bound to grow. The IPCC Special Report on sinks, due after 2000 (see Section 11.3), provides a key opportunity to explore some of these issues in more detail. A number of analyses have become available post-Kyoto that will inform the future debate.[25] As a result of the growing awareness of potential conflict between the Kyoto Protocol and the mentioned international agreements, COP 4 also decided "that issues of mutual concern to the [UNFCCC] and the Convention on Biological Diversity should be taken up by the subsidiary bodies" in 1999.[26] The COPs of the conventions on biodiversity and desertification are also in the process of addressing the relationship with the FCCC.

While discussions have begun, it is far from obvious in which way Parties may resolve the relevant issues. Avoiding conflicts and enhancing complementary implementation on the ground will require substantive integration, as well as procedural steps to be taken. As far as procedure is concerned, a good start would be enhanced institutionalised co-operation between the respective Secretariats.

22 UNU 1998, p. 42; see also Tarasofsky 1999; Brown 1999.
23 Ibid.; Downes 1999.
24 See Brown 1999.
25 E.g. Brown 1999; Gillespie 1998b; WBGU 1998.
26 FCCC/CP/1998/16/Add.1, p. 71.

24. The Landscape of International Climate Politics at the Turn of the Century

The future of the international co-operation on climate change depends on whether and when the Kyoto Protocol will be ratified and implemented by the major players. This, in turn, is heavily influenced by both their strategic considerations and their domestic climate politics. In the emerging global civil society, domestic and international climate policies are furthermore affected by non-governmental forces. The following sections draw a picture of the international landscape of climate politics and the prospects of the Kyoto Protocol as a part of this landscape. For this purpose, the post-Kyoto situation on the international level, in the major countries and regions and in respect to non-governmental developments at the turn of the century is taken under review.

24.1 The Post-Kyoto International Process: Buenos Aires and Beyond

In the immediate period following Kyoto, exhaustion prevailed and international climate policy was characterised by a low level of activity. However, the "prompt start" decision passed at Kyoto mandated the subsidiary bodies and COP 4 to make/prepare decisions, particularly regarding the Kyoto Mechanisms.[1] Consequently, the mechanisms (Joint Implementation, CDM, Emissions Trading) and also the issue of sinks have become the focus of climate diplomacy, which experienced another milestone with COP 4 in Buenos Aires in November 1998. Furthermore, the participation of developing countries has remained high on the international agenda. Overall progress, however, has been slow.

24.1.1 The Kyoto Mechanisms and Sinks

The Kyoto Mechanisms are at the centre of attention in the post-Kyoto process. There are many questions which remained open in the Protocol concerning the following list of topics: the institutional organisation, the problems of monitoring and verification, the mechanisms for determining a breach and for the enforcement of the rules, the methods of measurement to be used to calculate the reductions achieved through project based instruments, and lastly, the eligibility of sink projects under the CDM. In addition to the many technical issues concerning the concrete terms of the mechanisms, there is the politically highly controversial question as to whether upper limits ("caps") should be set on the use of the Kyoto

1 Decision 1/CP.3 in FCCC/CP/1997/7/Add.1.

Mechanisms in order to ensure that buying emission allowances abroad will remain "supplemental" (as stipulated by the Protocol).[2]

The political "camps" confronting each other on the design of the Kyoto Mechanisms have remained very much unchanged since Kyoto. On the one hand, the US and other members of the "umbrella group" (Australia, Canada, Iceland, Japan, New Zealand, Norway, Russian Federation, Ukraine, and the US), which formed as a result of Kyoto, prefer as little regulation as possible so as not to constrain the availability of the mechanisms and, above all, Emissions Trading. The chief negotiator of the US at COP 4 in Buenos Aires (2–13 November 1998), Stuart Eizenstat, emphasised the earnestness of the US opposition to any "caps" on the use of the mechanisms by quoting from Winston Churchill's famous World War II dictum: he vowed that the USA would "fight on the beaches and in the trenches" against such a limit.

On the other hand, the EU still favours stringent guidelines for the calculation of project-based (CDM, JI) emission reductions and verification. In 1999, the Union proposed a rather complicated cap formula and was then accused by the US of trying to "rewrite the Protocol" (see Section 15.3.3).[3] However, the EU has become increasingly divided internally on these issues.

Whereas the rules for Emissions Trading may only be required prior to the commitment period in the year 2008, the CDM, according to Article 12 of the Protocol, can be used to earn "certified emission reductions" from the year 2000 onwards. This will require binding regulations so as to prevent each country from following self-made rules for the calculation of emission reductions. In reality, the elaboration of all rules concerning flexibility instruments have been linked: virtually all industrialised countries have made the resolution of the related issues a precondition for their ratification of the Kyoto Protocol. As a result, the bundling of all issues related to the Kyoto Mechanisms was included in the Buenos Aires Plan of Action adopted at COP 4 after protracted negotiations.[4]

The work programme on mechanisms of the Kyoto Protocol,[5] which forms part of this action plan, specifies that detailed rules for all mechanisms should be adopted in one package at COP 6, scheduled to take place in late 2000 or early 2001 (probably in the Netherlands). The extent of the regulations to be established is immense, as is made apparent by the long catalogue of open questions attached to the agreement. This catalogue is a shopping list containing nearly all items

2 The relevant issues, including the "hot air" problem posed by the fact that especially Russia and the Ukraine have a large potential for selling emission allowances as a result of the economic decline in the 1990s, are spelled out in more detail in the respective Chapters dealing with each Mechanism; see Chapters 13 (JI), 14 (CDM) and 15 (Emissions Trading).

3 "Clinton accuses EU of trying to rewrite Global Warming Pact", Wall Street Journal, 18 May 1999; for the EU cap proposal see Council Conclusions on a Community Strategy on Climate Change, EU Doc. 8346/99, 18 May 1999.

4 See Decision 1/CP.4 in FCCC/CP/1998/16/Add.1.

5 See Decision 7/CP.4 in FCCC/CP/1998/16/Add.1.

mentioned by the different players and reflects the players' inability to close the gap between their differing positions.

By bundling the rules of all three Kyoto Mechanisms, industrialised countries successfully put pressure on developing countries to agree not only on the rules of the CDM, for which many had shown a particular interest, but also on Emissions Trading and Joint Implementation between industrialised countries. By doing so, there is the danger that the environmental effectiveness of the instruments will fall by the wayside.[6]

Another potential threat to the effectiveness of the regime would be a hasty expansion of the sink categories to be taken into account under the Kyoto Protocol (see Section 11.3). After lengthy discussions in 1998, Parties at COP 4 linked both the treatment of forest sinks and the inclusion of further categories of sinks to a special report by the IPCC to be adopted after the year 2000.[7] Although the issue was thus effectively postponed, some Parties, most importantly the US and other non-EU industrialised countries, tried to push the debate at the political level during the Subsidiary Body meetings in Bonn in June 1999.[8] These are attempts to create facts (i.e. an expansion of the sink categories under the Protocol) before the IPCC special report becomes available. The European Union has shown little interest to fend off these attempts.

24.1.2 Developing Country Participation

Developing country participation in the Kyoto Protocol (see Chapter 17), although not on the official agenda, has remained one of the most prominent issues throughout the post-Kyoto discussions. The issue attained utmost prominence in the corridors and backrooms of COP 4 in Buenos Aires and was partly responsible for the sluggish progress at the conference. At its root, the conflict is spurred by continuing calls from the US and other non-EU industrialised countries for the "meaningful participation" of developing countries in combating climate change. In response, China and India have made the per capita distribution of emission rights one of their central demands. The result has been deadlock.

The event that triggered and enhanced the debate at COP 4 in Buenos Aires was the welcoming address to the conference by Argentina's President Menem. Withdrawing from the consensus of the developing countries, he announced the adoption of voluntary quantitative commitments for his country by COP 5 in 1999. This was coupled with a call for equal access to all Kyoto Mechanisms – a clear reference to Emissions Trading.[9] If Argentina was to stay below its voluntary target, it would be able to sell the excess emission allowances on the international market.

---

6 Parties will have to assess the pilot phase on Activities Implemented Jointly (AIJ) in 1999; see Decision 6/CP.4 in FCCC/CP/1998/16/Add.1.

7 Decision 9/CP.4 in FCCC/CP/1998/16/Add.1; see also Section 11.3.

8 On these meetings see Earth Negotiations Bulletin, Vol. 12, No. 110, 14 June 1999.

9 See Annex I of FCCC/CP/1998/16.

To many observers, this kind of voluntary limits to emissions carries the danger of further dilution of the industrialised countries' reduction obligations. A high Argentinean voluntary target, based upon a (strategically increased) political business-as-usual energy scenario, could introduce a large quantity of excess emission allowances into the trading system (see also Chapter 17).

Post-Kyoto, some progress was achieved with respect to strengthening the transfer of financial resources, technology and know-how. Decisions on the transfer of technology[10] and the financial mechanism[11] were passed as part of the Buenos Aires Plan of Action. The GEF was entrusted with the operation of the financial mechanism, which will be reviewed every four years. The GEF will also begin financing measures designed to assist vulnerable developing countries in adapting to climate change impacts, like e.g. the rising sea level. At the same time, compensation for potentially negative effects of climate protection, in form of a reduced demand for fossil fuels and raw materials, will remain on the agenda of the Convention organs as well (see Chapter 17).[12]

24.1.3 Non-compliance Procedure and Other Matters

The number of other topics on the agenda of international climate policy has certainly not decreased post-Kyoto. Rather, the regime is further broadening its scope. Some activities of the pre-Kyoto phase have been continued, like the elaboration of a non-compliance procedure and the development of policies and measures (PAMs) for the mitigation of climate change. Regarding the former, COP 4 established a joint working group on compliance under the SBI and the SBSTA, which is to report to COP 5 and which might result in the elaboration of a non-compliance procedure by COP 6 (see Section 16.3). With respect to PAMs, Parties at COP 4 mandated the Secretariat to report on relevant "best practices" to SBSTA 11 in late 1999 and to organise a workshop on the matter in 2000 (see Chapter 10). Overall, the issue has not enjoyed high prominence post-Kyoto.

In addition, COP 4 endorsed the work programme on methodologies, (review of) national communications and emission inventories elaborated at the sessions of the subsidiary bodies in the summer of 1998.[13] These items belong to the preparations for the first session of the COP/MOP of the Protocol. The relationship to other international agreements, most prominently the Montreal Protocol with respect to the fluorinated GHGs and the Biodiversity Convention and forest-related agreements with respect to sinks, has also moved up the international agenda (see Chapter 23).[14]

---

10 Decision 4/CP.4 in FCCC/CP/1998/16/Add.1.

11 Decisions 2 and 3/CP.4 in FCCC/CP/1998/16/Add.1.

12 Decision 5/CP.4 in FCCC/CP/1998/16/Add.1.

13 Decision 8/CP.4 in FCCC/CP/1998/16/Add.1; on the latter aspect see also FCCC/SBSTA/1998/9 and FCCC/SBI/1998/7.

14 Decision 13/CP.4 in FCCC/CP/1998/16/Add.1.

In another field, the laggards clearly picked up points. The first review of the "adequacy of commitments" of industrialised countries under the FCCC at COP 1 in Berlin resulted in the Berlin Mandate and eventually in the Kyoto Protocol (see Chapter 4). In view of the importance of such provisions, Article 4.2(d) of the Convention mandated a second review by the 31 December 1998 at the latest. This, however, did not take place in Buenos Aires. This was mainly the result of the deadlock between industrialised and developing countries. Some industrialised countries demanded the inclusion of developing countries' emissions in the review. As a countermove, developing countries demanded the elaboration of detailed criteria in order to assess whether industrialised countries were complying with their obligations. Noteworthy and particularly worrying in this context is the fact that the Parties to the Convention seem to be willing to disregard concrete deadlines established in the treaty.

Equally serious was the inability of the Parties at COP 4 to agree upon a date for a third review. Negotiations took place and, according to the last draft, a third review date was foreseen for the end of 2002 at the latest. In this case, a renewed decision on the adequacy of commitments could have been based upon the Third Assessment Report of the IPCC due in 2001. Since the Parties reached an agreement on the submission of a third national communication for industrialised countries by the end of 2001 at COP 4,[15] a third (or rather: second) evaluation of the adequacy of commitments in 2002 could also be based on better national emission data.

24.2 The European Union and the Applicant Countries

Post-Kyoto, the EU has continued to strive for international leadership. In particular, it has attempted to maintain the pressure on the US and other members of the umbrella group to adopt an environmentally effective design for the Kyoto Mechanisms. In 1999, the EU presented a concrete, rather complex formula for determining a cap on the use of the Kyoto Mechanisms (see Section 15.3.3) in order to ensure that industrialised countries take effective domestic action.

Furthermore, the EU has made some effort to react to criticism of its inward focus during the Kyoto negotiations (see Section 21.1). As part of this effort, it strengthened its ties with CEITs and developing countries.[16] In this regard, the European Commission has embarked upon an effort to devise a strategy for the collaboration with developing countries.

The closer ties to Central and Eastern European CEITs are in part a consequence of the process of EU enlargement. This is expected to lead to the accession of the Czech Republic, Estonia, Hungary, Poland, and Slovenia early in the 21st century. More CEITs will follow in subsequent years. As part of the enlargement

15 Decision 11/CP.4 in FCCC/CP/1998/16/Add.1.

16 See also European Commission 1998b.

process, the accession countries increasingly adapt their national legislation and administrative structures to the EU requirements. As a consequence, the common interests and linkage with the EU have been growing. As a first positive result, the EU and several CEITs (as well as Switzerland) have presented a number of common positions on relevant topics in the post-Kyoto process.[17]

Under the British EU Presidency, the EU reached an agreement in June 1998 on an internal burden-sharing. This will form the basis of its declaration on joint fulfilment under Article 4 of the Kyoto Protocol upon its ratification (see Chapter 12). The total GHG emission reduction by 2010 is thus set to reach 8%, the combined EU commitment under the Kyoto Protocol (see Table 12.1, Section 12.3).

Despite this agreement, the EU has not yet taken the necessary steps for an early ratification of the Kyoto Protocol.[18] This appears to be mainly due to strategic considerations. First of all, given the unfinished business of Kyoto, early ratification would mean subscribing to obligations which are partly unknown. In addition, the EU might wish to keep pressure on the US and the "umbrella group" (see below) by threatening with non-ratification. Furthermore, the Union appears to be speculating internally on the accession of the first Eastern European countries. If this happens prior to ratification, this would enable the EU to incorporate some of the "hot air" available from these countries into its bubble agreement (see Chapter 12). There are, however, some good reasons that speak in favour of early ratification: the pressure on Member States to finally implement effective policies and measures to reduce GHG emissions on the national and European levels would increase. Such leadership could prove that effective climate policy is possible. It would also expose the laggards to the world public, as it would be obvious who was preventing the Kyoto Protocol from entering into force (see also Chapter 25).

Without ratification of the Kyoto Protocol, the EU has made little progress in developing its policies for limiting GHG emissions.[19] The main post-Kyoto achievement has been an agreement with European car manufacturers in 1998 to limit average CO_2 emissions of new cars to 140 mg per kilometre by 2008. This represents a 25% reduction from current levels and could contribute about one-sixth to achieving the Kyoto target of the EU.[20] Negotiations with non-European car manufacturers are underway. In addition, the Council started a process in 1998 to strengthen the integration of environmental and climate policy considerations into other policy areas. Strategy papers on several relevant policy fields (energy, agriculture, transport, industry, etc.) have been or are being prepared.[21] Moreover, the entry into force of the Amsterdam Treaty in mid-1999 may improve the prospects for environmental policy making at the European level, since it strengthens

---

17 See e.g. FCCC/CP/1998/MISC.7; FCCC/SB/1998/MISC.1/Add.3/Rev.1 and Add.6.

18 See the reminder in European Commission 1999, p. 11.

19 See WWF 1998.

20 EWWE, Vol. 7, No. 15, 7 August 1998, pp. 3–5.

21 See ENDS Daily, 23 June 1999.

the rather "green" European Parliament.[22] However, the concrete impact of these factors (policy integration, Amsterdam Treaty) remains unclear. According to the European Commission's estimates of May 1999, EU total GHG emissions are still expected to increase by some 6% from the 1990 level by 2010 if further measures are not taken.[23]

In contrast to the mentioned positive developments, internal divergence with respect to the Kyoto Mechanisms has become more and more obvious as some Member States increasingly explore the potential of the mechanisms. For example, the Netherlands are planning to realise 50% (compared to a business as usual scenario) of their target under the burden-sharing agreement of 1999 by drawing on Emissions Trading, JI and the CDM.[24] In addition, progress on a number of internal policy initiatives has been lacking. As of mid-1999, a long-awaited proposal for legislation supporting the feed-in of electricity produced by renewable energy sources had still not even been tabled by the European Commission. Negotiations on a proposal for the taxation of energy products appeared to be deadlocked as well.[25] In addition, the EU has been slow to develop policies and measures for the non-CO_2 GHGs. Regarding the fluorinated gases (HFCs, PFCs, SF_6), the Union had not elaborated a strategy until mid-1999.[26]

This lack of progress contrasts with a number of analyses, which show that cost-effective options exist to implement the EU target of an emission reduction of 8% by 2010. According to the Commission's estimates after Kyoto, two-thirds of this target could be attained at a cost of less than 5 Euro per ton CO_2 equivalent.[27] The measures needed to reach that target have to be adopted by the EU and its Member States rather soon, if they are to have effect in time.

To summarise, early ratification by the EU and Central and Eastern European CEITs thus seems unlikely, but there is little doubt that these countries are willing to become Parties to the Kyoto Protocol. Eventually, ratification can be expected in the first years of the 21st century. It remains to be seen, however, whether the EU will be able to realise its leadership potential by developing its own policies and measures and by establishing stable links and coalitions with other Parties and developing countries in particular (see Chapter 25).[28]

---

22 See Bär/Kraemer 1998; whether the new rules for a Common Foreign and Security Policy, that are also part of the Amsterdam Treaty, will improve the role of the EU in climate diplomacy remains to be seen.

23 See European Commission 1999, p. 3.

24 ENDS Daily, 23 June 1999; see also the website of the Dutch Environment Ministry (VROM) at <http://www.minvrom.nl/milieu/broeikaseffect/f.htm?41901.htm> as of 24 June 1999.

25 ENDS Daily, 26 May 1999.

26 See European Commission 1999.

27 See European Commission 1999, p. 10.

28 See, generally, Gupta/Grubb 1999.

24.3 The "Umbrella Group"

In the post-Kyoto period, the "umbrella group" has become the successor of JUSSCANNZ in international climate politics. The members of the group are Australia, Canada, Iceland, Japan, New Zealand, Norway, the Russian Federation, Ukraine, and the US. In addition to the core members of JUSSCANNZ, it thus includes the two CEITs with the highest GHG emissions and the greatest interest in "hot air" trading. The umbrella group encompasses all major GHG emitters among the industrialised countries except the EU. The exact agreements between the members of this group are a matter of speculation. However, US Chief negotiator Stuart Eizenstat indicated in May 1998 that the group had achieved a "conceptual understanding ... to trade emissions rights with each other" [29]

Little has changed in *US climate politics* since Kyoto. The US administration has continued to make the "meaningful participation of key developing countries" a precondition for considering the ratification of the Kyoto Protocol. At the same time, it has worked forcefully during international negotiations to prevent any quantitative restrictions on the use of the Kyoto Mechanisms. Domestically, Congress has discussed several relevant legislative climate policy initiatives that are, however, not expected to lead to a major revision of the trend of rising GHG emissions. To the contrary, the Senate Budget Committee deleted climate-friendly tax incentives and research grants in 1998, and a number of Republicans pledged to block any administration budgetary or regulatory moves towards implementing the US's Kyoto target.[30] As in the Kyoto process, the "Carbon Club" has retained considerable influence and has worked to prevent effective action in the US (see below).

US GHG emissions are projected to increase by 23% from 1990 levels by 2010.[31] The longer the internal US climate policies remain in stalemate and the longer legislative action is postponed, the more dependent the US will become on the Kyoto Mechanisms for reaching its target of minus 7%. At the turn of the century, it may already be too late for the US to execute its commitment by domestic action alone.

However, a number of developments point in a more positive direction. First, it has become increasingly evident that the climate sceptics represent an infinitely small minority among scientists. With the news on extreme weather events and rising global temperatures increasing in frequency, their influence on public policy making in the US is bound to decrease. Second, growing sectors of the business

---

29 Stuart Eizenstat, Statement before the House International Relations Committee, Washington, D.C., 13 May 1998, as released by the Bureau of Oceans and International Environmental and Scientific Affairs (on file with the authors); on the difficulties of entering into a bubble for non-EU countries see Chapter 12.

30 "Global-Warming Debate Gets No Consensus in Industry", Wall Street Journal, 16 April 1998.

31 FCCC/CP/1998/11/Add.1.

community, including those formerly associated with the "Carbon Club", appear to have accepted the need for action (see below). After ratification of the Protocol by other major trading partners of the US, opposition against the treaty by some sectors may wane because of different standards on global markets.[32] Furthermore, President Clinton has ordered all federal government agencies to improve energy efficiency in government buildings by 35% from 1985 levels and to reduce GHG emissions from energy use in government buildings to 30% below 1990 levels by the year 2010.[33]

The presidential and congressional elections in November 2000 could be an important event, since they provide an opportunity for changing the internal balance in favour of climate policies. Early in the 21st century, the composition of the US Senate that needs to approve ratification could become more favourable. Should Argentina and Kazakhstan (and possibly other developing countries) follow through with their announcement to take on quantitative commitments to limit GHG emissions and should the CDM become operational, the US demand for "meaningful participation" of developing countries might be satisfied sufficiently. As typical of US politics, public opinion will have a decisive say in the end. Overall, ratification prospects are very uncertain in the US, but a window of opportunity may open up early in the 21st century.

Japan has a strong diplomatic interest in participating constructively in international co-operation devoted to the Kyoto Protocol. In the end, the agreement was reached in Kyoto, and Japan would certainly not like to prevent its entry into force. Regarding domestic implementation, Japan has looked intensively into its options for reaching the Kyoto target of minus 6%. Already in 1998, it started exploring possibilities for JI projects in Russia (see Chapter 13). Japan has developed a strategy for meeting its Kyoto target. This strategy aims, among other things, at a stabilisation of CO_2 emissions and intends the use of JI and Emissions Trading to contribute 1.8% of the total 6% reduction of GHG emissions. Japan has passed some legislation to this end, including a law for the promotion of measures to prevent global warming (October 1998), and plans to expand the use of nuclear power greatly in order to achieve its target.[34] As it is unlikely that Japan will politically be in a position to fully achieve its target domestically, its eventual decision on whether or not to ratify the Kyoto Protocol will also depend upon the outcome of the international negotiations on the Kyoto Mechanisms. Assuming that these will allow Japan to make use of JI and Emissions Trading to some extent, Japan can be

---

32 See Smeloff 1998, p. 67. For example, Volkswagen decided to give up production of the old "beetle" after it became clear that it would not be able to comply with high Californian environmental standards.

33 Global Environmental Change Report, Vol. XI, No. 11, 11 June 1999, p. 3.

34 "Analyses on Japan's Post-Kyoto Policy Measures to Achieve the Kyoto Target", 31 May 1999, Bonn (on file with authors); see also "Law Concerning the Promotion of Measures to Cope with Global Warming", in: Japan Environment Quarterly, Vol. 3, No. 4, December 1998.

expected to join the Protocol early in the next century (if others, e.g. the EU, do so as well).

The ratification and implementation prospects for the Kyoto Protocol in *Russia* are most uncertain. Even before Kyoto, the Russian government did not possess much of a climate policy. This situation was aggravated by the financial crisis in the autumn of 1998. At the turn of the century, the economic and political prospects of Russia are uncertain as well. The economic recovery that has been predicted almost every year since the beginning of the 1990's remains elusive. Politically, it is not clear who will succeed the current President Yeltsin in 2000. With climate protection interests meagrely represented in Russian politics, much will depend on the international discussions on the design of Emissions Trading, as this will be decisive for Russia's ability to sell its excess emission allowances ("hot air"). The establishment of an Emissions Trading system would provide a major incentive for Russia to join the Kyoto Protocol.

The ratification of *other non-EU industrialised OECD countries* is not as decisive for the entry into force of the Kyoto Protocol, since their share of the total Annex I CO_2 emissions is relatively small. Depending upon which of the big players ratify, the smaller emitters could, however, tip the scales for reaching the necessary 55% of the total Annex I CO_2 emissions in 1990 for the Protocol to enter into force. No uniform development is visible in that respect. Norway and Switzerland reached their objectives in Kyoto to a large extent and have a history of constructive participation in international efforts. They may thus be expected to become members of the accord eventually, assuming, once again, a satisfying conclusion to the talks on the Kyoto Mechanisms. The situation in other countries (Australia, Canada, New Zealand) is more uncertain. In the end, these countries may ratify so as to be "in the boat", but are likely to allow others to launch it. Overall, these countries have taken little domestic action to revert GHG emission trends in an attempt to bring them closer to their Kyoto targets.

24.4 Developing Countries

Because a minimum of 55 ratifications are required, developing countries must be included in the game if the Kyoto Protocol is to enter into force. In several aspects, their involvement is decisive for the future success of the international cooperation within the framework of the Protocol. In the short and medium term, their constructive input is needed to reach agreement on the Kyoto Mechanisms, in particular Emissions Trading and the CDM. In the long term, they need to limit and reduce their own GHG emissions (see Chapter 17).

In the follow-up of Kyoto, *OPEC countries* have focused their efforts on compensation for countries dependent upon fossil fuel exports. They still have the potential and capability to slow down the process, although the Kyoto process has shown that their influence is limited when the other big players are determined to continue. If the Kyoto Protocol enters into force, they can be expected to join the club, if only to be able to influence the future game. After all, these countries had the political insight to join with all other countries in adopting the Protocol. Although they would qualify as prime candidates among developing countries in terms of per-capita emissions and per-capita income, Saudi Arabia and Kuwait are unlikely to commit to any emission limitations. They can also be expected to oppose action by other countries, although complying with OPEC production quotas will be far more important for world oil prices than any climate policy in the foreseeable future.

Post-Kyoto, *AOSIS* has been torn between two partially competing objectives. On the one hand, its members have pursued their obvious interest in maximising support for countries particularly vulnerable to climate change by raising resources for adaptation through the CDM (see Chapter 14). On the other hand, they have continued to support a further elaboration of the Kyoto Protocol intended to ensure the effectiveness of its mechanisms in combating climate change. Finding a balance between these two objectives represents a considerable challenge, since maximising the use of the CDM (and thus the share of proceeds available for adaptation) may relieve industrialised countries of the need to change their own emission trajectories. Overall, AOSIS will remain the moral conscience of the international climate negotiations. Indeed, AOSIS members were among the first to sign the Protocol, and, as of 16 June 1999, seven of the 10 countries that had ratified the Protocol were AOSIS members.

The positions of many other members of *G-77 and China* on the Kyoto Protocol are even more ambiguous. Little effort has been made to ratify the Protocol in most of these countries. However, interest in doing so (in order to be part of the club) will grow once entry into force becomes likely. Besides, the positions of G-77 countries have increasingly drifted apart within the boundaries of the extreme positions taken by AOSIS and OPEC. The growing internal divergence was made most evident by Argentina's declaration at COP 4 (followed by a similar statement from Kazakhstan) that it would announce a voluntary quantitative commitment at COP 5 in 1999 (see Chapter 17). As this willingness is closely related to the interest in taking part in Emissions Trading, Argentina also has an interest in the establishment of an Emissions Trading system. This does not necessarily mean the end of G-77 solidarity (which has been announced many times in the past). But other developing countries, India and China in particular, have voiced principal concerns about the allocation and distribution of emission rights under the Protocol. These concerns have to be taken into account when seeking their approval of Emissions Trading. A constructive dialogue on the allocation of emission rights and equity, if complemented by domestic action of industrialised countries, might motivate developing countries to participate in a discussion about their own future commitments (see Chapter 25 below).

24.5 Non-governmental Developments

At first glance, post-Kyoto business has continued much the same way as it had done before. Thus, in the immediate aftermath of Kyoto, the Carbon Club launched a campaign in the US to obstruct the Kyoto Protocol. It appealed to the US Congress not to sign the Protocol and elaborated plans to "spend millions of dollars to convince the public that the environmental accord is based on shaky science".[35] The fossil fuel lobby continued its work in both international climate diplomacy and on the domestic front to prevent effective action and ratification of the Protocol. All in all, corporate attitudes in the US "remain far more reticent"[36] than in other parts of the industrialised world.

However, a number of developments nourish the expectation that the balance of power in the business community might shift towards the more moderate and progressive forces. In April 1998, the oil giant Shell followed the example of BP and left the Global Climate Coalition.[37] Afterwards, a number of transatlantic mergers in the business community exposed the respective US branches to increased pressure from their more moderate European partners. For example, BP Amoco, which emerged in 1999, took on the former progressive BP stance.[38] BP chairman John Browne has, before and after Kyoto, consistently stressed the need for the company to take a proactive stance on climate change.[39] In addition to their strategic re-positioning in 1997 by investing heavily in solar energy (see Section 6.5), Shell and BP have declared plans to reduce their own GHG emissions by 10% from 1990 levels by 2002 (Shell) and by 2005 (BP). To do so, BP will use an internal emissions trading programme. At the same time, the progressive lobby groups within business, like Cogen, the Business Council for a Sustainable Energy Future and its European counterpart (e^5), have consolidated and strengthened their influence. In summary, "various companies have undertaken initiatives ranging from the purely voluntary to agreements negotiated with governments under the threat of more onerous regulation".[40]

Environmental NGOs have continued to work on the international process and the domestic implementation in industrialised countries. US groups in particular have shifted their focus from the international to the domestic level, since only congressional support will lead to the ratification of the Protocol. Co-ordinating within the framework of the Climate Action Network (CAN), environmental

---

35 "Industrial Group Plans to Battle Climate Treaty", The New York Times, 26 April 1998.

36 Grubb et al. 1999, p. 261.

37 See Dow Jones Newswires, 21 April 1998, "Shell oil withdraws from powerful U.S. energy lobby group"; Greenpeace Press Release, "Shell pulls out of U.S. Anti-Climate lobby group", 21 April 1998, available at <http://www.greenpeace.org/pressreleases/ 1998apr21.html>.

38 "BP Clashes with Esso over Action on Global Warming", The Guardian, 15 April 1999.

39 See Smeloff 1998, p. 67.

40 Grubb et al. 1999, p. 258.

NGOs have worked for early ratification of the Protocol and an environmentally effective design of the Kyoto Mechanisms on the international level. Environmental NGOs have faced two particular problems in the post-Kyoto period. First, the importance of climate change has generally declined on public agendas in industrialised countries. Second, NGOs have had difficulties conveying clear messages to the public because of the increasing complexity of the matter (Kyoto Mechanisms, sinks). With the adoption of the Kyoto Protocol, the environmental lobby has lost its focal point of the last years, which reinforced these problems (see below).

24.6 Conclusion: The State of Climate Policy and Some Underlying Causes

The international landscape of climate politics has remained relatively stable in the post-Kyoto period. As we enter the 21st century, the roles between the big players are still very similar to what they have been over the past years. Going beyond governments, civil society is characterised by contradictory trends. On the one hand, the importance of climate change has diminished on the public agenda in many industrialised countries (despite a continuous high public awareness). On the other hand, business has increasingly accepted the need to act, and big companies have strategically re-oriented themselves towards a more climate-friendly development. However, GHG emissions have continued to grow unabatedly in almost all countries.

At the international level, the Buenos Aires Plan of Action set the stage for the future international process under the FCCC and its Kyoto Protocol. This international process has lost momentum in the post-Kyoto period. In fact, progress is not only slow, it is nearly non-existent. Rather, various laggards appear to grasp every opportunity to torpedo the Kyoto agreement and retreat behind it.

Several underlying trends help to explain this situation. First, domestic implementation of the Protocol, which is the key to progressive positions on the international level, has remained, with few exceptions (see above), superficial at best. Second, the domestic political conflicts of the USA have been transferred to the international level. The call for "meaningful participation" of developing countries in the Kyoto-Protocol, formulated by the US Senate, was taken over by the US administration in order to facilitate the ratification of the Kyoto Protocol. The resulting demand for "voluntary commitments" of developing countries catalysed the formation of an insurmountable block, made up of the oil-producing countries, China, India and other developing countries, and led to stalemate in the international talks after Kyoto.

Third, the institutional capacity of the COP and the COP/MOP of the Protocol to deal with these problems appears to be woefully inadequate. The Convention process still lacks an agreement on voting in the Rules of Procedure. This structural deficit has been incorporated in the Kyoto Protocol and will likely further stifle progress. The consensus requirement will lead to even more decisions based on the lowest common denominator, as it grants individual countries and small groups of countries representing special interests, like OPEC or Australia, undue influence. As a result, more "loopholes" concerning the industrialised countries' obligations can be expected, while the more progressive forces, like the EU and AOSIS, are impeded in showing leadership.

Fourth, the negotiations after Kyoto are characterised by a record complexity. By bringing in several "flexibility instruments" and by including GHG sinks in the Kyoto Protocol, JUSSCANNZ managed to overload the process with a multitude of technical problems. Post-Kyoto, virtually no one involved in the negotiations is capable of grasping the overall picture of the climate negotiating process.

As a consequence, representatives of civil society also find themselves caught up in the complexity trap. Not only are they trapped in the microcosm of technical details, where they are compelled to become increasingly specialised, but they are also deprived of their role as intermediaries to the media. The big headlines are missing. After the acceptance of the Kyoto Protocol with its binding targets for reducing GHG emissions, they lack a concrete and compelling objective that could be presented to the public. Without a clear message, however, NGOs have difficulties raising public support and exerting pressure on governments.

Under these circumstances, the prospects for the entry into force and implementation of the Kyoto Protocol at the turn of the century are still very much uncertain. If domestic action on climate change in industrialised countries and entry into force are not achieved by 2003/2004, saving the Kyoto Protocol from complete failure will be a giant's task, possibly an unattainable one.[41] In this case, it will be impossible to reach the targets established by the Kyoto Protocol, in particular for the US and Japan, since reverting the emission trend takes time. Even if the window of opportunity is grasped, use of the Kyoto Mechanisms will be urgently needed, especially by the US and Japan (possibly by some European countries as well), to comply with the targets. While the resolution of the relevant pending issues on the international level is therefore of great importance, the major objectives of the Kyoto Protocol remains taking effective domestic action.

At the turn of the millennium, international climate policy has arrived at a critical juncture. The international process is clearly lacking fresh momentum, vision and leadership. After a decade of international discussions, the time is ripe for a leadership initiative based on strong domestic implementation. As things stand, there is only one industrialised player who could provide such leadership at the turn of the millennium: the European Union (see Chapter 25).

41 See also Grubb et al. 1999, pp. 253 et seq.

25. A View from the Anthill: Towards a Leadership Initiative on Climate Change

Following the third Conference of the Parties (COP 3) at Kyoto in November 1997, the Kyoto Protocol was regarded with a certain awe and was characterised as a "koan" in the Zen tradition.[1] Since then, the international process has become entrenched in battles of retreat, most notably with regards to the concrete design of the various flexibility instruments, the so-called "Kyoto Mechanisms" (see Section 24.1).

Time, however, is running and each year the implementation of the Kyoto targets is delayed renders their realisation more unlikely. According to projections of the International Energy Agency, CO_2 emissions in all major industrialised countries will rise considerably by 2010 if no additional measures are taken (see Chapter 22). Drastic policy changes are thus required soon in all major industrialised countries in order to achieve the Kyoto targets. If this policy change is not implemented, the urgent re-orientation of industrial societies towards a climate-friendly economy with low fossil fuel consumption will be delayed. A stagnant climate policy on the part of the Western industrialised countries, upon whom the world takes example, would serve to reinforce the traditional development path now followed by the developing countries – with a profound impact on future global emissions.

International climate policy at the turn of the millennium is clearly lacking momentum and initiative. A vicious cycle of inaction has emerged. While the more progressive industrialised countries (essentially a number of EU Member States) wait for the US to take on its responsibility and join forces with them, the US and other laggard countries like Australia are striving to delay action further in order to gain more concessions. The later agreement is reached and action starts, the better their case for watering down existing obligations, as these may not realistically be within reach anymore.

25.1 The Rationale of a Leadership Initiative on Climate Change

Although the EU has remained committed rhetorically to leadership on climate change post-Kyoto, its deeds hardly do justice to its words. The post-Kyoto strategy of the Union of binding the US into an agreement at all costs has proven ineffective. In contrast, experience with other regimes has shown that leadership can be exerted most effectively if a combination of three components is applied. First, the leader must make use of general political and economic weight ("structural leadership"). Second, leadership in international negotiations benefits

1 Earth Negotiations Bulletin, Vol. 12, No. 76, 13 December 1997, p. 14.

clearly from the skilful building of coalitions ("instrumental leadership"). Third, and perhaps most importantly, credible leadership is most effectively advanced by demonstrating solutions to others ("directional leadership"). Domestic action, i.e. "leadership by example", is a decisive element of the latter aspect.[2]

The emergence of the Montreal Protocol on Substances that Deplete the Ozone Layer in the mid-1980s is a prime example of strong environmental leadership on the part of the US (an unlikely possibility in the 1990s).[3] Similar leadership strategies proved highly successful in the negotiation of the Land Mines Treaty. A pioneering group assuming leadership on climate change would receive massive support from non-governmental organisations, as could be observed during the establishment of an International Criminal Court in Rome in June 1998.

Given the reluctance shown by the US and Japan, the EU is the only candidate that might reasonably be expected to grasp a Leadership Initiative on climate change. Led by the Union, a coalition that actively and publicly pursues climate protection could be formed on the international level. With some prospects for success, the EU could aim at winning support on different issues from a number of countries with economies in transition (CEITs), from committed developing countries and, possibly, from Japan as the host of COP 3. In this manner, the critical mass of countries necessary to breath new life into international climate policy could be reached.

The initiative would need to be pursued within the framework of the climate regime but independent of the co-operation of those unwilling to proceed.[4] As stated before, independent action is required, since simply waiting for the laggards would certainly paralyse the process. At the same time, however, it is important that a leadership group remains part of the climate regime, as any devaluation of the existing regime could severely jeopardise the climate process as such. As fragile as this process may seem, it is the only credible and viable one for the intergovernmental negotiating process at this moment and for a considerable time to come. The FCCC and the Kyoto Protocol have a considerable future potential, as they provide much of the structure needed for achieving effective international co-operation for protecting of the Earth's climate. The climate regime furthermore enjoys wide public attention and approval. Weakening the FCCC and the Kyoto Protocol could thus play into the hands of those inimical to the concept of restraints on the use of fossil fuels.

A Leadership Initiative should help give momentum to the international climate process through concurrent action on the international and domestic levels. Should it adopt this role, Europe would need to stop staring at the US. Many consider US participation to be vital, a result of fears concerning competitive economic disad-

---

2 See Gupta/Grubb 1999.

3 Benedick 1998a; Flavin 1998.

4 This is different from the approach proposed by Christopher Flavin, who suggests the formation of a leadership group of committed countries outside of the climate regime, joined by regional governments, cities and companies; see Flavin 1998.

vantages. Yet, by exploiting the available low-cost potentials for reducing GHG emissions and by investing in an ecological modernisation that will eventually lead to economic benefits, there is much room for acting without the US.[5] It is time to realise that the EU and others have been waiting for the US for nearly a decade now. Recent rumours suggest that COP 6 might be postponed until early 2001 in order to wait for the outcome of the next presidential elections in the US in November 2000. Climate protection was postponed in a similar way by the US presidential elections in 1992 and in 1996. Although the candidate supported by environmentalists won both times, the US has not altered its position much. The time has come for the EU to take the lead.

In many respects, the Union has a large potential for undertaking a renewed Leadership Initiative. It is without doubt one of the most influential players in international affairs, climate policy in particular, and has command over significant political and economic resources (structural leadership). It could use these resources to put pressure on its negotiating partners and to build effective coalitions. Its diverse diplomatic capabilities are enhanced by the combined experience of the Member States in foreign relations (instrumental leadership). The UK and France, but also Germany and other EU members, have long-established close relations with many parts of the world.[6] If these advantages were to be combined, the EU could initiate a strong leadership coalition on climate change. Last but not least, both the diverse experiences with implementation in the Member States and the existing system for co-ordinating policies on the European level give the Union exceptional tools for taking domestic action, and thus for leading by example and increasing the pressure on the US and others (directional leadership).

The envisioned Leadership Initiative might focus on the following core elements. First, it should aim for the *early ratification* of the Kyoto Protocol. This should be top priority. Second, the initiative should introduce measures for *domestic implementation* of the Kyoto obligations and make a concurrent effort to co-ordinate such measures internationally. Third, since the involvement of *developing countries* in the climate process is of utmost importance for the medium and long term effectiveness of climate protection and for the future development of the climate regime, special efforts need to be undertaken to integrate these countries into the formation of international climate policy. Each of these elements is explored in the following sections.

25.2 The First Element of a Leadership Initiative: Early Ratification

Ratification prospects and the entry into force of the Protocol after Kyoto are rather bleak (see Chapter 24). Since almost all of those AOSIS countries that are Parties to the Convention will join, the 55 ratifications required should not hinder

5 About two-thirds of the EU target could be met at low costs of up to 5 Euro per tonne of CO_2 equivalent, see European Commission 1999, p. 10.

6 See European Commission 1998b.

the entry into force of the Protocol. A serious obstacle is the requirement that the ratifying Annex I Parties represent at least 55% of the total CO_2 emissions of 1990 allocated to Annex I Parties (see Section 20.1). Therefore, at least two of the three main emitters (the US, EU and Russia) must ratify. First and foremost, this would require an internal strategy for a synchronised ratification of the Union and its Member States, as proposed by the Commission in mid-1999.[7]

Furthermore, the Union could attempt to break Russia and Japan (and the Ukraine) out of the "Umbrella Group" and win the support of other countries with economies in transition. The Umbrella Group was formed in Kyoto and comprises the core JUSSCANNZ countries (US, Japan, Canada, Australia, Norway, New Zealand) plus Russia and the Ukraine. United by the common goal to design the Kyoto Mechanisms as flexibly as possible, this group has effectively played the counterpart to the EU in the post-Kyoto negotiations to date.

Regarding climate change, it is perhaps Russia's major interest to sell as much of its hot air, i.e. its excess emission allowances (see Chapter 15), as possible. Since selling this hot air requires the entry into force of the Protocol and the trading regime, Russia has a strong interest in this. While it would like to have the US as a possible buyer on board, it would be more profitable for Russia to sell at least *some* of its emission allowances to interested European countries (and Japan) than to sell nothing at all, as would be the case should the Protocol not enter into force.

The Union could exploit this situation, but this might require a more proactive and flexible stance on the Kyoto Mechanisms. For example, the EU's request for a quantitative cap on the use of these Mechanisms is a major obstacle for an alliance between the EU and Russia (and Ukraine). The EU was already unsuccessful in pushing through this request at Kyoto.[8] The extent to which the EU's inflexibility on this matter delays entry into force of the Protocol, the pressure to allow the "loopholes" to be widened (most importantly the inclusion of further sink categories) increases. As time proceeds without the advent of decisive domestic implementation on the horizon, it becomes more and more difficult for many industrialised countries to meet their targets. This is also the case for a number of EU Member States, which intend to rely heavily on Emissions Trading, JI and the CDM for the fulfilment of their obligations.[9]

This is not to say that the EU must show utmost flexibility so as to achieve this aim. On the contrary, clear definitions, strict monitoring and reporting and the effective enforcement of the rules of the regime are of prime importance for the environmental and economic effectiveness of the mechanisms. It is essential that

---

7 See European Commission 1999, p. 11.

8 The US therefore accused the EU of trying to "rewrite the Kyoto Protocol"; see the quote by James Foley of the State Department in Global Environmental Change Report, Vol. XI, No. 10 of 28 May 1999, p. 1; see also "Clinton accuses EU of Trying to Rewrite Global Warming Pact", Wall Street Journal, 18 May 1999.

9 The Netherlands, for example, plan to achieve 50% of its target compared to the business-as-usual scenario with the Kyoto Mechanisms, see ENDS Daily, 23 June 1999.

the EU places emphasis on these points. Furthermore, an inflationary addition of new categories of sinks under the Protocol, which are only quantifiable in the rarest of cases, might jeopardise the basis of the Kyoto Protocol. The Union would earn much credit and would put its opponents on the defensive if it would insist upon a sound design for the Kyoto Mechanisms and for the rules on sinks.

However, increased flexibility regarding the "supplementarity" could pave the way for the formation of a strategic alliance with Russia and Japan that would secure the emission percentages required to enable the entry into force of the Protocol. Japan has a profound diplomatic interest in the success of the Kyoto Protocol which will provide a further incentive to join this new alliance. It should also be possible to win the support of eastern European CEITs, especially the candidates for accession to the EU, as these have tended to align themselves with the EU during the 1990s. In such a scenario, it might also be possible for this group of countries to agree upon a fee for all activities under the Kyoto Mechanisms (particularly on Emissions Trading), as is the case for the CDM. This would level the playing field between the different mechanisms and thus increase the chances of winning the approval of many developing countries that are interested in the CDM. It would also raise the transaction costs of Emissions Trading and thereby provide domestic measures with a comparative advantage (see Section 15.3.3).

This alliance would end the stifling confrontation between the blocks (EU – JUSSCANNZ; EU – Umbrella Group) by building new coalitions. It would have substantial weight in the elaboration of the Kyoto Mechanisms and could achieve an early entry into force of the Protocol after COP 6, during the first year of the new century. This would increase the pressure on others to join. The US would not be in such an advantageous position anymore to influence the decisions if it stayed outside the Protocol. It would end the *de-facto* veto power of the US in the international process and it would demonstrate again that the world can act independently of the last remaining superpower, as it has done in other cases.

25.3 The Second Element of a Leadership Initiative: Measures for Domestic Implementation and their Co-ordination

Domestic implementation is possibly the most fundamental basis for leadership on climate change. With all its weaknesses during the negotiations of the Kyoto Protocol, EU leadership was most credible when it started to lay the ground for domestic implementation (see Section 6.1). Post-Kyoto, however, this has not been followed up upon. Due to a lack of political will, the fear of comparative disadvantages on global markets, serious resistance on the part of some industrial sectors and a lack of public pressure, comprehensive domestic policies and measures for an effective climate policy are still the exception. A second pillar of a Leadership Initiative could therefore be the implementation of policies and measures among a larger group of countries to begin the long process of structural decarbonisation of the economy.

Even before Kyoto, a number of studies had provided evidence of the vast potential of low- and no-cost options for reducing GHG emissions in the EU and elsewhere (see Chapters 2 and 6). This has been confirmed and reinforced post Kyoto. An analysis of the European Commission in 1999 concluded that the EU can achieve two-thirds of its target of minus 8% for the first commitment period by implementing low-cost measures of up to 5 Euro per tonne of CO_2 equivalent.[10] Given this potential, it is not the economic costs but political opposition that prevents implementation of the Kyoto target in the EU.

Within the framework of a Leadership Initiative, such impediments could be removed by common political will of the EU Member States and other allies. Remaining concerns about competitiveness could be minimised by a co-ordination of policies, for example with Japan. The EU and its Member States have a wealth of experience in policy co-ordination which they can easily employ for this purpose, especially since preparatory work has been done by the European Commission and others in past years (see e.g. Section 6.1).

Experience with the rather unsuccessful attempts of the EU to establish binding policies and measures in the Protocol (see Chapter 10) suggests that, first, emphasis should not be placed on binding measures but on a transparent process of co-ordination subject to a high degree of public scrutiny. Second, due to the benefits that can be expected from co-ordination, the initiative should concentrate on a limited set of measures upon which agreement can most likely be reached. A leadership group of committed countries could thus emerge and demonstrate that ecological modernisation and an efficient economy are not only compatible, but also mutually reinforcing targets.

Fiscal policy is one of the most important tools of climate policy. Even minor tax differentiations can be very effective, as the experience with the phasing out of leaded petrol in Europe shows.[11] A number of countries (especially in Europe) have already successfully implemented energy/carbon taxes. This policy instrument does have a direct impact on the competitive situation of various businesses, and many of the relevant countries have therefore granted tax exemptions of various degrees to relevant industries. Because of this competitiveness aspect, the case for a European and international co-ordination of related efforts is strong. Taxation on a global level, however, appears to be politically unfeasible, largely due to fierce US opposition.[12] Japan is generally more favourable of taxation and could be an ally for the co-ordinated national implementation of carbon/energy taxes.[13]

---

10 European Commission 1999, p. 10.

11 See European Commission 1999, p. 7.

12 See for example the call by Cooper 1998 and Schneider 1998.

13 According to a recent poll released by the Japanese Environment Agency, the majority of Japanese firms support a tax on fossil fuels to curb CO_2 emissions, <http://biz.yahoo.com/rf/990527/u.html> as of 27 May 1999.

Furthermore, structural decarbonisation in the next century demands the fast development of low- or no-carbon energy sources.[14] A second focus of the leadership group might therefore be a commitment for a large-scale co-ordinated effort in research and development for renewable energy sources and for the efficient use of energy.[15] Co-ordination of such efforts certainly has a significant potential for producing synergies. Relying partly on existing international institutions and research networks, such a co-ordinated R&D strategy, should bring about substantial progress within the next 10 years, if funded appropriately.

To this end, resources could become available if, third, international agreement on guidelines for dismantling climate adverse subsidies could be reached. Numerous studies have shown that vast amounts of financial resources are flowing into carbon intensive and unsustainable industries and activities.[16] The rationale for such subsidies is frequently the protection of relevant industries from international competition. Internationally co-ordinated action could thus help address this concern. Such an initiative could be linked to efforts within the World Trade Organization (see Chapter 23). At least part of those subsidies could then be diverted into R&D and into (transitory) support for the development and introduction of renewable energy sources into the market.

Fourth, there is a rationale for co-ordinating efforts to strengthen the energy efficiency standards existing in all major industrialised countries. Improving the efficiency of energy use is not only of major strategic importance for a significant reduction of global GHG emissions. Co-ordination of efficiency standards (for household appliances, etc.) should also be of interest to most industrialised countries as it facilitates international trade by removing or avoiding potential trade barriers (e.g. where meeting efficiency requirements is a precondition for being allowed into foreign markets). Finally, reaching agreement on climate-friendly public procurement and emission reduction measures for the public sector[17] should be politically feasible as well. The public sector accounts for a large share of domestic demand and can thus exercise formidable influence on the market.[18]

These activities of the Leadership Initiative for the co-ordination of domestic measures could both decisively assist the members in implementing their obligations under the Kyoto Protocol and begin a process of learning. It should be complemented by strong public statements for ratification and implementation by the group of Parties. Such a strategy was highly successful in the ozone negotiations,

---

14 See Jacoby et al. 1998, p. 61.

15 See also the Battelle Global Energy Technology Strategy Project to Address Climate Change, <http://gtsp.pnl.gov/ge/home.nsf/webpage/> for a technology-based initiative.

16 See OECD 1995; Moor/Calami 1997; more specifically on climate see Koplow/Martin 1998.

17 See "Clinton orders government to reduce energy use and emissions", Global Environmental Change Report, Vol. XI, No. 11, 11 June 1999, p. 3.

18 Benedick 1998a, p. 20; see also the Climate Technology Initiative of the IEA: "Enhancing Markets for Climate Friendly Technologies: Leadership Through Government Purchasing"; June 1998, available at <http://www.iea.org/climate.htm>.

where various leading countries, in the absence of a consensual agreement on further measures, annexed resolutions and declarations with this content to the reports on the Meetings of the Parties. The majority of these demands were satisfied in the next round of negotiations.[19]

There are special advantages of proceeding in a co-ordinated manner in the mentioned areas. In addition, and irrespective of broader international co-ordination, there is ample room for unilateral action to implement the Kyoto commitments in the EU and elsewhere. Such unilateral efforts will be indispensable to successfully implement the Protocol and to underpin leadership on climate change.

25.4 The Third Element of a Leadership Initiative: Enhanced Involvement of Developing Countries

For a long time, indeed since the historical alliance between the EU and a great number of the developing countries at COP 1 in Berlin (see Chapter 4), the EU has neglected its traditionally close relations with developing countries. Increased participation of developing countries in the climate regime might therefore be a third goal of the Leadership Initiative. Although the demand of the US for "meaningful participation" is inappropriate and even damaging at the present stage of development of the international climate regime, the long term sustainability of our planet will require the substantial involvement of those countries in the not-too-distant future (see Chapter 17). Because the forceful demands of JUSSCANNZ before and after Kyoto have increased the sensitivity of developing countries, any such diplomatic effort needs to be conducted with great care if it is to succeed.

Developing countries' needs and interests should be addressed first. A primary interest of many developing countries lies with adaptation, since they are particularly vulnerable to changes in the global and regional climate. For this reason, an adaptation component was part of the Clean Development Fund proposed by Brazil in the run-up to Kyoto.[20] After Kyoto, the attendants of the African Ministerial Conference on Environment (AMCEN) in October 1998 recommended the establishment of an "adaptation fund" and a "seed fund".[21] At present, limited funds are available through the GEF for adaptation. Article 12.8 of the Kyoto Protocol furthermore requires that "a share of the proceeds" from CDM activities is to be used to assist developing countries to meet the costs of adaptation (see Chapter 14). However, these resources are rather limited, and will continue to be so even after the CDM becomes operational.

A first emphasis of the Leadership Initiative might therefore be on an adaptation strategy between the EU and some key developing countries. This should include both the financing of studies related to impacts and adaptation and the making available of additional resources for adaptation, e.g. through the estab-

---

19 See for example Ott 1998a, pp. 200 et seq.

20 See FCCC/AGBM/1997/MISC.1/Add.3.

21 See UNEP press release 1998/11.

lishment of an adaptation fund.[22] The resources needed to finance an adaptation strategy could be generated by a fee (similar to the one already applicable under the CDM) being raised on activities under all three mechanisms (see above), which could be used, at least partly, for such a fund. As a beneficial side-effect of this initiative, the true costs of adapting to climate change might become visible. This would put cost estimates for mitigating climate change into perspective.

Such an initiative on adaptation would generate a lot of support inside and outside the climate regime. It would demonstrate that the special situation of developing countries is being taken into account, without raising fears of a hidden agenda. And it would free AOSIS countries from their dilemma to advocate a strong and reliable design of the CDM and, at the same time, push for as high revenues as possible that could be used for adaptation (see Chapter 14).

A second short term component of the Leadership Initiative could consist of a diplomatic collaborative effort to design the CDM in an environmentally effective and economically efficient manner. The issues to be addressed are not only those pertaining to the proper functioning of this instrument, like verifiable and strict baselines, comparable methodologies for verification etc., but also those that are primarily emphasised by developing countries, like the "financial additionality" with regard to other sources of funding (Official Development Assistance) and the integration of sustainable development objectives.[23] A truly collaborative effort to design the CDM in a mutually beneficial manner could serve as a confidence-building exercise.

Third, the Leadership Initiative could begin a dialogue with developing countries on the fair and equitable allocation of "emission rights". The allocation of "assigned amounts of emissions" on the basis of 1990 emissions ("grandfathering") in the Kyoto Protocol was perceived by many to be inequitable. A number of alternative proposals have been made for the distribution of emission rights, including the pre-Kyoto Brazilian proposal to allocate emission rights on the basis of historic emissions ("effective emissions"). Most prominent is probably the proposal to aim at convergence towards equal per-capita emissions, as advocated by India at COP 1 in Berlin.[24] Others have proposed to take GDP per capita and other factors into account to allocate emissions allowances (see Chapter 17).

This is an extremely sensitive issue. Emphasis should therefore be placed on building confidence and common understanding through a discussion process

---

22 See also Humphreys, Stephen/Sokona, Youba/Thomas, Jean-Philippe: "Equity in the CDM", ENDA TM, Dakar, <http://www.enda.sn/energie/cdmequity.htm> as of 15 October 1998; Mathur, Ajay: "Climate Change: Post-Kyoto Perspectives from the South", TERI, New Delhi, <http://www.teriin.org/climate/cp-4/contents.html> as of 7 July 1999.

23 Humphreys et al., see footnote above.

24 See "Contraction and Convergence: A Global Solution to a Global Problem", Global Commons Institute, 18/07/1997 <http://www.gci.org.uk/contconv/cc.html> as of 9 June 1999; see also Friends of the Earth International 1998; INFRAS AG/TERI 1997; Kinzig/Kammen 1998; Baumert et al. 1999.

separate from negotiations. Such a discrete process might also take tension out of the negotiations if the EU (and its allies) would carefully avoid putting any pressure on their partners.[25] Apart from addressing the difficult issue of allocation of emission allowances between industrialised and developing countries, these discussions might also pave the way for a more constructive approach towards differentiation between developing countries.

The problem is partly caused by the binary classification of Parties as Annex I/non-Annex I countries in the Convention. The new Annex B of the Kyoto Protocol perpetuates this artificial division. The dividing line in economic and ecological terms, in contrast, runs through these categories. Some of those countries labelled as developing countries are more affluent than their counterparts: the GDP per capita of Singapore, South Korea[26] and Israel is equal or higher than that of some members of the European Union, such as Greece, Spain and Portugal.[27]

A process for the differentiation of commitments for developing countries appears therefore reasonable. The Kyoto Protocol itself recognises the need for differentiation through the specific targets listed in Annex B and through the "joint fulfilment" mechanism under Article 4, allowing the EU Member States to reach their reduction target via an internal burden-sharing (see Chapter 12). During the negotiations on commitments in the AGBM, a number of countries sought to establish differentiated targets based on a variety of possible criteria.[28] A considerable number of Latin American and AOSIS countries are open to some form of specific commitments for developing countries.[29]

The failure of industrialised countries to respond adequately to the challenge of climate change so far has not enhanced the willingness of developing countries to contribute to mitigation efforts in the near future.[30] Nevertheless, major developing countries have effectively de-coupled economic growth and emissions of GHGs to

25 By presenting its "Annex X" before Kyoto, the EU did give the impression that it was open for non-OECD developing countries, which severely strained its relations with these countries.

26 OECD member since 1996.

27 In comparison with recent OECD countries like Mexico, Turkey and Hungary their figures are up to nine times as high; see Fischer 1998; CIA World-Factbook 1998 <http://www.odci.gov/cia/publications/factbook> as of 6 April 1999.

28 Their proposals were issued as FCCC/AGBM/MISC.3 and MISC.3/Add.1 to 3.

29 For example Argentina, Kazakhstan and Costa Rica have already announced their willingness to take on binding commitments. South Korea is considering this as well, see "Towards Global Participation", Presentation by Mr. Raeckwon Chung on behalf of the Korean Ministry of Foreign Affairs and Trade at the OECD/IEA Climate Change Forum in Paris, 10 March 1999 (on file with the authors).

30 In the United States, net emissions of all GHG increased by 21.54% from 1990 to 1997 (EPA draft US GHG inventories, 3 February 99, <http://www.epa.gov/globalwarming/inventory/1999inv.html> as of 9 June 1999); of the 24 countries listed in Annex II, only seven are expected to emit CO_2 at or below 1990 levels by the end of the millennium (FCCC/CP/1998/11/Add.2).

a larger extent than many industrialised countries.[31] Under these circumstances, a carefully crafted effort is needed in order to begin a long term discussion process aimed at developing a process for the participation of developing countries that is considered fair by everyone involved.

25.5 Conclusion

After a decade of talk, it is high time to enter the new millennium with renewed momentum and to drive international climate policy towards action to reduce GHG emissions. The EU is the only industrialised player in sight with the ability to undertake a Leadership Initiative aimed at (1) bringing the Kyoto Protocol into force, (2) initiating broad domestic implementation of the Kyoto targets and (3) starting an international process of discussion with the long term goal of involving developing countries more closely and more equitably in the international efforts.

If the envisioned Leadership Initiative can be established successfully, it will open up a vast array of future opportunities. It will set into motion a societal learning process in which it will become apparent that well-being can indeed be improved upon without burning ever increasing amounts of fossil fuels. This could decisively facilitate the next steps of the international process. In this respect, quantified targets for industrialised countries for the second commitment period from 2013–2017 will be on the international agenda in the middle of the next decade. A substantial strengthening of existing commitments will be required in order to direct industrialised countries towards the long term goal of an emission reduction in the order of 80%. If an equitable scheme can be devised, commitments by the biggest emitters amongst developing countries could also become a reality in the second decade of the 21st century. After nearly 10 years of development, the Leadership Initiative within the climate regime would thus provide the opportunity for dynamic progress towards protecting the Earth's climate for the benefit of humankind.

31 See Reid/Goldemberg 1997; 1998.

Appendix

Kyoto Protocol to the United Nations Framework Convention on Climate Change

The Parties to this Protocol,

Being Parties to the United Nations Framework Convention on Climate Change, hereinafter referred to as "the Convention",

In pursuit of the ultimate objective of the Convention as stated in its Article 2,

Recalling the provisions of the Convention,

Being guided by Article 3 of the Convention,

Pursuant to the Berlin Mandate adopted by decision 1/CP.1 of the Conference of the Parties to the Convention at its first session,

Have agreed as follows:

Article 1

For the purposes of this Protocol, the definitions contained in Article 1 of the Convention shall apply. In addition:

1. "Conference of the Parties" means the Conference of the Parties to the Convention.
2. "Convention" means the United Nations Framework Convention on Climate Change, adopted in New York on 9 May 1992.
3. "Intergovernmental Panel on Climate Change" means the Intergovernmental Panel on Climate Change established in 1988 jointly by the World Meteorological Organization and the United Nations Environment Programme.
4. "Montreal Protocol" means the Montreal Protocol on Substances that Deplete the Ozone Layer, adopted in Montreal on 16 September 1987 and as subsequently adjusted and amended.
5. "Parties present and voting" means Parties present and casting an affirmative or negative vote.
6. "Party" means, unless the context otherwise indicates, a Party to this Protocol.
7. "Party included in Annex I" means a Party included in Annex I to the Convention, as may be amended, or a Party which has made a notification under Article 4, paragraph 2(g), of the Convention.

Article 2

1. Each Party included in Annex I, in achieving its quantified emission limitation and reduction commitments under Article 3, in order to promote sustainable development, shall:
   (a) Implement and/or further elaborate policies and measures in accordance with its national circumstances, such as:
      (i) Enhancement of energy efficiency in relevant sectors of the national economy;
      (ii) Protection and enhancement of sinks and reservoirs of greenhouse gases not controlled by the Montreal Protocol, taking into account its commitments under relevant international environmental agreements; promotion of sustainable forest management practices, afforestation and reforestation;
      (iii) Promotion of sustainable forms of agriculture in light of climate change considerations;
      (iv) Research on, and promotion, development and increased use of, new and renewable forms of energy, of carbon dioxide sequestration technologies and of advanced and innovative environmentally sound technologies;
      (v) Progressive reduction or phasing out of market imperfections, fiscal incentives, tax and duty exemptions and subsidies in all greenhouse gas emitting sectors that run counter to the objective of the Convention and application of market instruments;
      (vi) Encouragement of appropriate reforms in relevant sectors aimed at promoting policies and measures which limit or reduce emissions of greenhouse gases not controlled by the Montreal Protocol;
      (vii) Measures to limit and/or reduce emissions of greenhouse gases not controlled by the Montreal Protocol in the transport sector;
      (viii) Limitation and/or reduction of methane emissions through recovery and use in waste management, as well as in the production, transport and distribution of energy;
   (b) Cooperate with other such Parties to enhance the individual and combined effectiveness of their policies and measures adopted under this Article, pursuant to Article 4, paragraph 2(e)(i), of the Convention. To this end, these Parties shall take steps to share their experience and exchange information on such policies and measures, including developing ways of improving their comparability, transparency and effectiveness. The Conference of the Parties serving as the meeting of the Parties to this Protocol shall, at its first session or as soon as practicable thereafter, consider ways to facilitate such cooperation, taking into account all relevant information.
2. The Parties included in Annex I shall pursue limitation or reduction of emissions of greenhouse gases not controlled by the Montreal Protocol from avia-

tion and marine bunker fuels, working through the International Civil Aviation Organization and the International Maritime Organization, respectively.

3. The Parties included in Annex I shall strive to implement policies and measures under this Article in such a way as to minimize adverse effects, including the adverse effects of climate change, effects on international trade, and social, environmental and economic impacts on other Parties, especially developing country Parties and in particular those identified in Article 4, paragraphs 8 and 9, of the Convention, taking into account Article 3 of the Convention. The Conference of the Parties serving as the meeting of the Parties to this Protocol may take further action, as appropriate, to promote the implementation of the provisions of this paragraph.
4. The Conference of the Parties serving as the meeting of the Parties to this Protocol, if it decides that it would be beneficial to coordinate any of the policies and measures in paragraph 1(a) above, taking into account different national circumstances and potential effects, shall consider ways and means to elaborate the coordination of such policies and measures.

Article 3

1. The Parties included in Annex I shall, individually or jointly, ensure that their aggregate anthropogenic carbon dioxide equivalent emissions of the greenhouse gases listed in Annex A do not exceed their assigned amounts, calculated pursuant to their quantified emission limitation and reduction commitments inscribed in Annex B and in accordance with the provisions of this Article, with a view to reducing their overall emissions of such gases by at least 5 per cent below 1990 levels in the commitment period 2008 to 2012.
2. Each Party included in Annex I shall, by 2005, have made demonstrable progress in achieving its commitments under this Protocol.
3. The net changes in greenhouse gas emissions by sources and removals by sinks resulting from direct human-induced land-use change and forestry activities, limited to afforestation, reforestation and deforestation since 1990, measured as verifiable changes in carbon stocks in each commitment period, shall be used to meet the commitments under this Article of each Party included in Annex I. The greenhouse gas emissions by sources and removals by sinks associated with those activities shall be reported in a transparent and verifiable manner and reviewed in accordance with Articles 7 and 8.
4. Prior to the first session of the Conference of the Parties serving as the meeting of the Parties to this Protocol, each Party included in Annex I shall provide, for consideration by the Subsidiary Body for Scientific and Technological Advice, data to establish its level of carbon stocks in 1990 and to enable an estimate to be made of its changes in carbon stocks in subsequent years. The Conference of the Parties serving as the meeting of the Parties to this Protocol shall, at its first session or as soon as practicable thereafter, decide upon modalities, rules and guidelines as to how, and which, additional human-

induced activities related to changes in greenhouse gas emissions by sources and removals by sinks in the agricultural soils and the land-use change and forestry categories shall be added to, or subtracted from, the assigned amounts for Parties included in Annex I, taking into account uncertainties, transparency in reporting, verifiability, the methodological work of the Intergovernmental Panel on Climate Change, the advice provided by the Subsidiary Body for Scientific and Technological Advice in accordance with Article 5 and the decisions of the Conference of the Parties. Such a decision shall apply in the second and subsequent commitment periods. A Party may choose to apply such a decision on these additional human-induced activities for its first commitment period, provided that these activities have taken place since 1990.

5. The Parties included in Annex I undergoing the process of transition to a market economy whose base year or period was established pursuant to decision 9/CP.2 of the Conference of the Parties at its second session shall use that base year or period for the implementation of their commitments under this Article. Any other Party included in Annex I undergoing the process of transition to a market economy which has not yet submitted its first national communication under Article 12 of the Convention may also notify the Conference of the Parties serving as the meeting of the Parties to this Protocol that it intends to use an historical base year or period other than 1990 for the implementation of its commitments under this Article. The Conference of the Parties serving as the meeting of the Parties to this Protocol shall decide on the acceptance of such notification.
6. Taking into account Article 4, paragraph 6, of the Convention, in the implementation of their commitments under this Protocol other than those under this Article, a certain degree of flexibility shall be allowed by the Conference of the Parties serving as the meeting of the Parties to this Protocol to the Parties included in Annex I undergoing the process of transition to a market economy.
7. In the first quantified emission limitation and reduction commitment period, from 2008 to 2012, the assigned amount for each Party included in Annex I shall be equal to the percentage inscribed for it in Annex B of its aggregate anthropogenic carbon dioxide equivalent emissions of the greenhouse gases listed in Annex A in 1990, or the base year or period determined in accordance with paragraph 5 above, multiplied by five. Those Parties included in Annex I for whom land-use change and forestry constituted a net source of greenhouse gas emissions in 1990 shall include in their 1990 emissions base year or period the aggregate anthropogenic carbon dioxide equivalent emissions by sources minus removals by sinks in 1990 from land-use change for the purposes of calculating their assigned amount.
8. Any Party included in Annex I may use 1995 as its base year for hydrofluorocarbons, perfluorocarbons and sulphur hexafluoride, for the purposes of the calculation referred to in paragraph 7 above.

9. Commitments for subsequent periods for Parties included in Annex I shall be established in amendments to Annex B to this Protocol, which shall be adopted in accordance with the provisions of Article 21, paragraph 7. The Conference of the Parties serving as the meeting of the Parties to this Protocol shall initiate the consideration of such commitments at least seven years before the end of the first commitment period referred to in paragraph 1 above.
10. Any emission reduction units, or any part of an assigned amount, which a Party acquires from another Party in accordance with the provisions of Article 6 or of Article 17 shall be added to the assigned amount for the acquiring Party.
11. Any emission reduction units, or any part of an assigned amount, which a Party transfers to another Party in accordance with the provisions of Article 6 or of Article 17 shall be subtracted from the assigned amount for the transferring Party.
12. Any certified emission reductions which a Party acquires from another Party in accordance with the provisions of Article 12 shall be added to the assigned amount for the acquiring Party.
13. If the emissions of a Party included in Annex I in a commitment period are less than its assigned amount under this Article, this difference shall, on request of that Party, be added to the assigned amount for that Party for subsequent commitment periods.
14. Each Party included in Annex I shall strive to implement the commitments mentioned in paragraph 1 above in such a way as to minimize adverse social, environmental and economic impacts on developing country Parties, particularly those identified in Article 4, paragraphs 8 and 9, of the Convention. In line with relevant decisions of the Conference of the Parties on the implementation of those paragraphs, the Conference of the Parties serving as the meeting of the Parties to this Protocol shall, at its first session, consider what actions are necessary to minimize the adverse effects of climate change and/or the impacts of response measures on Parties referred to in those paragraphs. Among the issues to be considered shall be the establishment of funding, insurance and transfer of technology.

Article 4

1. Any Parties included in Annex I that have reached an agreement to fulfil their commitments under Article 3 jointly, shall be deemed to have met those commitments provided that their total combined aggregate anthropogenic carbon dioxide equivalent emissions of the greenhouse gases listed in Annex A do not exceed their assigned amounts calculated pursuant to their quantified emission limitation and reduction commitments inscribed in Annex B and in accordance with the provisions of Article 3. The respective emission level allocated to each of the Parties to the agreement shall be set out in that agreement.

2. The Parties to any such agreement shall notify the secretariat of the terms of the agreement on the date of deposit of their instruments of ratification, acceptance or approval of this Protocol, or accession thereto. The secretariat shall in turn inform the Parties and signatories to the Convention of the terms of the agreement.
3. Any such agreement shall remain in operation for the duration of the commitment period specified in Article 3, paragraph 7.
4. If Parties acting jointly do so in the framework of, and together with, a regional economic integration organization, any alteration in the composition of the organization after adoption of this Protocol shall not affect existing commitments under this Protocol. Any alteration in the composition of the organization shall only apply for the purposes of those commitments under Article 3 that are adopted subsequent to that alteration.
5. In the event of failure by the Parties to such an agreement to achieve their total combined level of emission reductions, each Party to that agreement shall be responsible for its own level of emissions set out in the agreement.
6. If Parties acting jointly do so in the framework of, and together with, a regional economic integration organization which is itself a Party to this Protocol, each member State of that regional economic integration organization individually, and together with the regional economic integration organization acting in accordance with Article 24, shall, in the event of failure to achieve the total combined level of emission reductions, be responsible for its level of emissions as notified in accordance with this Article.

Article 5

1. Each Party included in Annex I shall have in place, no later than one year prior to the start of the first commitment period, a national system for the estimation of anthropogenic emissions by sources and removals by sinks of all greenhouse gases not controlled by the Montreal Protocol. Guidelines for such national systems, which shall incorporate the methodologies specified in paragraph 2 below, shall be decided upon by the Conference of the Parties serving as the meeting of the Parties to this Protocol at its first session.
2. Methodologies for estimating anthropogenic emissions by sources and removals by sinks of all greenhouse gases not controlled by the Montreal Protocol shall be those accepted by the Intergovernmental Panel on Climate Change and agreed upon by the Conference of the Parties at its third session. Where such methodologies are not used, appropriate adjustments shall be applied according to methodologies agreed upon by the Conference of the Parties serving as the meeting of the Parties to this Protocol at its first session. Based on the work of, inter alia, the Intergovernmental Panel on Climate Change and advice provided by the Subsidiary Body for Scientific and Technological Advice, the Conference of the Parties serving as the meeting of the Parties to this Protocol shall regularly review and, as appropriate, revise such method-

ologies and adjustments, taking fully into account any relevant decisions by the Conference of the Parties. Any revision to methodologies or adjustments shall be used only for the purposes of ascertaining compliance with commitments under Article 3 in respect of any commitment period adopted subsequent to that revision.

3. The global warming potentials used to calculate the carbon dioxide equivalence of anthropogenic emissions by sources and removals by sinks of greenhouse gases listed in Annex A shall be those accepted by the Intergovernmental Panel on Climate Change and agreed upon by the Conference of the Parties at its third session. Based on the work of, inter alia, the Intergovernmental Panel on Climate Change and advice provided by the Subsidiary Body for Scientific and Technological Advice, the Conference of the Parties serving as the meeting of the Parties to this Protocol shall regularly review and, as appropriate, revise the global warming potential of each such greenhouse gas, taking fully into account any relevant decisions by the Conference of the Parties. Any revision to a global warming potential shall apply only to commitments under Article 3 in respect of any commitment period adopted subsequent to that revision.

Article 6

1. For the purpose of meeting its commitments under Article 3, any Party included in Annex I may transfer to, or acquire from, any other such Party emission reduction units resulting from projects aimed at reducing anthropogenic emissions by sources or enhancing anthropogenic removals by sinks of greenhouse gases in any sector of the economy, provided that:
   (a) Any such project has the approval of the Parties involved;
   (b) Any such project provides a reduction in emissions by sources, or an enhancement of removals by sinks, that is additional to any that would otherwise occur;
   (c) It does not acquire any emission reduction units if it is not in compliance with its obligations under Articles 5 and 7; and
   (d) The acquisition of emission reduction units shall be supplemental to domestic actions for the purposes of meeting commitments under Article 3.
2. The Conference of the Parties serving as the meeting of the Parties to this Protocol may, at its first session or as soon as practicable thereafter, further elaborate guidelines for the implementation of this Article, including for verification and reporting.
3. A Party included in Annex I may authorize legal entities to participate, under its responsibility, in actions leading to the generation, transfer or acquisition under this Article of emission reduction units.
4. If a question of implementation by a Party included in Annex I of the requirements referred to in this Article is identified in accordance with the relevant

provisions of Article 8, transfers and acquisitions of emission reduction units may continue to be made after the question has been identified, provided that any such units may not be used by a Party to meet its commitments under Article 3 until any issue of compliance is resolved.

Article 7

1. Each Party included in Annex I shall incorporate in its annual inventory of anthropogenic emissions by sources and removals by sinks of greenhouse gases not controlled by the Montreal Protocol, submitted in accordance with the relevant decisions of the Conference of the Parties, the necessary supplementary information for the purposes of ensuring compliance with Article 3, to be determined in accordance with paragraph 4 below.
2. Each Party included in Annex I shall incorporate in its national communication, submitted under Article 12 of the Convention, the supplementary information necessary to demonstrate compliance with its commitments under this Protocol, to be determined in accordance with paragraph 4 below.
3. Each Party included in Annex I shall submit the information required under paragraph 1 above annually, beginning with the first inventory due under the Convention for the first year of the commitment period after this Protocol has entered into force for that Party. Each such Party shall submit the information required under paragraph 2 above as part of the first national communication due under the Convention after this Protocol has entered into force for it and after the adoption of guidelines as provided for in paragraph 4 below. The frequency of subsequent submission of information required under this Article shall be determined by the Conference of the Parties serving as the meeting of the Parties to this Protocol, taking into account any timetable for the submission of national communications decided upon by the Conference of the Parties.
4. The Conference of the Parties serving as the meeting of the Parties to this Protocol shall adopt at its first session, and review periodically thereafter, guidelines for the preparation of the information required under this Article, taking into account guidelines for the preparation of national communications by Parties included in Annex I adopted by the Conference of the Parties. The Conference of the Parties serving as the meeting of the Parties to this Protocol shall also, prior to the first commitment period, decide upon modalities for the accounting of assigned amounts.

Article 8

1. The information submitted under Article 7 by each Party included in Annex I shall be reviewed by expert review teams pursuant to the relevant decisions of the Conference of the Parties and in accordance with guidelines adopted for this purpose by the Conference of the Parties serving as the meeting of the

Parties to this Protocol under paragraph 4 below. The information submitted under Article 7, paragraph 1, by each Party included in Annex I shall be reviewed as part of the annual compilation and accounting of emissions inventories and assigned amounts. Additionally, the information submitted under Article 7, paragraph 2, by each Party included in Annex I shall be reviewed as part of the review of communications.

2. Expert review teams shall be coordinated by the secretariat and shall be composed of experts selected from those nominated by Parties to the Convention and, as appropriate, by intergovernmental organizations, in accordance with guidance provided for this purpose by the Conference of the Parties.
3. The review process shall provide a thorough and comprehensive technical assessment of all aspects of the implementation by a Party of this Protocol. The expert review teams shall prepare a report to the Conference of the Parties serving as the meeting of the Parties to this Protocol, assessing the implementation of the commitments of the Party and identifying any potential problems in, and factors influencing, the fulfilment of commitments. Such reports shall be circulated by the secretariat to all Parties to the Convention. The secretariat shall list those questions of implementation indicated in such reports for further consideration by the Conference of the Parties serving as the meeting of the Parties to this Protocol.
4. The Conference of the Parties serving as the meeting of the Parties to this Protocol shall adopt at its first session, and review periodically thereafter, guidelines for the review of implementation of this Protocol by expert review teams taking into account the relevant decisions of the Conference of the Parties.
5. The Conference of the Parties serving as the meeting of the Parties to this Protocol shall, with the assistance of the Subsidiary Body for Implementation and, as appropriate, the Subsidiary Body for Scientific and Technological Advice, consider:
   (a) The information submitted by Parties under Article 7 and the reports of the expert reviews thereon conducted under this Article; and
   (b) Those questions of implementation listed by the secretariat under paragraph 3 above, as well as any questions raised by Parties.
6. Pursuant to its consideration of the information referred to in paragraph 5 above, the Conference of the Parties serving as the meeting of the Parties to this Protocol shall take decisions on any matter required for the implementation of this Protocol.

Article 9

1. The Conference of the Parties serving as the meeting of the Parties to this Protocol shall periodically review this Protocol in the light of the best available scientific information and assessments on climate change and its impacts, as well as relevant technical, social and economic information. Such reviews

shall be coordinated with pertinent reviews under the Convention, in particular those required by Article 4, paragraph 2(d), and Article 7, paragraph 2(a), of the Convention. Based on these reviews, the Conference of the Parties serving as the meeting of the Parties to this Protocol shall take appropriate action.

2. The first review shall take place at the second session of the Conference of the Parties serving as the meeting of the Parties to this Protocol. Further reviews shall take place at regular intervals and in a timely manner.

Article 10

All Parties, taking into account their common but differentiated responsibilities and their specific national and regional development priorities, objectives and circumstances, without introducing any new commitments for Parties not included in Annex I, but reaffirming existing commitments under Article 4, paragraph 1, of the Convention, and continuing to advance the implementation of these commitments in order to achieve sustainable development, taking into account Article 4, paragraphs 3, 5 and 7, of the Convention, shall:

(a) Formulate, where relevant and to the extent possible, cost-effective national and, where appropriate, regional programmes to improve the quality of local emission factors, activity data and/or models which reflect the socio-economic conditions of each Party for the preparation and periodic updating of national inventories of anthropogenic emissions by sources and removals by sinks of all greenhouse gases not controlled by the Montreal Protocol, using comparable methodologies to be agreed upon by the Conference of the Parties, and consistent with the guidelines for the preparation of national communications adopted by the Conference of the Parties;

(b) Formulate, implement, publish and regularly update national and, where appropriate, regional programmes containing measures to mitigate climate change and measures to facilitate adequate adaptation to climate change:

(i) Such programmes would, inter alia, concern the energy, transport and industry sectors as well as agriculture, forestry and waste management. Furthermore, adaptation technologies and methods for improving spatial planning would improve adaptation to climate change; and

(ii) Parties included in Annex I shall submit information on action under this Protocol, including national programmes, in accordance with Article 7; and other Parties shall seek to include in their national communications, as appropriate, information on programmes which contain measures that the Party believes contribute to addressing climate change and its adverse impacts, including the abatement of increases in greenhouse gas emissions, and enhancement of and removals by sinks, capacity building and adaptation measures;

(c) Cooperate in the promotion of effective modalities for the development, application and diffusion of, and take all practicable steps to promote, facilitate and finance, as appropriate, the transfer of, or access to, environmentally sound technologies, know-how, practices and processes pertinent to climate change, in particular to developing countries, including the formulation of policies and programmes for the effective transfer of environmentally sound technologies that are publicly owned or in the public domain and the creation of an enabling environment for the private sector, to promote and enhance the transfer of, and access to, environmentally sound technologies;

(d) Cooperate in scientific and technical research and promote the maintenance and the development of systematic observation systems and development of data archives to reduce uncertainties related to the climate system, the adverse impacts of climate change and the economic and social consequences of various response strategies, and promote the development and strengthening of endogenous capacities and capabilities to participate in international and intergovernmental efforts, programmes and networks on research and systematic observation, taking into account Article 5 of the Convention;

(e) Cooperate in and promote at the international level, and, where appropriate, using existing bodies, the development and implementation of education and training programmes, including the strengthening of national capacity building, in particular human and institutional capacities and the exchange or secondment of personnel to train experts in this field, in particular for developing countries, and facilitate at the national level public awareness of, and public access to information on, climate change. Suitable modalities should be developed to implement these activities through the relevant bodies of the Convention, taking into account Article 6 of the Convention;

(f) Include in their national communications information on programmes and activities undertaken pursuant to this Article in accordance with relevant decisions of the Conference of the Parties; and

(g) Give full consideration, in implementing the commitments under this Article, to Article 4, paragraph 8, of the Convention.

Article 11

1. In the implementation of Article 10, Parties shall take into account the provisions of Article 4, paragraphs 4, 5, 7, 8 and 9, of the Convention.
2. In the context of the implementation of Article 4, paragraph 1, of the Convention, in accordance with the provisions of Article 4, paragraph 3, and Article 11 of the Convention, and through the entity or entities entrusted with the

operation of the financial mechanism of the Convention, the developed country Parties and other developed Parties included in Annex II to the Convention shall:

(a) Provide new and additional financial resources to meet the agreed full costs incurred by developing country Parties in advancing the implementation of existing commitments under Article 4, paragraph 1(a), of the Convention that are covered in Article 10, subparagraph (a); and

(b) Also provide such financial resources, including for the transfer of technology, needed by the developing country Parties to meet the agreed full incremental costs of advancing the implementation of existing commitments under Article 4, paragraph 1, of the Convention that are covered by Article 10 and that are agreed between a developing country Party and the international entity or entities referred to in Article 11 of the Convention, in accordance with that Article.

The implementation of these existing commitments shall take into account the need for adequacy and predictability in the flow of funds and the importance of appropriate burden sharing among developed country Parties. The guidance to the entity or entities entrusted with the operation of the financial mechanism of the Convention in relevant decisions of the Conference of the Parties, including those agreed before the adoption of this Protocol, shall apply mutatis mutandis to the provisions of this paragraph.

3. The developed country Parties and other developed Parties in Annex II to the Convention may also provide, and developing country Parties avail themselves of, financial resources for the implementation of Article 10, through bilateral, regional and other multilateral channels.

Article 12

1. A clean development mechanism is hereby defined.
2. The purpose of the clean development mechanism shall be to assist Parties not included in Annex I in achieving sustainable development and in contributing to the ultimate objective of the Convention, and to assist Parties included in Annex I in achieving compliance with their quantified emission limitation and reduction commitments under Article 3.
3. Under the clean development mechanism:
   (a) Parties not included in Annex I will benefit from project activities resulting in certified emission reductions; and
   (b) Parties included in Annex I may use the certified emission reductions accruing from such project activities to contribute to compliance with part of their quantified emission limitation and reduction commitments under Article 3, as determined by the Conference of the Parties serving as the meeting of the Parties to this Protocol.
4. The clean development mechanism shall be subject to the authority and guidance of the Conference of the Parties serving as the meeting of the Parties to

this Protocol and be supervised by an executive board of the clean development mechanism.

5. Emission reductions resulting from each project activity shall be certified by operational entities to be designated by the Conference of the Parties serving as the meeting of the Parties to this Protocol, on the basis of:
   (a) Voluntary participation approved by each Party involved;
   (b) Real, measurable, and long-term benefits related to the mitigation of climate change; and
   (c) Reductions in emissions that are additional to any that would occur in the absence of the certified project activity.
6. The clean development mechanism shall assist in arranging funding of certified project activities as necessary.
7. The Conference of the Parties serving as the meeting of the Parties to this Protocol shall, at its first session, elaborate modalities and procedures with the objective of ensuring transparency, efficiency and accountability through independent auditing and verification of project activities.
8. The Conference of the Parties serving as the meeting of the Parties to this Protocol shall ensure that a share of the proceeds from certified project activities is used to cover administrative expenses as well as to assist developing country Parties that are particularly vulnerable to the adverse effects of climate change to meet the costs of adaptation.
9. Participation under the clean development mechanism, including in activities mentioned in paragraph 3(a) above and in the acquisition of certified emission reductions, may involve private and/or public entities, and is to be subject to whatever guidance may be provided by the executive board of the clean development mechanism.
10. Certified emission reductions obtained during the period from the year 2000 up to the beginning of the first commitment period can be used to assist in achieving compliance in the first commitment period.

Article 13

1. The Conference of the Parties, the supreme body of the Convention, shall serve as the meeting of the Parties to this Protocol.
2. Parties to the Convention that are not Parties to this Protocol may participate as observers in the proceedings of any session of the Conference of the Parties serving as the meeting of the Parties to this Protocol. When the Conference of the Parties serves as the meeting of the Parties to this Protocol, decisions under this Protocol shall be taken only by those that are Parties to this Protocol.
3. When the Conference of the Parties serves as the meeting of the Parties to this Protocol, any member of the Bureau of the Conference of the Parties representing a Party to the Convention but, at that time, not a Party to this Protocol,

shall be replaced by an additional member to be elected by and from amongst the Parties to this Protocol.

4. The Conference of the Parties serving as the meeting of the Parties to this Protocol shall keep under regular review the implementation of this Protocol and shall make, within its mandate, the decisions necessary to promote its effective implementation. It shall perform the functions assigned to it by this Protocol and shall:
   (a) Assess, on the basis of all information made available to it in accordance with the provisions of this Protocol, the implementation of this Protocol by the Parties, the overall effects of the measures taken pursuant to this Protocol, in particular environmental, economic and social effects as well as their cumulative impacts and the extent to which progress towards the objective of the Convention is being achieved;
   (b) Periodically examine the obligations of the Parties under this Protocol, giving due consideration to any reviews required by Article 4, paragraph 2(d), and Article 7, paragraph 2, of the Convention, in the light of the objective of the Convention, the experience gained in its implementation and the evolution of scientific and technological knowledge, and in this respect consider and adopt regular reports on the implementation of this Protocol;
   (c) Promote and facilitate the exchange of information on measures adopted by the Parties to address climate change and its effects, taking into account the differing circumstances, responsibilities and capabilities of the Parties and their respective commitments under this Protocol;
   (d) Facilitate, at the request of two or more Parties, the coordination of measures adopted by them to address climate change and its effects, taking into account the differing circumstances, responsibilities and capabilities of the Parties and their respective commitments under this Protocol;
   (e) Promote and guide, in accordance with the objective of the Convention and the provisions of this Protocol, and taking fully into account the relevant decisions by the Conference of the Parties, the development and periodic refinement of comparable methodologies for the effective implementation of this Protocol, to be agreed on by the Conference of the Parties serving as the meeting of the Parties to this Protocol;
   (f) Make recommendations on any matters necessary for the implementation of this Protocol;
   (g) Seek to mobilize additional financial resources in accordance with Article 11, paragraph 2;
   (h) Establish such subsidiary bodies as are deemed necessary for the implementation of this Protocol;
   (i) Seek and utilize, where appropriate, the services and cooperation of, and information provided by, competent international organizations and intergovernmental and non-governmental bodies; and

(j) Exercise such other functions as may be required for the implementation of this Protocol, and consider any assignment resulting from a decision by the Conference of the Parties.

5. The rules of procedure of the Conference of the Parties and financial procedures applied under the Convention shall be applied mutatis mutandis under this Protocol, except as may be otherwise decided by consensus by the Conference of the Parties serving as the meeting of the Parties to this Protocol.
6. The first session of the Conference of the Parties serving as the meeting of the Parties to this Protocol shall be convened by the secretariat in conjunction with the first session of the Conference of the Parties that is scheduled after the date of the entry into force of this Protocol. Subsequent ordinary sessions of the Conference of the Parties serving as the meeting of the Parties to this Protocol shall be held every year and in conjunction with ordinary sessions of the Conference of the Parties, unless otherwise decided by the Conference of the Parties serving as the meeting of the Parties to this Protocol.
7. Extraordinary sessions of the Conference of the Parties serving as the meeting of the Parties to this Protocol shall be held at such other times as may be deemed necessary by the Conference of the Parties serving as the meeting of the Parties to this Protocol, or at the written request of any Party, provided that, within six months of the request being communicated to the Parties by the secretariat, it is supported by at least one third of the Parties.
8. The United Nations, its specialized agencies and the International Atomic Energy Agency, as well as any State member thereof or observers thereto not party to the Convention, may be represented at sessions of the Conference of the Parties serving as the meeting of the Parties to this Protocol as observers. Any body or agency, whether national or international, governmental or non-governmental, which is qualified in matters covered by this Protocol and which has informed the secretariat of its wish to be represented at a session of the Conference of the Parties serving as the meeting of the Parties to this Protocol as an observer, may be so admitted unless at least one third of the Parties present object. The admission and participation of observers shall be subject to the rules of procedure, as referred to in paragraph 5 above.

Article 14

1. The secretariat established by Article 8 of the Convention shall serve as the secretariat of this Protocol.
2. Article 8, paragraph 2, of the Convention on the functions of the secretariat, and Article 8, paragraph 3, of the Convention on arrangements made for the functioning of the secretariat, shall apply mutatis mutandis to this Protocol. The secretariat shall, in addition, exercise the functions assigned to it under this Protocol.

Article 15

1. The Subsidiary Body for Scientific and Technological Advice and the Subsidiary Body for Implementation established by Articles 9 and 10 of the Convention shall serve as, respectively, the Subsidiary Body for Scientific and Technological Advice and the Subsidiary Body for Implementation of this Protocol. The provisions relating to the functioning of these two bodies under the Convention shall apply mutatis mutandis to this Protocol. Sessions of the meetings of the Subsidiary Body for Scientific and Technological Advice and the Subsidiary Body for Implementation of this Protocol shall be held in conjunction with the meetings of, respectively, the Subsidiary Body for Scientific and Technological Advice and the Subsidiary Body for Implementation of the Convention.
2. Parties to the Convention that are not Parties to this Protocol may participate as observers in the proceedings of any session of the subsidiary bodies. When the subsidiary bodies serve as the subsidiary bodies of this Protocol, decisions under this Protocol shall be taken only by those that are Parties to this Protocol.
3. When the subsidiary bodies established by Articles 9 and 10 of the Convention exercise their functions with regard to matters concerning this Protocol, any member of the Bureaux of those subsidiary bodies representing a Party to the Convention but, at that time, not a party to this Protocol, shall be replaced by an additional member to be elected by and from amongst the Parties to this Protocol.

Article 16

The Conference of the Parties serving as the meeting of the Parties to this Protocol shall, as soon as practicable, consider the application to this Protocol of, and modify as appropriate, the multilateral consultative process referred to in Article 13 of the Convention, in the light of any relevant decisions that may be taken by the Conference of the Parties. Any multilateral consultative process that may be applied to this Protocol shall operate without prejudice to the procedures and mechanisms established in accordance with Article 18.

Article 17

The Conference of the Parties shall define the relevant principles, modalities, rules and guidelines, in particular for verification, reporting and accountability for emissions trading. The Parties included in Annex B may participate in emissions trading for the purposes of fulfilling their commitments under Article 3. Any such trading shall be supplemental to domestic actions for the purpose of meeting quantified emission limitation and reduction commitments under that Article.

Article 18

The Conference of the Parties serving as the meeting of the Parties to this Protocol shall, at its first session, approve appropriate and effective procedures and mechanisms to determine and to address cases of non-compliance with the provisions of this Protocol, including through the development of an indicative list of consequences, taking into account the cause, type, degree and frequency of non-compliance. Any procedures and mechanisms under this Article entailing binding consequences shall be adopted by means of an amendment to this Protocol.

Article 19

The provisions of Article 14 of the Convention on settlement of disputes shall apply mutatis mutandis to this Protocol.

Article 20

1. Any Party may propose amendments to this Protocol.
2. Amendments to this Protocol shall be adopted at an ordinary session of the Conference of the Parties serving as the meeting of the Parties to this Protocol. The text of any proposed amendment to this Protocol shall be communicated to the Parties by the secretariat at least six months before the meeting at which it is proposed for adoption. The secretariat shall also communicate the text of any proposed amendments to the Parties and signatories to the Convention and, for information, to the Depositary.
3. The Parties shall make every effort to reach agreement on any proposed amendment to this Protocol by consensus. If all efforts at consensus have been exhausted, and no agreement reached, the amendment shall as a last resort be adopted by a three-fourths majority vote of the Parties present and voting at the meeting. The adopted amendment shall be communicated by the secretariat to the Depositary, who shall circulate it to all Parties for their acceptance.
4. Instruments of acceptance in respect of an amendment shall be deposited with the Depositary. An amendment adopted in accordance with paragraph 3 above shall enter into force for those Parties having accepted it on the ninetieth day after the date of receipt by the Depositary of an instrument of acceptance by at least three fourths of the Parties to this Protocol.
5. The amendment shall enter into force for any other Party on the ninetieth day after the date on which that Party deposits with the Depositary its instrument of acceptance of the said amendment.

Article 21

1. Annexes to this Protocol shall form an integral part thereof and, unless otherwise expressly provided, a reference to this Protocol constitutes at the same time a reference to any annexes thereto. Any annexes adopted after the entry

into force of this Protocol shall be restricted to lists, forms and any other material of a descriptive nature that is of a scientific, technical, procedural or administrative character.

2. Any Party may make proposals for an annex to this Protocol and may propose amendments to annexes to this Protocol.
3. Annexes to this Protocol and amendments to annexes to this Protocol shall be adopted at an ordinary session of the Conference of the Parties serving as the meeting of the Parties to this Protocol. The text of any proposed annex or amendment to an annex shall be communicated to the Parties by the secretariat at least six months before the meeting at which it is proposed for adoption. The secretariat shall also communicate the text of any proposed annex or amendment to an annex to the Parties and signatories to the Convention and, for information, to the Depositary.
4. The Parties shall make every effort to reach agreement on any proposed annex or amendment to an annex by consensus. If all efforts at consensus have been exhausted, and no agreement reached, the annex or amendment to an annex shall as a last resort be adopted by a three-fourths majority vote of the Parties present and voting at the meeting. The adopted annex or amendment to an annex shall be communicated by the secretariat to the Depositary, who shall circulate it to all Parties for their acceptance.
5. An annex, or amendment to an annex other than Annex A or B, that has been adopted in accordance with paragraphs 3 and 4 above shall enter into force for all Parties to this Protocol six months after the date of the communication by the Depositary to such Parties of the adoption of the annex or adoption of the amendment to the annex, except for those Parties that have notified the Depositary, in writing, within that period of their non-acceptance of the annex or amendment to the annex. The annex or amendment to an annex shall enter into force for Parties which withdraw their notification of non-acceptance on the ninetieth day after the date on which withdrawal of such notification has been received by the Depositary.
6. If the adoption of an annex or an amendment to an annex involves an amendment to this Protocol, that annex or amendment to an annex shall not enter into force until such time as the amendment to this Protocol enters into force.
7. Amendments to Annexes A and B to this Protocol shall be adopted and enter into force in accordance with the procedure set out in Article 20, provided that any amendment to Annex B shall be adopted only with the written consent of the Party concerned.

Article 22

1. Each Party shall have one vote, except as provided for in paragraph 2 below.
2. Regional economic integration organizations, in matters within their competence, shall exercise their right to vote with a number of votes equal to the number of their member States that are Parties to this Protocol. Such an

organization shall not exercise its right to vote if any of its member States exercises its right, and vice versa.

Article 23

The Secretary-General of the United Nations shall be the Depositary of this Protocol.

Article 24

1. This Protocol shall be open for signature and subject to ratification, acceptance or approval by States and regional economic integration organizations which are Parties to the Convention. It shall be open for signature at United Nations Headquarters in New York from 16 March 1998 to 15 March 1999. This Protocol shall be open for accession from the day after the date on which it is closed for signature. Instruments of ratification, acceptance, approval or accession shall be deposited with the Depositary.
2. Any regional economic integration organization which becomes a Party to this Protocol without any of its member States being a Party shall be bound by all the obligations under this Protocol. In the case of such organizations, one or more of whose member States is a Party to this Protocol, the organization and its member States shall decide on their respective responsibilities for the performance of their obligations under this Protocol. In such cases, the organization and the member States shall not be entitled to exercise rights under this Protocol concurrently.
3. In their instruments of ratification, acceptance, approval or accession, regional economic integration organizations shall declare the extent of their competence with respect to the matters governed by this Protocol. These organizations shall also inform the Depositary, who shall in turn inform the Parties, of any substantial modification in the extent of their competence.

Article 25

1. This Protocol shall enter into force on the ninetieth day after the date on which not less than 55 Parties to the Convention, incorporating Parties included in Annex I which accounted in total for at least 55 per cent of the total carbon dioxide emissions for 1990 of the Parties included in Annex I, have deposited their instruments of ratification, acceptance, approval or accession.
2. For the purposes of this Article, "the total carbon dioxide emissions for 1990 of the Parties included in Annex I" means the amount communicated on or before the date of adoption of this Protocol by the Parties included in Annex I in their first national communications submitted in accordance with Article 12 of the Convention.

3. For each State or regional economic integration organization that ratifies, accepts or approves this Protocol or accedes thereto after the conditions set out in paragraph 1 above for entry into force have been fulfilled, this Protocol shall enter into force on the ninetieth day following the date of deposit of its instrument of ratification, acceptance, approval or accession.
4. For the purposes of this Article, any instrument deposited by a regional economic integration organization shall not be counted as additional to those deposited by States members of the organization.

Article 26

No reservations may be made to this Protocol.

Article 27

1. At any time after three years from the date on which this Protocol has entered into force for a Party, that Party may withdraw from this Protocol by giving written notification to the Depositary.
2. Any such withdrawal shall take effect upon expiry of one year from the date of receipt by the Depositary of the notification of withdrawal, or on such later date as may be specified in the notification of withdrawal.
3. Any Party that withdraws from the Convention shall be considered as also having withdrawn from this Protocol.

Article 28

The original of this Protocol, of which the Arabic, Chinese, English, French, Russian and Spanish texts are equally authentic, shall be deposited with the Secretary-General of the United Nations.

DONE at Kyoto this eleventh day of December one thousand nine hundred and ninety-seven.

IN WITNESS WHEREOF the undersigned, being duly authorized to that effect, have affixed their signatures to this Protocol on the dates indicated.

Annex A

Greenhouse gases
Carbon dioxide (CO_2)
Methane (CH_4)
Nitrous oxide (N_2O)
Hydrofluorocarbons (HFCs)
Perfluorocarbons (PFCs)
Sulphur hexafluoride (SF_6)

Sectors/source categories
Energy
  Fuel combustion
    Energy industries
    Manufacturing industries and construction
    Transport
    Other sectors
    Other
  Fugitive emissions from fuels
    Solid fuels
    Oil and natural gas
    Other

Industrial processes
  Mineral products
  Chemical industry
  Metal production
  Other production
  Production of halocarbons and sulphur hexafluoride
  Consumption of halocarbons and sulphur hexafluoride
  Other

Solvent and other product use

Agriculture
  Enteric fermentation
  Manure management
  Rice cultivation
  Agricultural soils
  Prescribed burning of savannas
  Field burning of agricultural residues
  Other

Waste
  Solid waste disposal on land
  Wastewater handling
  Waste incineration
  Other

Annex B

Party	Quantified emission limitation or reduction commitment (percentage of base year or period)
Australia	108
Austria	92
Belgium	92
Bulgaria[a]	92
Canada	94
Croatia[a]	95
Czech Republic[a]	92
Denmark	92
Estonia[a]	92
European Community	92
Finland	92
France	92
Germany	92
Greece	92
Hungary[a]	94
Iceland	110
Ireland	92
Italy	92
Japan	94
Latvia[a]	92
Liechtenstein	92
Lithuania[a]	92
Luxembourg	92
Monaco	92
Netherlands	92
New Zealand	100
Norway	101
Poland[a]	94
Portugal	92
Romania[a]	92
Russian Federation[a]	100
Slovakia[a]	92
Slovenia[a]	92
Spain	92
Sweden	92
Switzerland	92
Ukraine[a]	100
United Kingdom of Great Britain and Northern Ireland	92
United States of America	93

[a] Countries that are undergoing the process of transition to a market economy.

Table Appendix 1: CO_2 Emissions of Annex I Parties in 1990

Party	Emissions (Gg CO_2)	Percentage
Australia	288,965	2.1
Austria	59,200	0.4
Belgium	113,405	0.8
Bulgaria	82,990	0.6
Canada	457,441	3.3
Czech Republic	169,514	1.2
Denmark	52,100	0.4
Estonia	37,797	0.3
Finland	53,900	0.4
France	366,536	2.7
Germany	1,012,443	7.4
Greece	82,100	0.6
Hungary	71,673	0.5
Iceland	2,172	0.0
Ireland	30,719	0.2
Italy	428,941	3.1
Japan	1,173,360	8.5
Latvia	22,976	0.2
Liechtenstein	208	0.0
Luxembourg	11,343	0.1
Monaco	71	0.0
Netherlands	167,600	1.2
New Zealand	25,530	0.2
Norway	35,533	0.3
Poland	414,930	3.0
Portugal	42,148	0.3
Romania	171,103	1.2
Russian Federation	2,388,720	17.4
Slovakia	58,278	0.4
Spain	260,654	1.9
Sweden	61,256	0.4
Switzerland	43,600	0.3
United Kingdom of Great Britain and Northern Ireland	584,078	4.3
United States of America	4,957,022	36.1
Total	**13,728,306**	**100.0**

Source: (FCCC/CP/98/7/Add.1)

Bibliography

ABARE 1997: *The Economic Impact of International Climate Change Policy* (Research Report 97.4). Canberra: ABARE.

AFEAS 1998: *Production, Sales and Atmospheric Release of Fluorocarbons through 1996.* Washington D. C.: AFEAS.

Agarwal H. and S. Narain 1991: *Global Warming in an Unequal World. A Case of Environmental Colonialism.* New Delhi: Centre for Science and Environment.

Alcamo, J. and E. Kreileman 1998: Emission Scenarios and Global Climate Protection, in: J. Alcamo, R. Leemans and G.J.J. Kreileman (eds.): *Global Climate Change Scenarios of the 21st Century. Results from the IMAGE 2.1. Model.* Oxford (UK): Pergamon/Elsevier Science, 163–192.

Alexy, Robert 1995: *Recht, Vernunft, Diskurs. Studien zur Rechtsphilosophie.* Frankfurt a.M: Suhrkamp.

Altvater, Elmar, Achim Brunnengräber and Heike Walk 1998: Substitut des Staatsvolkes: Nicht-Regierungsorganisationen als neue Akteure transnationaler (Umwelt-) Politik. *Ökologisches Wirtschaften*, No. 2 (1998), 23-25.

Arrhenius, Svante 1896: On the Influence of Carbonic Acid in the Air Upon the Temperature of the Ground, in: *Philosophical Magazine* 41, No. 251 (April 1896), 237–277.

Australia 1997: *Australia's Second National Report under the United Nations Framework Convention on Climate Change.* Canberra: Department of the Environment.

Bail, Christoph 1998: Das Klimaschutzregime nach Kyoto. *Europäische Zeitschrift für Wirtschaftsrecht*, Vol. 9, No. 15, 457–464.

Bals, Christoph 1997: *Klimaskeptiker im Treibhaus. Die wissenschaftliche Debatte um den globalen Treibhauswandel.* Bonn: Forum Umwelt & Entwicklung.

Bär, Stefani and R. Andreas Kraemer 1998: European Environmental Policy After Amsterdam. *Journal of Environmental Law*, Vol. 10, No. 2, 315–330.

Barratt-Brown, Elizabeth 1991: Building a Monitoring and Compliance Regime Under the Montreal Protocol. *Yale Journal of International Law*, No. 16 (1991), 519–570.

Baumert, Kevin A., Ruchi Bhandari and Nancy Kete 1999: *What might a Developing Country Climate Commitment Look Like?* Washington, D.C.: World Resources Institute.

Belliveau, Michael 1998: *Smoke Mirrors. Will Global Pollution Trading Save the Climate or Promote Injustice and Fraud?* San Francisco: Transnational Resource & Action Center.

Benedick, Richard Elliot 1998a: *Ozone Diplomacy. New Directions in Safeguarding the Planet.* Second edition, Cambridge (Massachusetts): Harvard University Press.

Benedick, Richard Elliot 1998b: Auf dem falschen Weg zum Klimaschutz? *Universitas*, November 1998, 1017–1031.

Berk, Marcel, Michel den Elzen and Bert Metz 1999: *Global Climate Protection and Equitable Burden Sharing – An Exploration of Some Options.* Paper presented at the EFIEA policy workshop on "Integrated climate policies in the European Environment, costs and opportunities", Milan, 4–6 March 1999.

Bernauer, Thomas 1995: The Effect of International Environmental Institutions: How We Might Learn More. *International Organization*, Vol. 49, No. 2, 351–377.

Beuermann, Christiane and Jill Jäger 1996: Climate Change Politics in Germany: How Long Will any Double Dividend Last? in: Timothy o'Riordan and Jill Jäger (eds.): *Politics of Climate Change – A European Perspective*. London/New York: Routledge, 186–227.

Beyerlin, Ulrich and Markus Ehrmann 1997: Fünf Jahre nach dem Erdgipfel von Rio: eine kritische Bestandsaufnahme der Sondergeneralversammlung vom Juni 1997. *Umweltpolitik und Recht (UPR)*, No. 9, 356–361.

Biermann, Frank 1997: Financing Environmental Policies in the South. Experiences from the Multilateral Ozone Fund. *International Environmental Affairs*, Vol. 9, No. 3, 179–218.

Blok, K., G.J.M. Phylipsen and J.W. Bode 1997: *The Triptique Approach. Burden Differentiation of CO_2 Emission Reduction Among European Union Member States.* Discussion paper for the informal workshop for the European Union Ad Hoc Group on Climate Change, Zeist, The Netherlands, 16–17 January 1997.

Bodansky, David 1993: The United Nations Framework Convention on Climate Change: A Commentary. *Yale Journal of International Law*, Vol. 18, 451–558.

Botcheva, Liliana 1996: Focus and Effectiveness of Environmental Activism in Eastern Europe: A Comparative Study of Environmental Movements in Bulgaria, Hungary, Slovakia, and Romania. *Journal of Environment and Development*, Vol. 5, No. 3, 292–308.

Brack, Duncan 1996: *International Trade and the Montreal Protocol*. London: Earthscan.

Brauch, Hans Günter (ed.) 1996: *Klimapolitik – naturwissenschaftliche Grundlagen, internationale Regimebildung und Konflikte, ökonomische Analysen sowie nationale Problemerkennung und Politikumsetzung*. Berlin/Heidelberg: Springer.

Breitmeier, Helmut 1996: *Wie entstehen globale Umweltregime? Der Konfliktaustrag zum Schutz der Ozonschicht und des globalen Klimas*. Opladen: Leske+Budrich.

Brown, Paige 1998: *Climate, Biodiversity, and Forests: Issues and Opportunities from the Kyoto Protocol*. Washington, D.C.: World Resources Institute.

Browne, John 1997: *Climate Change: The New Agenda* (A presentation to Stanford University, California, 19 May 1997). London: British Petroleum Company.

Brownlie, Ian 1990: *Principles of Public International Law*. Oxford: Oxford University Press.

Burtraw, Dallas 1999: Cost savings, market performance and economic benefits of the US Acid Rain Program, in: Steve Sorell and Jim Skea (eds.) 1999: *Pollution for Sale: Emissions Trading and Joint Implementation*. Northampton (Massachusetts): Edward Elgar, 43–62.

Cameron, James and Matthias Buck 1998a: *The Clean Development Mechanism and Investment Opportunities for the Private Sector*. Paper, presented at UNEP, Fourth International Roundtable on Finance and the Environment, Cambridge/UK, 17–18 September 1998.

Cameron, James and Matthias Buck 1998b: *Designing the 'Clean Development Mechanism'*. FIELD Working Paper, presented to Aspen Global Forum in October 1998.

Campbell, Laura B. 1998: Emission Trading, Joint Implementation and the Clean Development Mechanism: The Role of Private Sector and Other Non-State Actors in Implementation, in: W. Bradnee Chambers (ed.). *Global Climate Governance: Inter-linkages Between the Kyoto Protocol and Other Multilateral Regimes*. Tokyo: United Nations University, Institute of Advanced Studies, 7-12.

Carius, Alexander and Kurt M. Lietzmann (eds.) 1999: *Environmental Change and Security. A European Perspective*. Berlin: Springer.

CEPS (Centre for European Policy Studies) 1997: *EU Climate Change Policy on the Road to Kyoto and Beyond*. Policy Conclusions and Background Report. Brussels: CEPS.

Center for Health and the Global Environment 1998: *Extreme Weather Events: The Health and Economic Cosequences of the 1997/98 El Niño and La Niña*. Havard: Havard Medical School.

Centre for Science and Environment 1998: *The Kyoto Protocol. What it says?* New Delhi (India): CSE.

Chambers, W. Bradnee (ed.) 1998a: *Global Climate Governance: Inter-linkages Between the Kyoto Protocol and Other Multilateral Regimes*. Tokyo: United Nations University, Institute of Advanced Studies.

Chambers, W. Bradnee 1998b: International Trade Law and the Kyoto Protocol, in: W. Bradnee Chambers (ed.). *Global Climate Governance: Inter-linkages Between the Kyoto Protocol and Other Multilateral Regimes*. Tokyo: United Nations University, Institute of Advanced Studies, 39–58.

Chartier, Denis and Jean-Paul Deléage 1998: The International Environmental NGOs: From the Revolutionary Alternative to the Pragmatism of Reform. *Environmental Politics*, Vol. 7, No. 3, 26–41.

Chayes, Abram and Eugene B. Skolnikoff 1992: *A Prompt Start: Implementing the Framework Convention on Climate Change*. Conference Report, Bellagio/Italy, 28–30 January 1992.

Chayes, Abram and Antonia Handler Chayes 1993: On compliance. *International Organization*, Vol. 47, No. 2, 175–205.

Chatterje, Pratap and Matthias Finger 1994: *The Earth Brokers. Power, Politics and Development*. London and New York: Routledge.

CAN-CEE/CNE (Climate Action Network – Central and Eastern Europe and Climate Network Europe) 1995: *Independent NGO Evaluations of National Plans for Climate Change Mitigation: Central and Eastern Europe*, First Review, January 1995.

CNE/USCAN (Climate Network Europe/United States Climate Action Network) 1995: *Independent NGO Evaluations of National Plans for Climate Change Mitigation: OECD Countries – Third Review*, January 1995.

CNE/USCAN (Climate Network Europe/United States Climate Action Network) 1997: *Independent NGO Evaluations of National Plans for Climate Change Mitigation: OECD Countries – Fifth Review*, September 1997.

CNE (Climate Network Europe) 1997: *Targeting Kyoto and Beyond. The Role of Europe*. Conference Proceedings, Bonn, 16/17 October 1997.

Coenen, Reinhard and Gerhard Sardemann 1998: Das Kyoto-Protokoll zum Schutz des Klimas – Erfolg oder Mißerfolg? (ITAS Arbeitsbericht 1/1998). Karlsruhe: Forschungszentrum Karlsruhe/ITAS.

Collier, Ute 1996: *Implementing a Climate Change Strategy in the European Union: Obstacles and Opportunities*. EUI Working Paper RSC No. 96/1. Badia Fiosolana: European University Institute.

Collier, Ute 1997: The EU and Climate Change Policy: The Struggle Over Policy Competences, in: Ute Collier (ed.): *Cases in Climate Change Policy: Political Reality in the European Union*. London: Earthscan, 43–64.

Contini, Paolo and Peter H. Sand 1972: Methods to expedite Environment Protection: International Ecostandards. *American Journal of International Law*, No. 66 (1972), 37–59.

Cook, Elizabeth (ed.) 1996: *Ozone Protection in the United States. Elements of Success*. Washington, D.C.: World Resources Institute.

Cooper, Richard N. 1998: Toward a Real Global Warming Treaty. *Foreign Affairs*, Vol. 77, No. 2, 66–79.

Corfee Morlot, Jan 1998: *Ensuring Compliance With a Global Climate Change Agreement*. OECD Information Paper: ENV/EPOC (98)5/REV1. Paris: OECD.

Craig, Paul and Graáinne de Búrca 1998: *EU Law, Text, Cases and Materials*. Oxford: Oxford University Press.

Dessus, Benjamin 1998: Equity, Sustainability and Solidarity Concerns, in: José Goldemberg (ed.): *Issues & Options: The Clean Development Mechanism*. UNDP.

DeSombre, Elizabeth R. and Joanne Kauffman 1996: The Montreal Protocol Multilateral Fund: Partial Success, in: Robert O. Keohane and Marc A. Levy (eds.): *Institutions for Environmental Aid. Pitfalls and Promise*. Cambridge (Massachusetts) and London: MIT Press.

DFAT/ABARE 1995: *Global Climate Change: Economic Dimensions of a Cooperative International Policy Response Beyond 2000*. Canberra: ABARE.

Downes, David R. 1999: Global Forest Policy and Selected International Instruments: a Preliminary Review, in: Richard G. Tarasofsky (ed.): *Assessing the International Forest Regime*.

Gland (Switzerland), Cambridge (UK) and Bonn (Germany): International Union for Conservation of Nature and Natural Ressources (IUCN).

Dudek, Daniel J. 1996: *Emission Budgets: Creating Rewards, Lowering Costs and Ensuring Results*. Springfield (Virginia): Proceedings Climate Change Analysis Workshop (6–7 June).

Dutschke, Michael 1998: *Nord-Süd-Kooperation in der Klimapolitik. Zwischen Umwelt und Entwicklung: Joint Implementation am Beispiel Costa Rica*. Berlin: Öko-Institut.

Dutschke, Michael and Axel Michaelowa 1998a: *Der Handel mit Emissionsrechten für Treibhausgase; Empfehlungen aus ökonomischer Sicht auf der Grundlage des Kyoto-Protokolls* (HWWA-Report 187). Hamburg: HWWA.

Dutschke, Michael and Axel Michaelowa 1998b: *Issues and open questions of Greenhouse Gas Emission Trading under the Kyoto Protocoll* (Executive Summary). Hamburg: HWWA.

Dworkin, Ronald 1972: *Taking Rights Seriously*. London: Duckworth.

Easterbrook, Gregg 1998: Hot and Not Bothered. *The New Republic*, 4 May, 1998, 20-22.

Ehrmann, Markus 1997: Die Globale Umweltfazilität (GEF). *Zeitschrift für ausländisches öffentliches Recht und Völkerrecht/ Heidelberg Journal of International Law*, Vol. 57, No. 2–3, 565–614.

Ehrmann, Markus and Sebastian Oberthür 1997: Spring in Climate Negotiations? Environmental Policy and Law, Vol. 27, No. 3, 192–196.

Ehrmann, Markus 1999: *Erfüllungskontrolle im Umweltvölkerrecht – Verfahren der Erfüllungskontrolle in der umweltvölkerrechtlichen Vertragspraxis*. PhD-thesis to be published late 1999.

EIA 1999: *International Energy Outlook*. Washington, D.C.: Energy Information Administration.

Eizenstat, Stuart 1998a: Stick with Kyoto. A Sound Start on Global Warming. *Foreign Affairs*, Vol. 77, No. 3, 119–121.

Ellerman, Denny and Richard Schmalensee, Paul L. Joskow, Juan Pablo Montero, Elizabeth M. Bailey 1999: Summary evaluation of the US SO_2 Emissions Trading Program as Implemented in 1995, in: Steve Sorell and Jim Skea (eds.) 1999: *Pollution for Sale: Emissions Trading and Joint Implementation*. Northampton (Massachusetts): Edward Elgar, 27–42.

Environmental Defense Fund 1997: *Emissions Budgets. Building an Effective International Greenhouse Gas Control System.* Washington, D.C.: EDF.

European Commission 1996a: *Communication from the Commission under the United Framework Convention on Climate Change* (Articel 4, 2, b, c, d and Articel 12), (COM(96) 217 final), Brussels.

European Commission 1996b: *Communication from the Commission: Strategy Paper for Reducing Methane Emissions* (COM(96) 557 final), Brussels.

European Commission 1996c: *European Energy to 2020*. Directorate General for Energy (DG XVII), Brussels.

European Commission 1996d: *Recommendation for a Decision of the Council* (SEK(96) 2002 final), Brussels.

European Commission 1997a: *Communication from the Commission: An Overall View of Energy Policy and Actions* (COM(97) 167 final), Brussels.

European Commission 1997b: *Communication from the Commission: The Energy Dimension of Climate Change* (COM(97) 196 final), Brussels.

European Commission 1997c: *Communication from the Commission: Climate Change – The EU Approach for Kyoto* (COM(97) 481 final), Brussels.

European Commission 1997d: *Proposal for a Council Decision Concerning a Multiannual Programme for the Promotion of Renewable Energy Sources in the Community* (Altener II) (COM (97) 87 final), Brussels.

European Commission 1997e: *Proposal for a Council Directive Restructuring the Community Framework for the Taxation of Energy Products* (COM(97) 30 final), Brussels.

European Commission 1997f: *Amended Proposal for a Council Directive to Introduce Rational Planning Techniques in the Electricity and Gas Sectors* (COM(97) 69 final), Brussels.

European Commission 1998a: *Communication from the Commission – Energy Efficiency in the European Union – Towards a Strategy for the Rational Use of Energy* (no number), Brussels.

European Commission 1998b: *Climate Change – Towards an EU Post-Kyoto Strategy. Commission Communication to the Council and the European Parliament.* (COM(98)353 final), Brussels.

European Commission 1999: *Preparing for Implementation of the Kyoto Protocol. Commission Communication to the Council and the Parliament* (COM(99)230 final), Brussels.

Fairman, David 1996: The Global Environment Facility: Haunted by the Shadow of the Future, in: Robert O. Keohane and Marc A. Levy (eds): *Institutions for Environmental Aid: Pitfalls and Promises*. Cambridge (Massachusetts): MIT Press, 55–88.

Fan, S., M. Floor, J. Mahlmann, S. Pacala, J. Sarmiento, T. Takashashi and P. Tans 1998: A Large Terrestrial Carbon Sink in North America Implied by Atmospheric and Oceanic Carbon Dioxide Data and Models. *Science*, No. 282, 442–446.

Filho, Gylvan Meira 1998: Ideas for Implementation, in: José Goldemberg (ed.): *Issues & Options: The Clean Development Mechanism*. UNDP.

Fischer 1998: *Fischer Weltalmanach 1999*. Frankfurt am Main: Fischer.

Flavin, Christopher 1998: Last Tango in Buenos Aires. *World•Watch*, November/December 1998, 11–18.

Friends of the Earth International 1998: *Carbon Justice! Developing an Equitable Global Solution to Stabilize Greenhouse Gas Concentrations at a Safe Level in Accordance With the Climate Convention* (Climate Change Briefing October 1998). Amsterdam.

Gallup 1992: *The Health of the Planet Survey. A Preliminary Report on Attitudes on the Environment and Economic Growth Measured by Surveys of Citizens in 22 Nations to Date*. Princeton, New York: The George H. Gallup International Institute.

GEF 1996: *Operational Strategy of the Global Environment Facility*. Washington, D.C.: GEF.

Gehring, Thomas 1990: International Environmental Regimes: Dynamic Sectoral Legal Systems. *Yearbook of International Environmental Law*, Vol. 1, 35–56.

Gehring, Thomas 1994: *Dynamic International Regimes. Institutions for International Environmental Governance*. Frankfurt am Main: Peter Lang.

Gehring, Thomas 1996: Arguing and Bargaining in internationalen Verhandlungen, in: Volker von Prittwitz (ed.): *Verhandeln und Argumentieren. Dialog, Interessen und Macht in der Umweltpolitik*. Opladen: Leske & Budrich, 207–238.

Gelbspan, Ross 1995: The Heat is On. *Harper Magazine*, December 1995, 31–37.

Gelbspan, Ross 1997: *The Heat is On. The High Stakes Battle over Earth's Threatened Climate*. Reading (Massachusetts).

Germany 1997: *Second Communication of the Federal Republic of Germany to the UNFCCC*, Bonn: Federal Ministry for the Environment, Nature Protection and Nuclear Safety.

Gillespie, Alexander 1998a: The Formation of Climate Change Policy in New Zealand 1992–1995: The Conflict Between International and Domestic Responses. *Environmental Politics*, Vol. 7, No. 3, 134–143.

Gillespie, Alexander 1998b: Sinks, Biodiversity & Forests: The Implications of the Kyoto Protocol Upon the Other Primary UNCED Instruments, in: Bradnee W. Chambers (ed.): *Global Climate Governance: Inter-linkages between the Kyoto Protocol and other Multilateral Regimes*. Tokyo: United Nations University, Institute of Advanced Studies, 117–139.

Global Commons Institute 1991: *A 'Comprehensible Approach' to Human Induced Climate Change, Recognizes the Natural Link Between 'Equity and Survival'*. London: GCI.

Goldberg, Donald M. 1998: *Carbon Conservation. Climate Change, Forests and the Clean Development Mechanism*. Washington, D:C.: Center for International Environmental Law (CIEL) and Centro de Derecho Ambiental y de los Recursos Naturales (CEDARENA).

Goldemberg, José (ed.) 1998: *Issues & Options: The Clean Development Mechanism*. UNDP.

Goldemberg, José and Walter Reid 1999: Greenhouse Gas Emissions and Developments: A Review of Lessons Learned, in: José Goldemberg and Walter Reid (eds.): *Promoting Develop-*

ment while Limiting Greenhouse Gas Emissions: Trends & Baselines. New York: UNDP, 1-13.

Gore, Albert 1992: *Earth in the Balance: Ecology and the Human Spirit*. New York: Plume Book.

Grant, Cedric 1995: Equity in International Relations: a Third World Perspective. *International Affairs*, Vol. 71, No. 3, 567–587.

Greenpeace International/Stockholm Environment Institute 1997: *Towards a fossil free energy future. The Next Energy Transition*. Boston.

Greenpeace 1997: *Towards the Kyoto Protocol. Update on Key Issues and Country Positions*. Bonn, 22–31 October 1997 (AGBM 8).

Greenpeace 1998: *Greenpeace Analysis of the Kyoto Protocol* (Greenpeace Briefing Paper). Bonn: UNFCCC Sessions of the subsidiary bodies June 2–12, 1998.

Grubb, Michael 1995a: From Rio to Kyoto via Berlin: Climate Change and the Prospects for International Action, in: Michael Grubb and Dean Anderson (eds.): *The Emerging International Regime for Climate Change. Structures and Options after Berlin*. London: The Royal Institute of International Affairs, 79–96.

Grubb, Michael 1995b: Seeking Fair Wheather: Ethics and the International Debate on Climate Change. *International Affairs*, Vol. 71, No. 3, 463–496.

Grubb, Michael 1996: The Rise of Climate Change, in: John V. Mitchell, Peter Beck and Michael Grubb (eds.): *The New Geopolitics of Energy*. London: The Royal Institute of International Affairs.

Grubb, Michael 1998: International Emissions Trading under the Kyoto Protocol: Core Issues in Implementation. *Review Of European Community & International Environmental Law*, Vol. 7, No. 2, 140–146.

Grubb, Michael, Christiaan Vrolijk and Duncan Brack 1999: *The Kyoto Protocol. A Guide and Assessment*. London: The Royal Institute of International Affairs, Earthscan.

Gupta, Joyeeta; Richard van der Wurff and Gerd Junne 1995: *International Policies to Address the Greenhouse Effect*. Amsterdam: University of Amsterdam – Deptartment of International Relations and Public International Law.

Gupta, Joyeeta 1997: *The Climate Change Convention: From Conflict to Consensus?* Dordrecht: Kluwer.

Gupta, Joyeeta 1998: Leadership in the Climate Regime: Inspiring the Commitments of Developing Countries in the Post-Kyoto Phase. *Review of European Community & International Environmental Law*, Vol. 7, No. 2, 180–190.

Gupta, Joyeeta and Michael Grubb (eds.) 1999: *Climate Change and the Leadership Paradigm: Options for the European Union*. Dordrecht: Kluwer (forthcoming).

Haas, Peter M. 1993: Epistemic Communities and the Dynamics of International Environmental Co-operation, in: Volker Rittberger (ed.): *Regime Theory and International Relations*. Oxford: Clarendon Press, 168–201.

Haigh, Nigel 1991: The European Community and International Environmental Policy. *International Environmental Affairs*, Vol. 3, No. 3, 163–180.

Haigh, Nigel 1996: Climate Change Policies and Politics in the European Community. A European Perspective, in: Timothy o'Riordan and Jill Jäger (eds.): *Politics of Climate Change*. London: Routledge, 155–184.

Hamilton, Clive 1997: *Climate Change Policies in Australia. A Briefing to a Meeting of the Ad Hoc Group on the Berlin Mandate* (Bonn (Germany), 5 August 1997). Lyneham ACT: The Australian Institute.

Hamwey, Robert 1998: *Can Attractive Models for Joint Implementation and the Clean Development Mechanism be Developed for Buenos Aires*? Geneva (Switzerland): International Academy of the Environment.

Handler Chayes, Antonia, Abram Chayes and Ronald B. Mitchell 1995: Active Compliance Management in Environmental Treaties; in: Winfried Lang (ed.): *Sustainable Development and International Law*. London/Dordrecht/Boston: Graham & Trotman/Martinus Nijhoff, 75–89.

Hansjürgens, Bernd 1998: Wie erfolgreich ist das neue Schwefeldioxid-Zertifikatesystem in den U.S.A.? – Erste Erfahrungen und Lehren für die Zukunft. *Zeitschrift für Umweltpolitik & Umweltrecht*, Vol. 21, No. 1, 1–32.

Hadley Centre 1998: *Climate Change and its Impacts*. Bracknell: UK Meterological Office.

Harris, Paul G. 1998: *Understanding America's Climate Change Policy: Realpolitik, Pluralism, and Ethic Norms*. OCEES Research Paper No 15. Oxford: Oxford Center for the Environment, Ethics & Society.

Hassing, Paul and Matthew S. Mendis 1998: Sustainable Development and Greenhouse Gas Reduction, in: José Goldemberg (ed.): *Issues & Options: The Clean Development Mechanism*. UNDP.

Heinegg, Wolff Heintschel von 1998: §22 EG im Verhältnis zu internationalen Organisationen und Einrichtungen (zugleich zur Vertragsschlußkompetenz der Gemeinschaft im Bereich des Umweltschutzes), in: Hans-Werner Rengeling (ed.): *Handbuch zum europäischen und deutschen Umwltrecht. Eine systematische Darstellung des europäischen Umweltrechts mit seinen Auswirkungen auf das deutsche Recht und mit rechtspolitischen Perspektiven*. Köln: Carl Heymanns, 648–692.

Hession, Martin and Richard Macory 1994: The Legal Framework of European Community Participation in International Environmental Agreements. *New Europe Law Review*, Vol. 2, No. 1, 61–136.

Hoof, G.J.H. van and K. de Vey Mestdagh 1984: Mechanisms of International Supervision; in: Peter van Dijk (gen.ed.): *Supervisory Mechanisms in International Economic Organisations*. Deventer: Kluwer.

Houghton, John T. 1997: *Global Warming: The Complete Briefing*. Second edition, Cambridge: Cambridge University Press.

Housman, Robert, Donald Goldberg, Brennan Van Dyke and Durwood Zaelke (eds.) 1995: *The Use of Trade Measures in Select Multilateral Environmental Agreements*. UNEP.

Hurrell, Andrew and Benedict Kingsbury 1992: The International Politics of the Environment: An Introduction, in: Andrew Hurrell and Benedict Kingsbury (eds.): *The International Politics of the Environment. Actors, Interests, and Institutions*. Oxford: Oxford University Press, 1–47.

IEA (International Energy Agency) 1996: *World Energy Outlook 1996 Edition*. Paris: IEA/OECD.

IEA (International Energy Agency) 1997: *CO_2 Emissions from Fuel Combustion. A New Basis for Comparing Emissions of a Major Greenhouse Gas*. Paris: IEA/OECD.

IEA (International Energy Agency) 1998: *World Energy Outlook 1998 Edition*. Paris: IEA/OECD.

INFRAS AG/TERI 1997: *Long term carbon emission targets aiming towards convergence*. New Delhi.

IPCC 1990: *Climate Change. The IPCC Scientific Assessment* (ed. by J. T. Houghton, G. J. Jenkins, J. J. Ephraums). Cambridge: Cambridge University Press.

IPCC 1992: *Climate Change 1992. The Supplementary Report to the IPCC Scientific Assessment* (ed. by J. T. Houghton et al.). Cambridge: Cambridge University Press.

IPCC 1994: *Radiative Forcing of Climate Change. The 1994 Report of the Scientific Assessment Working Group of IPCC* (ed. by Bert Bolin, J. T. Houghton, L. Gylvan Meira Filho). WMO/UNEP (without place).

IPCC 1995: *IPCC Second Assessment: Climate Change 1995. A Report of the Intergovernmental Panel on Climate Change*. WMO/UNEP (without place).

IPCC 1996a: *Climate Change 1995. The Science of Climate Change*. Contribution of Working Group I to the Second Assessment Report of the Intergovernmental Panel on Climate Change (edited by J.T. Houghton, L.G. Meira Filho, B.A. Callander, N. Harris, A. Kattenberg, K. Maskell). Cambridge: Cambridge University Press.

IPCC 1996b: *Climate Change 1995. Impacts, Adaptations and Mitigation of Climate Change: Scientific-Technical Analyses*. Contribution of Working Group II to the Second Assessment Report of the Intergovernmental Panel on Climate Change (edited by Robert T. Watson, Marufu C. Zinyowera, Richard H. Moss, David J. Dokken). Cambridge: Cambridge University Press.

IPCC 1996c: *Climate Change 1995. Economic and Social Dimensions of Climate Change*. Contribution of Working Group III to the Second Assessment Report of the Intergovernmental Panel on Climate Change (edited by James P. Bruce, Hoesung Lee, Erik F. Haites). Cambridge: Cambridge University Press.

IPCC 1997: *Stabilization of Atmospheric Greenhouse Gases: Physical, Biological and Socio-economic Implications* (edited by John T. Houghton, L. Gylvan Meira Filho, David J. Griggs, Kathy Maskell). WMO/UNEP (without place).

IPCC 1998: IPCC/OECD/IEA *Programme on National Greenhouse Gas Inventories: Managing Uncertainty in National Greenhouse Gas Inventories* (Meeting Report). Paris.

IPCC 1999: *Aviation and the Global Atmosphere. A Special Report of IPCC Working Groups II and III* (edited by Joyce E. Penner, David H. Lister, David J. Griggs, David J. Dokken, Mack McFarland). Cambridge: Cambridge University Press.

IPCC/TEAP 1999: *Report of the Joint IPCC/TEAP Meeting on Options for the Limitation of Emissions of HFCs and PFCs* (26–28 May 1999). Petten (The Netherlands).

Jacoby, Henry D., Ronald G. Prinn and Richard Schmalensee 1998: Kyoto's Unfinished Business. *Foreign Affairs*, Vol. 77, No. 4, 54–66.

Jaeger, Jill and H. L. Ferguson (World Meteorological Organisation) 1991: *Climate Change: Science, Impacts and Policy: Proceedings of the Second World Climate Conference*. Cambridge: Cambridge University Press.

Jaeger, Jill and Reinhard Loske 1994: *Options for the Further Development of the Commitments Within the Framework Convention on Climate Change* (Wuppertal Paper No. 22, August 1994). Wuppertal: Wuppertal Institute.

Jancar-Webster, B. (ed.) 1993: *Environmental Action in Eastern Europe*. London: M. E. Sharpe.

Janett, Daniel 1996: Allianzsysteme von Nicht-Regierungsorganisationen in der Klimapolitik. Zwischenbericht aus einer Fallstudie zum Berliner Klimagipfel. *Forschungsjournal Neue Soziale Bewegungen*, Vol. 9, No. 2, 83–89.

Jänicke, Martin and Helmut Weidner (eds.) 1997: *National Environmental Policies. A Comparative Study of Capacity-Building*. Berlin: Springer.

Japan 1997: Japan's Second National Communication under the United Nations Framework Convention on Climate Change. Tokyo: The Government of Japan.

Jefferson, Michael 1996: *Climate Change 1995, The Intergovernmental Panel on Climate Change Second Assessment Report Reviewed*.

Jepma, Catrinus J. (ed.) 1995: *The Feasibility of Joint Implementation*. Dordrecht: Kluwer.

Kamieniecki, Sheldon (ed.) 1993: *Environmental Politics in the International Arena. Movements, Parties, Organizations, and Policy*. Albany: SUNY Press.

Kassler, Peter and Matthew Paterson 1997: *Energy Exporters and Climate Change. Energy and Environmental Programme*. London: The Royal Institute of International Affairs.

Kelly, P. Mick 1990: *The Climate in Crisis: Halting Global Warming*. Climatic Research Unit/School of Environmental Sciences, University of East Anglia/Norwich, prepared for Greenpeace International, Amsterdam.

Kinley, R.J. 1994: The Communication and Review of Information under the United Nations Framework Convention on Climate Change, in: W. Katscher, G. Stein, J. Lanchbery and J. Salt (eds.): *Greenhouse Gas Verification – Why, How and How Much? Proceedings of a workshop*. Konferenzen des Forschungszentrums Jülich, Vol. 14, 141–146.

Kinzig, Ann P. and Daniel M. Kammen 1998: National Trajectories of Carbon Emissions: Analysis of Proposals to Foster the Transition to Low-carbon Eonomies. *Global Environmental Change*, Vol. 8, No. 3, 183–208.

Klaassen, Ger 1999: Emissions trading in the European Union: Practice and Prospects. in: Steve Sorell and Jim Skea (eds.) 1999: *Pollution for Sale: Emissions Trading and Joint Implementation*. Northampton (Massachusetts): Edward Elgar, 83–100.

Koplow, Douglas and Aaron Martin 1998: *Fueling Global Warming. Federal Subsidies to Oil in the United States* (a report for Greenpeace).

Koskenniemi, Martti 1992: Breach of Treaty or Non-Compliance? Reflections on the Enforcement of the Montreal Protocol. *Yearbook of International Environmental Law*, Vol. 3, 122–162.

Krägenow, Timm 1995: *Verhandlungspoker um Klimaschutz. Beobachtungen und Ergebnisse der Vertragsstaatenkonferenz zur Klimarahmenkonvention in Berlin*. Freiburg i.B.: Öko-Institut.

Krämer, Ludwig 1998: *E.C. Treaty and Environment Law*. London: Sweet & Maxwell.

Lanchbery, John and David Victor 1995: The Role of Science in the Global Climate Negotiations, in: Helge Ole Bergesen and Georg Parman (eds.): *Green Globe Yearbook of International Cooperation on Environment and Development*. Oxford: Oxford University Press, 29–39.

Lanchbery, John 1997: What to Expect from Kyoto. *Environment*, Vol. 39, No. 9, 4–11.

Lanchbery, John 1998: Verifying Compliance with the Kyoto Protocol. *Review of European Community & International Environmental Law*, Vol. 7, No. 2, 170–175.

Lashof, Daniel A., Benjamin J. DeAngelo, Scott R. Saleska and John Harte 1997: Terrestrial Ecosystem Feedbacks to Global Climate Change. *Annual Review of Energy and Environment*, Vol. 22 (1997), 75–118.

Levy, Marc A. et al. 1995: The Study of International Regimes. *European Journal of International Relations*, Vol. 1, 267–330.

Liberatore, Angela 1995: Arguments, Assumptions and the Choice of Policy Instruments. The Case of the Debate on the CO_2/Energy Tax in the European Community", in: Bruno Dente (ed.): *Environmental Policy in Search of New Instruments*. Dordrecht-Boston-London: Kluwer, 55–71.

Loske, Reinhard and Sebastian Oberthür 1994: Joint Implementation under the Climate Change Convention. *International Environmental Affairs*, Vol. 6, No. 1, 45–58.

Loske, Reinhard 1996: *Klimapolitik. Im Spannungsfeld von Kurzzeitinteressen und Langzeiterfordernissen*. Marburg: Metropolis.

Loske, Reinhard and Hermann E. Ott 1998: Reflections on Kyoto – "Success or Failure"? *Oxford Energy Forum*, No. 33 (May 1998), 6–8.

Luhmann, Hans-Jochen, Hermann E. Ott, Christiane Beuermann, Manfred Fischdick, Peter Hennicke and Liesbeth Bakker 1997: *Simulation von Joint Implementation innerhalb der Klimarahmenkonvention anhand ausgewählter Projekte*. Berlin: Erich Schmidt.

Macrory, Richard and Martin Hession 1996: The European Community and Climate Change: The Role of Law and Legal Competence, in: Tim o'Riordan and Jill Jäger (eds.): *Politics of Climate Change. A European Perspective*. London/New York: Routledge, 106–154.

Michaelis, Laurie 1997a: *Policies and Measures for Common Action – Special Issues in Carbon/Energy Taxation: Marine Bunker Fuel Charges*. Working Paper No. 11 of the Annex I Expert Group on the UN FCCC. Paris: OECD/IEA.

Michaelis, Laurie 1997b: *Policies and Measures for Common Action – Special Issues in Carbon/Energy Taxation: Carbon Charges on Aviation Fuels*. Working Paper No. 12 of the Annex I Expert Group on the UN FCCC. Paris: OECD/IEA.

Michaelowa, Axel 1998: Impact of Interest Groups on EU Climate Policy. *European Environment*, Vol. 8, 152–160.

Michaelowa, Axel and Michael Dutschke 1999: Economic and Political Aspects of Baselines in the CDM Context, in: José Goldemberg and Walter Reid (eds.): *Promoting Development while Limiting Greenhouse Gas Emissions: Trends & Baselines*. New York: UNDP, 115-133.

Mintzer, Irving M. and Amber J. Leonhard 1994: *Negotiating Climate Change: The Inside Story of the Rio Convention*. Cambridge: Cambridge University Press.

Missfeldt, Fanny 1998: Flexibility Mechanisms: Which Path to Take after Kyoto? *Review of European Community & International Environmental Law*, Vol. 7, No. 2, 128–139.

Moltke, Konrad von 1993: U.S. Foreign Environmental Policy: A New Era in Transatlantic Relations?, in: Reinhard Loske (ed.), *The Future of Environmental Policy in Transatlantic Relations*. Conference Report, Wuppertal: Wuppertal Institute for Climate, Environment & Energy 1997, 26–43.

Moltke, Konrad von and Atiq Rahman 1996: External Perspectives on Climate Change: A View from the United States and the Third World, in: Timothy o'Riordan and Jill Jäger (eds.): *Politics of Climate Change – A European Perspective*. London/New York: Routledge, 330–345.

Moltke, Konrad von 1997: Institutional Interactions: The Structure of Regimes for Trade and the Environment, in: Oran R. Young (ed.): *Global Governance. Drawing Insights from the Environmental Experience*. Cambridge (Massachusetts) and London: MIT Press, 247–272.

Moor, André de and Peter Calami 1997: *Subsidizing Unsustainable Development. Undermining the Earth with Public Funds* (commissioned by the Earth Council).

Müller, Benito 1998: *Justice in Global Warming Negotiations. How to Obtain a Procedurally Fair Compromise*. Oxford: Oxford Institute for Energy Studies.

Müller-Kraenner, Sascha 1998: Zur Umsetzung und Weiterentwicklung des Kioto-Protokolles. *Zeitschrift für Umweltrecht*, Vol. 9, No. 3, 113–115.

Müller-Rommel, Ferdinand 1993: *Grüne Parteien in Westeuropa – Entwicklungsphasen und Erfolgsbedingungen*. Opladen: Westdeutscher Verlag.

Mullins, Fiona 1997: *Lessons from Existing Systems for International Greenhouse Gas Emissions Trading*. Paris: OECD.

Mullins, Fiona and Richard Baron 1997: *International GHG Emission Trading. "Policies and Measures for Common Action"* (Working Paper 9). Paris: OECD.

Multilateral Fund 1998: *Policies, Procedures, Guidelines and Criteria* (as at March 1998). Multilateral Funds for the Implementation of the Montreal Protocol.

Munich Re 1998: *Topics 1998*. Munich.

NATO/CCMS 1999: *Environment & Security in an International Context. Final Report March 1999* (edited by Kurt M. Lietzmann and Gary D. Vest). Berlin.

Nordic Council 1997: *Criteria and Perspectives for Joint Implementation. Ten Nordic Projects in Eastern Europe*. Copenhagen: Nordic Council of Ministers.

Oberthür, Sebastian 1993: *Politik im Treibhaus. Die Entstehung des internationalen Klimaschutzregimes*. Berlin: Edition Sigma.

Oberthür, Sebastian and Hermann Ott 1995: UN/Convention on Climate Change: The First Conference of the Parties. *Environmental Policy and Law*, Vol. 25, No. 4/5, 144–156.

Oberthür, Sebastian 1996a: Sign of Progress. *Environmental Policy and Law*, Vol. 26, No. 4, 158–160.

Oberthür, Sebastian 1996b: The Second Conference of the Parties. *Environmental Policy and Law*, Vol. 26, No. 5, 195–201.

Oberthür, Sebastian 1997 *Umweltschutz durch internationale Regime. Interessen, Verhandlungsprozesse, Wirkungen*. Opladen: Leske & Budrich.

Oberthür, Sebastian and Stefanie Bär 1998: *Untersuchung zur Umsetzung eines Protokolls zur Klimarahmenkonvention: Auswirkungen von Aktivitäten und Regelungen der EU auf die Emission klimawirksamer Gase*, Berlin: Ecologic.

Oberthür, Sebastian 1999a:The EU as an International Actor: The Case of the Protection of the Ozone Layer. *Journal of Common Market Studies* (forthcoming).

Oberthür, Sebastian 1999b: *Linkages between the Montreal and the Kyoto Protocols*. Paper prepared for the International Conference on Synergies and Coordination between Multilateral Environmental Agreements, Tokyo, 14–16 July 1999.

Oberthür, Sebastian 1999c: The EU in International Environmental Regimes and the Energy Charter Treaty. The Existing Experience, in: Joyeeta Gupta and Michael Grubb (eds.): *Climate Change and the Leadership Paradigm: Options for the European Union*. Dordrecht: Kluwer (forthcoming).

OECD 1992: *Climate Change: Designing a Tradeable Permit System*. Paris: OECD.

OECD 1993a: *Indicators for the Integration of Environmental Concerns into Energy Policies*. Environmental Monographs No. 79. Paris: OECD.

OECD 1993b: *Indicators for the Integration of Environmental Concerns into Transport Policies*. Environmental Monographs No. 80. Paris: OECD.

OECD 1995: *Subsidies and the Environment – Exploring the Linkages*. Paris: OECD.

OECD 1997: *Economic Globalisation and the Environment*. Paris: OECD.

OECD 1999: *Experience with Emission Baselines under the AIJ Pilot Phase*. OECD Information Paper. Paris: OECD.

O'Riordan, Timothy and Jill Jäger 1996: *Politics of Climate Change – A European Perpective*. London/ New York: Routledge.

Ott, Hermann E. 1991: The New Montreal Protocol: A Small Step for the Protection of the Ozone Layer, a Big Step for International Law and Relations. *Verfassung und Recht in Übersee/Law and Politics in Africa, Asia and Latin America*, No. 24 (1991), 188–208.

Ott, Hermann E. 1994: Tenth Session of the INC/FCCC: Results and Options for the First Conference of the Parties. *Environmental Law Network International Newsletter*, No. 2/1994, 3–7.

Ott, Hermann 1996a: Völkerrechtliche Aspekte der Klimarahmenkonvention, in: Hans Günter Brauch (ed.): *Klimapolitik – naturwissenschaftliche Grundlagen, internationale Regimebildung und Konflikte, ökonomische Analysen sowie nationale Problemerkennung und Politikumsetzung*. Berlin/Heidelberg: Springer, 61–74.

Ott, Hermann E. 1996b: Elements of a Supervisory Procedure for the Climate Regime. *Zeitschrift für ausländisches öffentliches Recht und Völkerrecht/Heidelberg Journal of International Law*, No. 56 (1996), 732–749.

Ott, Hermann 1997: Outline of EU Climate Policy in the FCCC – Explaining the "EU-bubble"; in: Japan Center of International and Comparative Environmental Law, UNFCCC Protocol Working Group 1997: *Climate Change and the Future of Mankind. Pre-COP3 International Symposium on Legal Strategies to Prevent Climate Change*, 13–14 September 1997, 1–7.

Ott, Hermann E. 1998a: *Umweltregime im Völkerrecht. Eine Untersuchung über neue Formen internationaler institutionalisierter Kooperation am Beispiel der Verträge zum Schutz der Ozonschicht und zur Kontrolle grenzüberschreitender Abfallverringerung*. Baden-Baden: Nomos.

Ott, Hermann E. 1998b: Operationalizing 'Joint Implementation'. Organizational and Institutional Aspects of a New Instrument in International Climate Policy. *Global Environmental Change*, Vol. 8, No. 1, 11–47.

Ott, Hermann E. 1998c: The Kyoto Protocol. Unfinished Business. *Environment*, Vol. 40, No. 6, 16–20, 41–45.

Ozone Secretariat 1996: *Handbook for the International Treaties for the Protection of the Ozone Layer, Fourth Edition (1996)*. Nairobi: Ozone Secretariat/UNEP.

Ozone Secretariat 1998: *1997 Update of the Handbook for the International Treaties for the Protection of the Ozone Layer*. Nairobi: Ozone Secretariat/UNEP.

Palmer, Geoffrey 1992: New Ways to Make International Environmental Law. *American Journal of International Law*, No. 86 (1992), 259–283.

Panayotou, Theodore 1998: Six Questions of Design and Governance, in: José Goldemberg (ed.): *Issues & Options: The Clean Development Mechanism*. UNDP.

Parson, Edward A. 1993: Protecting the Ozone Layer, in: Peter M. Haas; Robert O. Keohane and Marc A. Levy (eds.): *Institutions for the Earth. Sources of Effective International Environmental Protection*. Cambridge (Massachusetts): MIT Press, 27–73.

Parson, Edward and Owen Greene 1995: The Complex Chemistry of the International Ozone Agreements. *Environment*. Vol. 37, No. 2, 17–20, 35–43.

Parson, Edward A. and Karen Fisher-Vanden 1997: *Joint Implementation and its alternatives: choosing systems to distribute global emissions abatement and finance*. Cambridge (Massachusetts): Center for Science and International Affairs.

Pearce, Fred 1998: Growing Pains. *The New Scientist*, 24 October 1998, 20–21.

Petschel-Held, G. and H.J. Schnellnhuber 1998: The Tolerable Windows Approach to Climate Control: Optimization, Risks and Perspectives, in: F.L.Toth (ed.): *Cost-Benefit Analyses of Climate Change. The broader Perspectives*. Basel: Birkhäuser.

Pinto, Frank, S.M. Si-Ahmed and Steve Gorman 1999: *Options for Reduction of Emissions of HFCs in Developing Countries*. Paper presented at the Joint IPCC/TEAP Expert Meeting on Options for the Limitation of Emissions HCFs and PFCs, The Netherlands, 26–28 May 1999.

Princen, Thomas and Finger, Matthias (eds.) 1994: *Environmental NGOs in World Politics. Linking the Local and the Global*. London: Routledge.

Prittwitz, Volker von 1984: *Umweltaußenpolitik. Grenzüberschreitende Luftverschmutzung in Europa*. Frankfurt a.M.: Campus.

Prittwitz, Volker von 1990: *Das Katastrophenparadox. Elemente einer Theorie der Umweltpolitik*. Opladen: Leske & Budrich.

Rahmstorf, Stefan 1997: Risk of sea-change in the Atlantic. *Nature*, Vol. 388, 28 August 1997, 825–826.

Ramakrishna, Kilaparti 1994: *The Financial Mechanism of the Framework Convention on Climate Change: Operational Issues*. Woods Hole (Massachusetts): Woods Hole Research Center.

Raufer, Roger K. 1996: Market-Based Pollution Control Regulation: Implementing Economic Theory in the Real World. *Environmental Policy & Law*, Vol. 26, No. 4, 177–184.

Rayner, Steve 1991: The Greenhouse Effect in the US: the Legacy of Energy Abundance, in: Michael Grubb (ed.): *Energy Policies and the Greenhouse Effect*. London.

Reid, Walter V. and José Goldemberg 1997: Are Developing Countries Already Doing as Much as Industrialized Countries to Slow Climate Change? *Climate Notes*, July 1997. Washington, D.C.: World Resources Institute.

Reid, Walter V. and José Goldemberg 1998: Developing Countries are Combating Climate Change. *Energy Policy*, Vol. 26, No. 3, 233–237.

Richardson, Dick and Chris Roots 1995: *The Development of Green Parties in Europe*. London and New York: Routledge.

Ringius, Lasse, Asbjørn Torvanger and Bjart Holtsmark 1998: *Can Multi-criteria Rules Fairly Distribute Climate Burdens? OECD Results from Three Burden Sharing Rules*. Oslo: CICERO.

Roland, Kjell and Torleif Haugland 1995: Joint Implementation - Difficult to Implement? in: Catrinus C. Jepma (ed.): *The Feasibility of Joint Implementation*. Dordrecht, Boston, London: Kluwer.

Rose, A., B. Stevens, J. Edmonds and M. Wise 1998: *International Equity and Differentiation in Global Warming Policy, Environmental and Resource Economics*. 12, 25–51.

Roson, R. and F. Bosello 1999: *Distributional Consequences of Alternative Emissions Trading Schemes*. Paper submitted for the EFIEA Workshop "Integrating climate policies in the European environment: costs and opportunities", Milan, 4–6 March.

Rowbotham, Elizabeth J. 1996: Legal Obligations and Uncertainties in the Climate Change Science and Politics, in: Timothy o'Riordan and Jill Jäger (eds.): *Politics of Climate Change: A European Perspective*. London/New York: Routledge, 32–50.

Rowlands, Ian H. 1995: *The Politics of Global Atmospheric Change*. Manchester/New York: Manchester University Press.

Salt, Julian E. 1998: Kyoto and the Insurance Industry: An Insider's Perspective. *Environmental Politics*, Vol. 7, No. 2, 160–165.

Sampson, Gary 1998: WTO Rules and Climate Change: The Need for Policy Coherence, in: Bradnee W. Chambers (ed.): *Global Climate Governance: Inter-linkages between the Kyoto Protocol and other Multilateral Regimes*. Tokyo: United Nations University, Institute of Advanced Studies, 29–38.

Sand, Peter H. 1990: *Lessons Learned in Global Environmental Governance*. Washington, D.C.: World Resources Institute.

Sand, Peter H. 1997: Das Washingtoner Artenschutzabkommen (CITES) von 1973, in: Thomas Gehring and Sebastian Oberthür (eds.): *Internationale Umweltregime: Umweltschutz durch Verhandlungen und Verträge*. Opladen: Leske & Budrich, 165-184.

Sands, Phillippe 1992: The United Nations Framework Convention on Climate Change. *Review of European Community & International Law*, No. 1 (1992), 270–277.

Sands, Phillippe 1994: *Greening International Law*. New York: The New Press.

Schärer, Bernd 1999: Tradable Emission Permits in German Clean Air Policy: Considerations on the Efficiency of Environmental Policy Instruments, in: Steve Sorell and Jim Skea (eds.) 1999: *Pollution for Sale: Emissions Trading and Joint Implementation*. Northampton (Massachusetts): Edward Elgar, 141–153.

Schellnhuber, Hans-Joachim and Ursula Fuentes 1997: Globaler Klimaschutz: Ziele, Hemmnisse und Chancen für ein Steuerungsproblem vom Multiakteur-Langfrist-Typus. *Zeitschrift für Angewandte Umweltforschung*, Vol. 10, No. 4, 441–447.

Schlamadinger, Bernhard and Gregg Marland 1998: *Some Technical Issues Regarding Land-use Change and Forestry in the Kyoto Protocol*. Oak Ridge (USA): Oak Ridge National Laboratory, Environmental Sciences Division.

Schneider, Stephen S. 1998: Kyoto Protocol: The Unfinished Agenda. *Climatic Change*, No. 39 (1998), 1–21.

Schmidt, Hilmar and Ingo Take 1997: *More Democratic and Better? The Contribution of NGOs to the Democratization of International Politics and the Solution of Global Problems* (Working Paper No. 7, World Society Research Group). Darmstadt: Technical University/Darmstadt and Johann Wolfgang Goethe University/Frankfurt.

Schreurs, Miranda A. 1996: *Domestic Institutions, International Agendas, and Global Environmental Protection in Japan and Germany*. Dissertation, Department of Political Science/University of Michigan.

Schreurs, Miranda 1997: A Political System's Capacity for Global Environmental Leadership: A Case Study of Japan, in: Lutz Mez und Helmut Weidner (eds.): *Umweltpolitik und Staatsversagen. Perspektiven und Grenzen der Umweltpolitikanalyse* (Festschrift für Martin Jänicke zum 60. Geburtstag). Berlin: Sigma, 323–331.

Simeonova, Katia and Fanny Mißfeldt 1997: *Emission Trends in Economies in Transition* (EEP Climate Change Briefing). London: The Royal Institute of International Affairs.

Simonis, Udo Ernst 1998: Das Kioto-Protokoll und seine Bewertung. *Spektrum der Wissenschaft*, March 1998, 96–103.

Sjöstedt 1998: The EU Negotiates Climate Change. External Performance and Internal Structural Change. *Cooperation and Conflict*, Vol. 33, No. 3, 227–256.

Smeloff, Edward A. 1998: Global Warming: The Kyoto Protocol and Beyond. *Environmental Policy and Law*, Vol. 28, No. 2, 63–68.

Sohn, Louis B. 1983: The Future of Dispute Settlement; in: R.St.J. MacDonald and D.M. Johnston (eds.): *The Structure and Process of International Law. Essays in Legal Philosophy, Doctrine and Theory*. The Hague: Martinus Nijhoff, 1121–1146.

Solomon, Barry D. 1995: Global CO_2 Emissions Trading: Early Lessons from the U.S. Acid Rain Program. *Climatic Change*, No. 30 (1995), 75–96.

Sorell, Steve and Jim Skea (eds.) 1999: *Pollution for Sale: Emissions Trading and Joint Implementation*. Northampton (Massachusetts): Edward Elgar.

Sprinz, Detlef and Urs Luterbacher 1996: *International Relations and Global Climate Change*. PIK Report No. 21. Potsdam: Potsdam Institute for Climate Impact Research.

Steward, Richard B., Jonathan B. Wiener and Philippe Sands 1996: *Legal Issues Presented by a Pilot International Greenhouse Gas Trading System*. Genf: UNCTAD, United Nations.

Sydnes, Anne Kristin 1996: Norwegian Climate Policy: Environmental Idealism and Economic Realism, in: Timothy o'Riordan and Jill Jäger (eds.): *Politics of climate change*. London/New York: Routledge, 268–297.

Széll, Patrick 1995: The Development of Multilateral Mechanisms for Monitoring Compliance, in: Winfried Lang (ed.): *Sustainable Development and International Law*. London/Amsterdam: Graham & Trotman/Martinus Nijhoff, 97–109.

Taalab, Azza 1998: *Rising Voices Against Global Warming*. Frankfurt a.M.: IZE.

Take, Ingo 1997: *NGOs: Protagonists of World Society? Strategies and Levels of NGO Influence on International Relations*. Working Paper No. 8. World Society Research Group. Darmstadt: Technical University Darmstadt/Johann Wolfgang Goethe University Frankfurt.

Tarasofsky, Richard G. 1998: Ensuring Compatibility between Multilateral Environmental Agreements and GATT/WTO. *Yearbook of International Environmental Law*, Vol. 7, 52–74.

Tarasofsky, Richard G. (ed.) 1999: *Assessing the International Forest Regime*. Gland (Switzerland), Cambridge (UK) and Bonn (Germany): International Union for Conservation of Nature and Natural Ressources (IUCN).

Taylor, Bron et al. 1993: Grass-Roots Resistance: The Emergence of Popular Environmental Movements in Less Affluent Countries, in: Sheldon Kamieniecki (ed.), *Environmental Politics in the International Arena: Movements, Parties, Organizations, and Policy*. Albany: SUNY Press, 69–90.

The Corner House 1997: *Climate and Equity After Kyoto*. London.

Tietenberg, Tom, Michael Grubb, Axel Michaelowa, Byron Shift and Zhong Xian Zhang 1998: *International Rules for Greenhouse Gas Emissions Trading*. Geneva: UNCTAD.

UNCTAD 1994: *Combating Global Warming. Possible Rules, Regulations and Administrative Arrangements for a Global Market in CO_2 Emission Entitlements*. Geneva: UNCTAD, United Nations.

UNEP/SEI 1996: *Regulations to Control Ozone-Depleting Substances: A Guidebook*. UNEP and Stockholm Environment Institute.

UNEP 1997: *Statement of Environmental Commitment by the Insurance Industry*.

UNEP 1998: *Scientific Assessment of Ozone Depletion: 1998*. World Meteorological Organization Global Ozone Research and Monitoring Project – Report No. 44: NOAA/NASA/UNEP/WMO/European Commission.

UNEP 1999: *Handbook on Data Reporting under the Montreal Protocol*. UNEP/ Multilateral Fund for the Implementation of the Montreal Protocol.

UN 1996: *Indicators of Sustainable Development. Framework and Methodologies*. New York: United Nations.

UNU 1998: *Global Climate Governance: A Report on the Inter-linkage between the Kyoto Protocol and other Multilateral Regimes*. Tokyo: The United Nations University (UNU) and Global Environment Information Centre.

U.S. Department Of Energy 1996: *Policies and Measures For Reducing Energy Related Greenhouse Gas Emissions. Lessons From Recent Literature*. Washington, D.C.

U.S. Department of State 1997: *Environmental Diplomacy. The Environment and U.S. Foreign Policy*. Washington, D.C.

Vari, A. and P. Tamas (eds.) 1993: *Environment and Democratic Transition: Policy and Politics in Central and Eastern Europe*. Boston: Kluwer.

Victor, David G. and Julian E. Salt 1994a: Keeping the Climate Treaty Relevant. *Nature*, Vol. 373, 280–282.

Victor, David G. and Julian E. Salt 1994b: From Rio to Berlin: Managing Climate Change. *Environment*, Vol. 36, No. 10, 6–15; 25–32.

Victor, David G. 1995: *Design Options for Article 13 of the Framework Convention on Climate Change: Lessons from the GATT Dispute Panel System*. Laxenburg (Austria): International Institute for Applied Systems Analysis.

Victor, David G. 1998: The Operation and Effectiveness of the Montreal Protocol's Non-Compliance Procedure, in: David G. Victor, Kal Raustiala and Eugene B. Skolnikoff (eds.): *The Implementation and Effectiveness of International Environmental Commitments: Theory and Practice*. Cambridge (Massachusetts): MIT Press, 137–176.

Victor, David G., Kal Raustiala and Eugene B. Skolnikoff (eds.) 1998: *The Implementation and Effectiveness of International Environmental Commitments: Theory and Practice*. Cambridge (Massachusetts): MIT Press.

Vine, Edward and Jayant Sathaye 1999: *Guidelines for the Monitoring, Evaluation, Reporting, Verification, and Certification of Energy-Efficiency Projects for Climate Change Mitigation*. Berkeley: Energy Analysis Department.

Vogel, David 1986: *National Styles of Regulation. Environmental Policy in Great Britain and the United States*. Ithaca and London: Cornell University Press.

Wang, Xueman 1998: Towards a System of Compliance: Designing a Mechanism for the Climate Change Convention. *Review of European Community & International Environmental Law*, Vol. 7, No. 2, 176–179.

Wapner, Paul 1996: *Environmental Activism and World Civic Politics*. Albany: SUNY Press.

Wapner, Paul 1997: Governance in Global Civil Society, in: Oran R. Young (ed.): *Global Governance. Drawing Insights from the Environmental Experience*. Cambridge (Massachusetts) and London : MIT Press, 65–84.

WBGU (German Advisory Council on Global Change) 1997: *Targets for Climate Protection, 1997. A Study for the Third Conference of the Parties to the Framework Convention on Climate Change in Kyoto* (September 1997). Bremerhaven: WBGU.

WBGU (German Advisory Council on Global Change) 1998: *Die Anrechnung biologischer Quellen und Senken im Kyoto-Protokoll: Fortschritt oder Rückschlag für den Umweltschutz?* (Sondergutachten 1998).

Weizsäcker, Ernst Ulrich von, Amory B. Lovins and L. Hunter Lovins 1997: *Factor Four. Doubling Wealth - Halving Resource Use*. London: Earthscan.

Weizsäcker, Ernst Ulrich von and Hermann E. Ott 1998a: Tax Bads, not Goods. *Our Planet*, Vol. 9, No. 6, 17–18.

Weizsäcker, Ernst Ulrich von and Hermann E. Ott 1998b: Das Klimaschutz-Protokoll: Auch ein Erfolg der Zivilgesellschaft *(The Kyoto Protocol: A success of civil society)*. Contribution to the second Environment Report 1998 of Asea Brown Boveri (ABB).

Werksman, Jacob and Farhana Yamin 1995: Carrying forward the Berlin Mandate: Protocol Negotiations and Implementing Activities Jointly, in: K. Ramakrishna, A. Deutz and L. Jacobsen (eds.): *The Ad Hoc Process to Strengthen the Framework Convention on Climate Change*. Woods Hole (Massachusetts): The Woods Hole Research Center.

Werksman, Jacob 1996: Compliance and Transition: Russia's Non-Compliance Tests the Ozone Regime. *Zeitschrift für ausländisches öffentliches Recht und Völkerrecht (Heidelberg Journal of International Law)*, No. 56 (1996), 750–773.

Werksman, Jakob 1998a: The Clean Development Mechanism: Unwrapping the 'Kyoto Surprise'. *Review of European Community & International Environmental Law*, Vol. 7, No. 2, 147–158.

Werksman, Jacob 1998b: *Responding to Non-Compliance under the Climate Change Regime*. OECD Information Paper October 1998. Paris: OECD.

Werksman, Jake and Claudia Santoro 1998: Investing in Sustainable Development: the Potential Interaction Between the Kyoto Protocol and a Multilateral Agreement on Investment, in: Bradnee W. Chambers (ed.): *Global Climate Governance: Inter-linkages between the Kyoto Protocol and other Multilateral Regimes*. Tokyo: United Nations University, Institute of Advanced Studies, 59–76.

World Bank 1999: *World Development Report 1998/99. Knowledge for Development*. Oxford: Oxford University Press.

World Energy Council (WEC) 1996: *Climate Change 1995 – The Intergovernmental Panel on Climate Change. Second Assessment Report Reviewed*. London: WEC.

WRI (World Resources Institute) 1996: *World Resources 1996–97*. Oxford: Oxford University Press.

WWF 1996: *Policies and Measures to Reduce CO_2 Emissions by Efficiency and Renewables. A Preliminary Survey for the Period to 2005* (edited by Kornelis Blok, Detlef van Vuuren, Ad van Wijk and Lars Hein). Utrecht (The Netherlands): Department of Science, Technology and Society, Utrecht University.

WWF 1997a: *Policies and Measures to Reduce CO_2 Emissions in the United States. An Analysis of Options for 2005 and 2010* (edited by Stephen Bernow, Wiliam Dougherty, Max Duckworth, Sivan Kartha, Michael Lazarus and Michael Ruth). Boston/Massachusetts: Tellus Institute and Stockholm Environment Institute.

WWF 1997b: *Key Technology Policies to Reduce CO_2 Emissions in Japan. An Indicative Survey for 2005 and 2010* (edited by Haruki Tsuchiya, Yuzuru Matsuoka, Ad J.M. van Wijk and G.J.M. Phylipsen).Tokyo: WWF.

WWF 1998: *A Review of the Stage of Implementation of European Union Policies and Measures for CO_2 Emission Reduction* (edited by Dian Phylipsen, Kornelis Blok and Chris Hendriks). Utrecht (Netherlands): WWF.

WWF 1999: *Integrating Climate Change and Forest Conservation Policies. The WWF Perspective.*

Yamin, Farhana 1995: Additional Commitments and Joint Implementation: the Post-Berlin Landscape, in: Michael Grubb and Dean Anderson (eds.): *The Emerging International Regime for Climate Change*. London: The Royal Institute of International Affairs, 59–66.

Yamin, Farhana 1997: *NGO Participation in the Convention on Biological Diversity: Insights and Recommendations for the UNFCCC Process*. London: Field Working Paper, 24 July 1997.

Yamin, Farhana 1998a: The Kyoto Protocol: Origins, Assessment and Future Challenges. *Review of European Community & International Environmental Law*, Vol. 7, No. 2, 113–127.

Yamin, Farhana 1998b: Unanswered Questions, in: José Goldemberg (ed.): *Issues & Options: The Clean Development Mechanism*. UNDP.

Young, Oran R. 1989: The Politics of International Regime Formation: Managing Natural Resources and the Environment. *International Organization*, Vol. 43, No. 3, 349–375.

Young, Oran R. 1991: Political Leadership and Regime Formation: On the Development of Institutions in International Society. *International Organization*, Vol. 45, No. 3, 281–308.

Young, Oran R. 1994: *International Governance. Protecting the Environment in a Stateless Society*. Ithaca: Cornell University Press.

Young, Oran R. 1996: Institutional Linkages in International Society: Polar Perspectives. *Global Governance*, Vol. 2, No. 1, 1–24.

Young, Oran R. 1999: *Institutional Dimensions of Global Environmental Change:Science Plan* (IHDP Report No. 9). Bonn: International Human Dimensions Programme on Global Environmental Change (IHDP).

Ziesing, Hans Joachim et al. 1998: *Ursachen der CO2-Entwicklung in Deutschland in den Jahren 1990 bis 1995*. Berlin: Umweltbundesamt.

Index

MIX
Papier aus verantwortungsvollen Quellen
Paper from responsible sources
FSC® C105338
FSC
www.fsc.org

If you have any concerns about our products,
you can contact us on
ProductSafety@springernature.com

In case Publisher is established outside the EU,
the EU authorized representative is:
Springer Nature Customer Service Center GmbH
Europaplatz 3, 69115 Heidelberg, Germany

Printed by Libri Plureos GmbH
in Hamburg, Germany